Surveillance and Management of Parasitic Diseases in China

Xiao-Nong Zhou
Editor

Surveillance and Management of Parasitic Diseases in China

Springer
PMPH PEOPLE'S MEDICAL PUBLISHING HOUSE

Editor
Xiao-Nong Zhou
National Institute for Parasitic Disease at Chinese Center for Disease Control and Prevention (Chinese Center for Tropical Diseases Research)
Shanghai, China

ISBN 978-981-97-1240-3 ISBN 978-981-97-1241-0 (eBook)
https://doi.org/10.1007/978-981-97-1241-0

Jointly published with People's Medical Publishing House, PR of China

This Springer imprint is published by the registered company Springer Nature Singapore Pte Ltd.
The registered company address is: 152 Beach Road, #21-01/04 Gateway East, Singapore 189721, Singapore

Preface

Parasitic diseases seriously endanger human health and hinder the socioeconomic development of endemic areas. With global economic integration and climate change, some parasitic diseases have been listed as emerging or reemerging infectious diseases as well as neglected tropical diseases by the World Health Organization (WHO). In February 2007 at the Bangkok Conference, Margaret Chan, Director-General of the WHO, pointed out that this is an "ancient, quiet, hidden and neglected disease" that is often prevalent in poor areas of developing countries. Disease and poverty form a vicious circle; that is, poverty occurs due to disease, while disease prevalence occurs because of poverty.

The main purpose of disease surveillance is to identify priority issues of public health; describe the characteristics of infectious diseases in terms of time, region, and population distributions; monitor and identify epidemic trends and developmental patterns to put forward hypotheses regarding the causes of disease or outbreaks; and then carry out epidemiological research on infectious diseases to provide a scientific basis for evaluating and formulating disease prevention and control countermeasures. It is helpful to formulate feasible plans of regional disease control and implementation in line with future socioeconomic development and health service needs. Surveillance is an important part of disease prevention and control programs and is one of the basic functions of the Centers for Disease Control and Prevention (CDC) at all levels. Since 2005, the Chinese Center for Disease Control and Prevention (China CDC) has established a unified and standardized national parasitic disease surveillance system in more than 150 counties of 20 provinces across China, covering schistosomiasis, malaria, soil-transmitted parasitic diseases, foodborne parasitic diseases, and others. In addition, sentinel surveillance sites have been set up in various provinces, forming a national parasitic disease-monitoring and evaluation network. This network provides a scientific basis for understanding the epidemic dynamics, transmission patterns, and influencing factors of parasitic diseases in China, as it performs early detection and guidance for epidemic treatment, provides disease prediction and early warning, and formulates disease prevention and control countermeasures to support parasitic disease prevention and control processes.

Parasitic disease surveillance in China has always been crucial in parasitic disease prevention and control programs and is indispensable in public health. Health workers at all levels should understand the working procedures, relevant knowledge, and surveillance techniques in parasitic disease surveillance and control fieldwork. Therefore, this book serves as a protocol for professionals, aiming at the prevention and control of parasitic diseases in the vast rural areas of China with a focus on the surveillance techniques that all professional health workers must perform in endemic areas. The Introduction chapter introduces the concept of parasitic disease surveillance and the classification and application of surveillance techniques. Other chapters present surveillance techniques for the regional distribution of pathogens, case-population distribution, key vectors, ecological environment, and early-warning and response techniques. The concepts, methods, operating procedures, data collection, results analysis, reports, and conclusions of each surveillance technique are described in detail. Some examples of and experience with the surveillance of parasitic diseases are included to guide health institutions and clinical service providers in surveillance of and response to common parasitic diseases.

Because of the tight schedule, there are inevitably omissions and even errors in this book. We sincerely hope that readers will point them out.

Shanghai, China Xiao-Nong Zhou

Acknowledgments

Special acknowledgment goes to Senior Researchers and Professors Jiang Zheng, Zhong-Xing Wu, Cui'e Chen, Wu-Cheng Zhang, Tian-Ping Wang, Li-Yong Wen, and You-Sheng Liang, who provided reviews and revisions that resulted in great improvements to the book. This book was funded by the National Major Project for Infectious Diseases (No. 2012ZX10004-220), the Public Welfare Health Industry Science and Technology Project (No. 201202019), and the China-UK Global Health Support Project (No. GHSP-CS-OP2-02).

Contents

Contributors

Document Drafting and Data Collection

Zhuo Chen National Institute of Parasitic Diseases at Chinese Center for Disease Control and Prevention (Chinese Center for Tropical Diseases Research), Shanghai, China

Yao Deng Jiangsu Institute of Parasitic Diseases, Wuxi, Jiangsu, China

Qing Fu National Institute of Parasitic Diseases at Chinese Center for Disease Control and Prevention (Chinese Center for Tropical Diseases Research), Shanghai, China

Qing-Biao Hong Jiangsu Institute of Parasitic Diseases, Wuxi, Jiangsu, China

Shi-Zhu Li National Institute of Parasitic Diseases at Chinese Center for Disease Control and Prevention (Chinese Center for Tropical Diseases Research), Shanghai, China

Ying-Jun Qian National Institute of Parasitic Diseases at Chinese Center for Disease Control and Prevention (Chinese Center for Tropical Diseases Research), Shanghai, China

Le-Ping Sun Jiangsu Institute of Parasitic Diseases, Wuxi, Jiangsu, China

Qiang Wang National Institute of Parasitic Diseases at Chinese Center for Disease Control and Prevention (Chinese Center for Tropical Diseases Research), Shanghai, China

Tian-Ping Wang Anhui Institute of Parasitic Diseases, Hefei, Anhui, China

Shang Xia National Institute of Parasitic Diseases at Chinese Center for Disease Control and Prevention (Chinese Center for Tropical Diseases Research), Shanghai, China

Jing Xu National Institute of Parasitic Diseases at Chinese Center for Disease Control and Prevention (Chinese Center for Tropical Diseases Research), Shanghai, China

Jun-Fang Xu Hubei Institute for Nationalities, Enshi, Hubei, China

Guo-Jing Yang Jiangsu Institute of Parasitic Diseases, Wuxi, Jiangsu, China

Kun Yang Jiangsu Institute of Parasitic Diseases, Wuxi, Jiangsu, China

Yi Zhang National Institute of Parasitic Diseases at Chinese Center for Disease Control and Prevention (Chinese Center for Tropical Diseases Research), Shanghai, China

Xiao-Nong Zhou National Institute of Parasitic Diseases at Chinese Center for Disease Control and Prevention (Chinese Center for Tropical Diseases Research), Shanghai, China

English Translation and Editing

An-Qi Chao School of Global Health, Chinese Center for Tropical Diseases Research, Shanghai Jiao Tong University School of Medicine, Shanghai, China

Xian-Fa Chen National Institute of Parasitic Diseases at Chinese Center for Disease Control and Prevention (Chinese Center for Tropical Diseases Research), Shanghai, China

Wei Ding National Institute of Parasitic Diseases at Chinese Center for Disease Control and Prevention (Chinese Center for Tropical Diseases Research), Shanghai, China

Qiu-Feng Lan National Institute of Parasitic Diseases at Chinese Center for Disease Control and Prevention (Chinese Center for Tropical Diseases Research), Shanghai, China

Xin-Chen Li School of Global Health, Chinese Center for Tropical Disease Research, Shanghai Jiao Tong University School of Medicine, Shanghai, China

Lu Liu National Institute of Parasitic Diseases at Chinese Center for Disease Control and Prevention (Chinese Center for Tropical Diseases Research), Shanghai, China

Shan Lv National Institute of Parasitic Diseases at Chinese Center for Disease Control and Prevention (Chinese Center for Tropical Diseases Research), Shanghai, China

Xue-Jiao Ma National Institute of Parasitic Diseases at Chinese Center for Disease Control and Prevention (Chinese Center for Tropical Diseases Research), Shanghai, China

Er-Ya Wan School of Global Health, Chinese Center for Tropical Diseases Research, Shanghai Jiao Tong University School of Medicine, Shanghai, China

Duo-Quan Wang National Institute of Parasitic Diseases at Chinese Center for Disease Control and Prevention (Chinese Center for Tropical Diseases Research), Shanghai, China

Chen-Sheng Zhang National Institute of Parasitic Diseases at Chinese Center for Disease Control and Prevention (Chinese Center for Tropical Diseases Research), Shanghai, China

Qi-Yu Zhang School of Global Health, Chinese Center for Tropical Diseases Research, Shanghai Jiao Tong University School of Medicine, Shanghai, China

Xiao-Nong Zhou National Institute of Parasitic Diseases at Chinese Center for Disease Control and Prevention (Chinese Center for Tropical Diseases Research), Shanghai, China

School of Global Health, Chinese Center for Tropical Diseases Research, Shanghai Jiao Tong University School of Medicine, Shanghai, China

Hainan Center for Tropical Diseases Research, Haikou, China

Abbreviations

ACTs	Artemisinin-based combination therapies
ANN	Artificial neural network
CBA	Cost-benefit analysis
CEA	Cost-effectiveness analysis
China CDC	Chinese Center for Disease Control and Prevention
CIE	Counterimmunoelectrophoresis
CLPT	Circum larval precipitation test for trichinella
CMR	Capture-mark-recapture
CT	Computed tomography
CUA	Cost-utility analysis
DALYs	Disability-adjusted life years
DW	Disability weight
EHEC	Enterohemorrhagic *E. coli*
ELISA	Enzyme-linked immunosorbent assay
EWMA	Exponentially weighted moving average
FBS	Family burden scale
GBD	Global burden of disease
GIS	Geographic information systems
GPS	Global positioning system
H5N1	Human avian influenza
IEST	Immunoenzymatic staining test
IFA	Indirect fluorescent antibody method
LQAS	Lot quality assurance sampling
MRI	Magnetic resonance imaging
NHC	National Health Commission
NIPD	National Institute of Parasitic Diseases at Chinese Center for Disease Control and Prevention
NMPA	National Medical Products Administration
PCR	Polymerase chain reaction
PRC	People's Republic of China
QALYs	Quality-adjusted life years
RAPD	Random amplified polymorphic DNA
RDT	Rapid diagnostic test
RFLP	Restriction fragment length polymorphism

RS	Remote sensing
SARS	Severe acute respiratory syndrome
SBI	Social burden index
SCN	Social contact networks
WHA	World Health Assembly
WB	Western blot
WHO	World Health Organization
YIPD	Yunnan Institute of Parasitic Diseases
YLD	Years lived with disability
YLL	Years of life lost

Introduction 1

Xiao-Nong Zhou

Parasitic diseases are characterized by being infectious and endemic and having natural focal and other transmission features with specific pathogens, including protozoa, helminths, and arthropods. When a pathogen is excreted by a source of infection and invades a new susceptible organism, it experiences the transmission process of staying in the organism while metastasizing to the outside world. The invasion of pathogens and the influencing factors involved in this process, as well as the health resources utilized in controlling transmission, are all covered by modern disease surveillance and are the key elements in strengthening monitoring capacity and improving innovative technologies.

As with all infectious diseases, the prevalence of parasitic diseases has three necessary factors, namely, sources of infection, transmission paths, and susceptible populations. However, different parasitic diseases have unique transmission and epidemic features. For example, foodborne parasitic diseases are prone to group outbreaks; zoonosis is associated with highly repetitive infections; soil-transmitted parasitic diseases are transmitted by multiple species of parasites; opportunistic parasites are prevalent in special populations; and vector-borne parasitic diseases feature natural foci. These epidemiological characteristics determine the scopes and levels of prevalence, making surveillance of parasitic diseases arduous and complicated. These characteristics also provide important scientific references for the determination of parasitological surveillance indicators.

Surveillance has always been performed throughout parasitic disease prevention and control. In the early stage of prevention and control, surveillance provides the basic information for determining the distribution of parasites and the prevalence of parasitic diseases. In the medium term, it offers important data for clarifying

X.-N. Zhou (✉)
National Institute for Parasitic Disease at Chinese Center for Disease Control and Prevention (Chinese Center for Tropical Diseases Research), Shanghai, China
e-mail: zhouxn@nipd.chinacdc.cn

X.-N. Zhou (ed.), *Surveillance and Management of Parasitic Diseases in China*,
https://doi.org/10.1007/978-981-97-1241-0_1

prevention and control measures, organizational priorities and resources. In the late stage, it provides important technical support for assessing the effectiveness of the national control program of parasitic diseases and consolidating the achievements from the program. Finally, when the parasitic disease is nearly eliminated or meets the criteria for elimination, surveillance can promptly provide technical support for detecting sources of residual infection or imported infection. The level of surveillance and the disease information obtained through monitoring directly influence the levels of parasitic disease prevention and control in a region and guide the formulation of control strategies to ensure the quality of strategy implementation. Therefore, surveillance of parasitic diseases is indispensable to grassroots prevention and control. Based on the epidemiological characteristics of local parasitic diseases, a program for the prevention and control of parasitic diseases in local regions should be well planned and organized, and primary health resources should be applied to continuously improve the capacity of surveillance. Prevention and control of parasitic diseases should be promoted by monitoring and evaluating the quality of the control program to provide scientific support for the elimination of parasitic diseases.

1.1 Brief Introduction

1.1.1 Basic Concepts

Surveillance refers to continuously and systematically collecting, analyzing, and interpreting information related to public health programs and assessments and using the data for timely disease prevention and control. Therefore, surveillance is crucial in the whole process of the disease control program.

The process of surveillance is summarized as collection, analysis, feedback, and utilization. The purpose of surveillance of parasitic diseases is to monitor the occurrence and development of diseases, analyze and provide feedback on valid information, and provide timely response and treatment. On the one hand, the results of surveillance data analysis can be used to understand the occurrence and epidemic or transmission patterns of parasitic diseases, such as determining the outbreak population group, the epidemic scope, and the spread to surrounding areas; on the other hand, the analysis results are fed back to the administrative department to directly guide prevention and treatment efforts, including the formulation of prevention and control plans, measure implementation, evaluation of the effectiveness of measures as the emergency response to outbreaks, assessment and inspection at different stages of the control program. Therefore, surveillance has always been necessary throughout the process of parasitic disease control program (Fig. 1.1).

In recent years, with the development of modern informatics technology, systematic surveillance data with long duration and statistical analysis or mathematical models can be used to predict the development trend of a disease and correctly predict and establish an early-warning mechanism to directly provide emergency response services. The early-warning mechanism is based on long-term

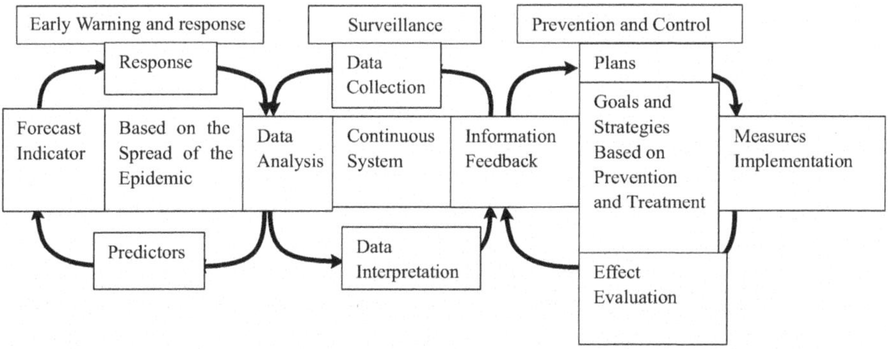

Fig. 1.1 Diagram of the relationships among functions, including prevention and control, early warning and response in surveillance of parasitic diseases

surveillance data and other basic information. It focuses on a series of epidemic indicators or a possible outbreak to predict in advance the risk level of the epidemic and the probability of occurrence as well as the epidemic trend on temporal and spatial levels to minimize the harm caused by the outbreak or occurrence of the epidemic. Based on the level of early warning, the response involves the local government utilizing resources in a timely manner according to contingency plans to control the epidemic spread and reduce the level of harm to ensure people's security. Surveillance, early warning, and response must be combined with a clear division of work and coherence in order to achieve the ideal effects in controlling the spread of parasitic diseases.

Therefore, the development and scientific application of surveillance techniques directly influence the surveillance quality as well as the effectiveness of the control program of parasitic diseases. It is also the basis for establishing an effective and rapid surveillance and early-warning mechanism. Only by establishing a scientific, efficient, and sensitive surveillance and early-warning response mechanism can we effectively prevent and control the spread of parasitic diseases to the maximum possible extent and minimize the incidence of parasitic diseases to protect the health of the population.

1.1.2 Development History

As early as the twentieth century, experts from several countries such as the United Kingdom, the United States, and Italy introduced the concept of surveillance and reported on notifiable infectious diseases. Systematic disease surveillance began in the Centers for Disease Control and Prevention of the United States in the late 1940s. Participants in the 21st World Health Assembly (WHA) in 1968 discussed national and international communicable disease surveillance. Many countries have carried out extensive surveillance since the 1970s to observe the epidemic situation of infectious diseases, then extended surveillance to chronic noncommunicable diseases and evaluated preventive measures and effects, and ultimately shifted from a purely biomedical approach to wider surveillance of creatures, psychology, and society.

Deaths and local population records in England and Wales in the sixteenth century were well documented. These demographic data were preserved and continued until the nineteenth century with continuous improvements of the statistics of lethal diseases and causes of death. These disease surveillance data mainly recorded the times and places of many diseases, such as acute diseases (at the early stage) and chronic diseases (at the late stage) and played a very important role in analyzing the impacts of the ecological environment and socioeconomic development on diseases. These analyses were employed to mobilize resources to promote local health and improve the health of the population. To stop maritime immigrants infected with cholera, chicken pox, plague, and yellow fever, the US Naval Hospital Service (the predecessor of the US Public Health Service) began registering and collecting the incidence of these diseases in 1878. The US Congress approved the weekly publication of these materials in 1893 and made the information publicly available until 1979. The US Congress passed a law in 1902 regarding reporting on a number of notifiable infectious diseases and announcing them in different forms. In 1961, the US Centers for Disease Control and Prevention was designated as a unit for collecting statistics on notifiable infectious diseases. To date, there are 60 kinds of notifiable diseases in the United States, and the Centers for Disease Control and Prevention of the United States are responsible for the statistical analysis of these diseases.

An integral public health information system for disease surveillance in China has gradually been formed, covering the period from the registration of the disease to standardized reporting. At present, China has established a system of complementarity between the disease surveillance system and the network of notifiable infectious diseases, covering 39 Class A, B, and C notifiable communicable diseases as well as 14 key diseases (including vectors) in 458 national-level sentinel sites in 31 provinces (cities and districts). This system helps explain national epidemic characteristics and patterns and develop scientific prevention and control measures, combining the routine reporting of notifiable infectious diseases and surveillance and the results of outbreak surveys with the use of epidemiological factor information acquired through the surveillance of vector-borne organisms, population immunization, etiology, and the environment. It compensates for the lack of surveillance of infectious diseases and emerging infectious diseases in addition to those that are notifiable and provides a strong support for the standardized management of disease prevention and control in China.

The development of disease surveillance has gone through three different stages: surveillance of diseases, epidemiology, and public health. Disease surveillance mainly includes systematic collection, sorting and analysis, feedback, and disease occurrence and development. In epidemiological surveillance, epidemic factors, including natural, social, and economic impacts, are considered in addition to disease surveillance. During the period of public health surveillance, health resources are covered in addition to the abovementioned disease and epidemiology surveillance. This is of great significance to the optimization of public health resources and capacity building to deal with sudden emergencies in a timely manner. Therefore, whether it is a matter of disease surveillance or public health, surveillance is an important foundation and source of information for the development, implementation, and evaluation of disease prevention and control strategies and measures.

With the global spread and prevalence of emerging infectious diseases and the impact of ecological and environmental factors, such as global climate change, on the transmission of infectious diseases, early warning and response to infectious diseases have received widespread attention in recent years. Corresponding surveillance and warning technologies have undergone increasingly rapid development. The WHO established an early-warning and response network in 2000 to guarantee that the global community can jointly resist the international transmission of infectious diseases and ensure that appropriate technical assistance reaches the affected countries promptly, which contributes to long-term epidemic prevention and capacity building.

1.2 Surveillance System

1.2.1 Components and Measures

1.2.1.1 Components of Surveillance

The control of diseases stems from the information generated by modern medicine's approach to the process of health and disease and depends on such feedback. Therefore, the surveillance of parasitic diseases includes four main basic processes: collection, collation and analysis, feedback, and information utilization:

1. Collection of information: Information comes mainly from routine surveillance or site surveillance. At present, the routine surveillance of parasitic diseases in China is still at the level of "control," that is, surveillance of infection sources, vectors (intermediate hosts), people in epidemic areas and epidemic stages, and impact factors in place of routine parasitic disease prevention and control measures, which consolidate the results of prevention efforts and prevent the recurrence of diseases. There are still some difficulties in conducting this work since the components of surveillance are varied and complicated. However, sentinel sites can compensate for the shortcomings of routine surveillance, such as a high false-positive rate and poor surveillance quality. These sites are low-cost, highly efficient, and serve an important purpose. According to the epidemiological characteristics of specific parasitic diseases and special requirements for prevention and control as well as the existing weak points of the system, representative surveillance sites can be chosen to serve certain specific purposes. For example, for malaria surveillance, sentinel sites can be established to collect epidemic information to correct malaria epidemic data reported by the National Public Health Information Coalition website, and surveillance sites can be set up for *P. falciparum* resistance in areas where falciparum malaria is endemic as well as for mosquito surveillance in typical regions. Moreover, an epidemiological sur-

vey of national parasitic diseases such as schistosomiasis is conducted every 10 years or so to enable the adjustment of the national plan for the prevention and control of parasitic diseases. Taking 1% of the population in an epidemic area as the sample and applying the stratified sampling method, we can calculate the national parasitic infection rate and quantitatively evaluate major epidemic factors and the economic benefits of prevention and control planning.

2. Collation and analysis of data: The collation and analysis of surveillance is an intermediate link in surveillance, which is important to ensure the quality and utilization of surveillance. Routine and sentinel surveillance together with thematic surveys must be conducted under the guidance of a unified science-based technological program that includes data collation and analysis after collection so that the collated and analyzed surveillance information supports the prevention and control of parasitic diseases. With the development of computer science and modern information technology, tools and software for organizing and analyzing all kinds of information have gradually been applied to surveillance, laying a foundation for improving the efficiency of these efforts.
3. Information feedback: The timeliness of data feedback is very important for the quality of parasitic disease control and revisions of prevention and control plans. Only when the data analysis results are fed back to decision-making departments (e.g., health administrative departments) on the front line of grassroots prevention and control or in a certain area can surveillance be effective in increasing the utilization of data. This is achieved mainly through an outbreak reporting system. At present, the epidemic reports of infectious diseases in China include the national important infectious disease reporting system, the public health emergency reporting system and various special disease reporting systems. These three types of reporting systems all cover different types of parasitic diseases. The parasitological reporting system started operation in 2010. These reporting systems can improve to ensure the reporting speed and correctness of information about an outbreak but cannot solve problems of omission, concealment, and false reporting. The ultimate solution to these problems involves strengthening the construction of the grassroots surveillance system to ensure the successful implementation of surveillance at the county, township, and village levels.
4. Employment of data: Employment of data should be fully used to promote the prevention and treatment of parasitic diseases. The data can be used in the following five main aspects:
   (a) Identifying the prevalence of parasitic diseases
   (b) Discovering new epidemic areas for parasitic diseases
   (c) Understanding the epidemic dynamics of parasitic diseases
   (d) Evaluating the effects of parasitic disease prevention and control strategies
   (e) Predicting and forecasting the epidemic trends of parasitic diseases

1.2.1.2 Surveillance Methods

After a long-term practice, parasitic disease surveillance refers to different surveillance techniques and their applications. The types of surveillance methods can be categorized as proactive or reactive surveillance: by surveillance period, as near-term or long-term surveillance; by surveillance object, as population-based, laboratory-based sample or hospital-based symptom surveillance; by surveillance form, as fixed or mobile surveillance sites; and by surveillance area, as surveillance in potential epidemic areas and imported infection sources (hosts). The surveillance methods adopted in the development of a parasitic disease surveillance program depend on the surveillance quality indicators, such as sensitivity, representativeness, and acceptability:

1. Proactive Surveillance and Reactive Surveillance: Proactive surveillance is the surveillance task performed by an official (or unit) on the surveillance objects at actively selected surveillance sites. It includes data collection, collation, and analysis, feedback, and utilization. Reactive surveillance is when an official (or unit) passively accepts the surveillance objects to carry out a survey. Therefore, the official is proactive, while the object is passive in proactive surveillance, and the official is passive, while the target is active in reactive surveillance. Both have advantages and disadvantages and can be used for different diseases or surveillance purposes. For example, proactive surveillance should be used to investigate epidemic false-negative rates because proactive surveillance is significantly better than reactive surveillance for such an investigation; a survey of malaria false negatives in 2000 showed that the actual incidences of malaria in Yunnan and Hainan Provinces were 18 times and 25 times higher than that of the reported rates, respectively. Therefore, the actual incidence of malaria can be determined only through proactive surveillance. Symptom surveillance is reactive surveillance. Such surveillance often focuses on patients with serious symptoms who must go to the hospital for diagnosis and treatment. For example, symptom surveillance of angiostrongyliasis is reactive. The disease is incurred by eating raw *Pomacea canaliculata* snails, but there is no peak incidence during the year, and few cases occur; therefore, reactive surveillance on the basis of symptoms can improve the case detection rate.
2. Near-Term and Long-Term Surveillance: The concept of near-term and long-term surveillance is often used to consolidate surveillance efforts to eliminate the risk of parasitic diseases. Generally, surveillance with a more recent starting point (often <10 years) is considered near-term surveillance, while surveillance over a longer period (>10 years) is called long-term surveillance and tends to occur in areas such as filariasis and schistosomiasis elimination. Post-disease surveillance depends largely on the sensitivity of the surveillance. In near-term surveillance, the main objective is to continue to eliminate sources of residual infection and to carry out surveillance aimed at preventing the importation of sources of infection. The strategy employed is mainly proactive surveillance supplemented by reactive surveillance. In long-term surveillance, the main objective is to carry out surveillance to prevent the importation of sources of infection. Therefore, the strategy used is reactive surveillance supplemented by proactive surveillance of the transmission factors.
3. Population Surveillance, Sample Surveillance, and Symptom Surveillance: According to different surveillance objects and methods, surveillance is divided into population surveillance, sample surveillance, and symptom surveillance.

Population surveillance, such as surveillance of the disease burdens of schistosomiasis and echinococcosis (B-ultrasonic examination), is carried out among large groups. In sample surveillance, such as surveillance of *Plasmodium* resistance and kala-azar, pathogenic and vector samples are collected in the field and brought to the laboratory for laboratory tests. Symptom surveillance, such as surveillance of angiostrongyliasis and malaria-fever patients, is based on hospital outpatient visits. For a particular parasitic disease, the three types of methods can be used interchangeably or simultaneously to serve different surveillance purposes and to improve quality indicators such as the sensitivity and acceptability of surveillance.

4. Fixed and Mobile Surveillance Sites: Both fixed and mobile surveillance sites are forms of sentinel surveillance. This method aims to compensate for deficiencies in the system of epidemic reporting and to improve surveillance sensitivity. Fixed surveillance sites may be set up for longitudinal and historical observation and mobile surveillance sites for lateral and current-status surveys. Fixed surveillance sites can systematically observe a parasitic infection in a fixed population over a long period, analyze the distribution and prevalence of diseases, propose influencing factors that affect the spread of diseases, and provide systematic and complete information for adjusting prevention and control strategies. However, fixed surveillance is often directly involved in the implementation of local prevention and control, which may affect the practical epidemic situation, lose the representativeness of surveillance, and cause problems of low positive predictive value and poor surveillance acceptability. These problems can be compensated by mobile surveillance sites. In particular, when using surveillance data to estimate the number of patients in a region, carefully adjust local prevention and control strategies, or observe the effectiveness of prevention and control, the two methods are often combined to ensure the effectiveness of scientific prevention and control. For example, when schistosomiasis in China was under the World Bank loan program (1992–2001) and filariasis was basically eradicated in the later stage of prevention and control, the combination of fixed and mobile surveillance sites was often used.
5. Surveillance in Potential Transmission Areas and Imported Infection Sources (Hosts): When disease control enters the phase of elimination or a particular risk factor arises, surveillance should be carried out in potential epidemic areas to prevent disease outbreaks or the importation of infection sources (hosts) to prevent the disease from spreading inward. For example, the potential spread of schistosomiasis as a result of global climate change or changes in the ecological environment of the Three Gorges has already been predicted. Therefore, routine surveillance of schistosomiasis in the predicted potential epidemic area is carried out to prevent the spread of the disease to nonendemic areas. The focus of malaria surveillance in the southern border area of Yunnan Province was caused by the imported cases. With the goal of eliminating malaria in China, surveillance of imported malaria cases could strengthen the prevention from the misdiagnosis of falciparum malaria cases and imported cases as a source of infection in local transmission.

1.2.2 System and Management

The overall objective of building the parasitological surveillance system is to control the epidemic and improve the capacity and effectiveness of parasitic disease prevention and control. A well-established surveillance system for parasitic diseases should have six subsystems and seven evaluation indicators (Fig. 1.2).

1.2.2.1 Surveillance System

1. Surveillance Network of Parasitic Diseases Based on County, Township, and Village Levels: At present, it is still difficult to establish a regular surveillance system for parasitic diseases in China. However, strengthening a representative surveillance site system is a low-cost, efficient, and feasible approach. According to information on epidemic or potential epidemic areas of various parasitic dis-

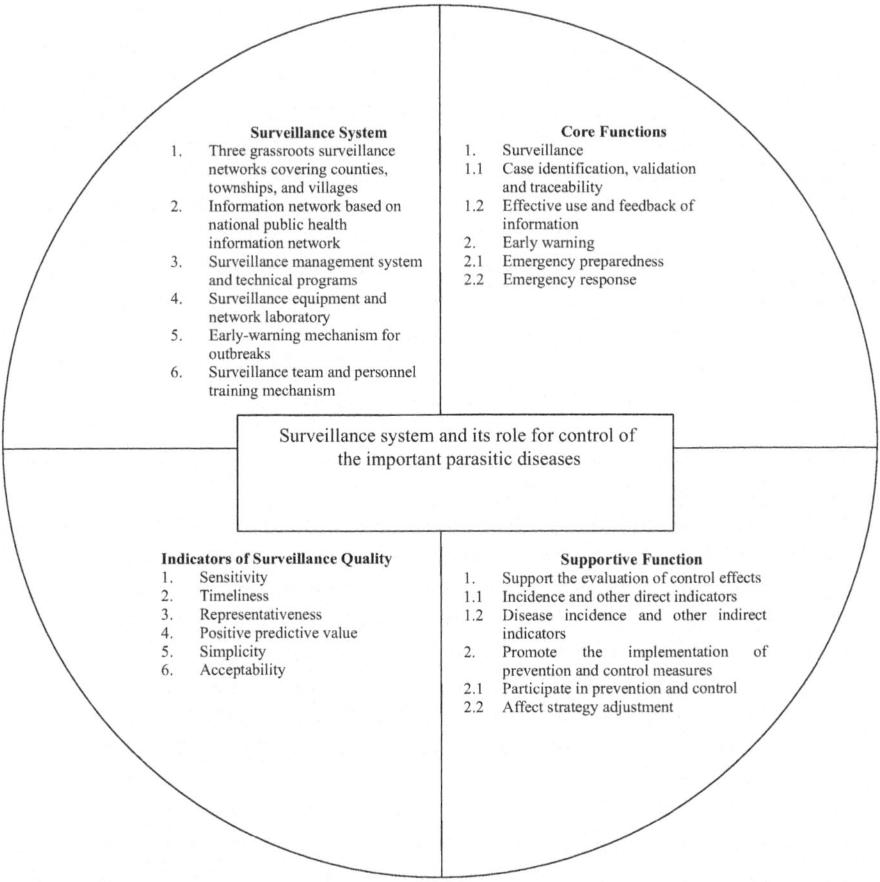

Fig. 1.2 Surveillance and early-warning system of parasitic diseases and framework of the working mechanism

eases, national or provincial surveillance sites have been set up in different counties, townships, and villages. Parasitic diseases that are being in the surveillance system include schistosomiasis, malaria, filariasis, soil-transmitted parasitic diseases, and foodborne parasitic diseases. Because of the different transmission characteristics of each parasitic disease, the ranges of surveillance sites also differ. According to the key task of surveillance, surveillance systems can be divided into three categories: population-based, laboratory-based, and hospital-based surveillance systems. The transmission of zoonotic parasitic diseases (e.g., schistosomiasis) is directly related to environmental and human factors. Therefore, surveillance sites should be set up within incorporated or unincorporated villages to establish a population-based surveillance system for epidemic analysis and data collection. Vector-borne parasitic diseases (such as malaria and filariasis) are transmitted by insects. The surveillance units are based on the flying range of insect vectors (in villages and towns), which reflects the law of disease transmission and is beneficial for the establishment of a laboratory-based surveillance system. Additionally, symptom surveillance centers have been established in county-level or township-level hospitals for foodborne parasitic diseases (e.g., angiostrongyliasis). When case information is registered, attention should be paid to details such as the village or street where a patient lives. A hospital-based surveillance system has been set up accordingly.

2. National-, Provincial-, and County-level Parasitic Disease Information Networks Based on the National Public Health Information Network: The successful operation of China's National Public Health Information Network guarantees quick access for key personnel to information on the epidemic status of parasitic diseases. However, due to the large number of parasitic diseases in China, some have not yet been included among the notifiable infectious diseases with statutorily required reporting. Therefore, on the basis of the National Public Health Information Network, a special information network on parasitic disease prevention and control has been set up to provide outbreak information, prevention and control information, resource collection, and analysis of different parasitic diseases and vectors (intermediate hosts).
3. Parasitic Disease Surveillance and Management System and Technical Protocols. The construction of a surveillance management system and technical protocols is one of the important components of software construction for the surveillance of parasitic diseases in China. Creating uniform surveillance standards and technical protocols and implementing them under a unified management system will help improve the level and efficiency of surveillance at all levels in China. This system construction is still a long-term task and should be adjusted in light of the pace of hardware construction for surveillance around the country.
4. Surveillance Equipment and Network Laboratory of Parasitic Diseases. The construction of surveillance equipment and a network laboratory is one of the important tasks of hardware construction in the surveillance of parasitic diseases in China. Surveillance equipment needs to be based on the needs of different parasitic diseases and the distribution of the diseases. The network laboratory will ensure the quality of routine surveillance; in particular, the diagnosis and differential diag-

nosis of certain rare parasitic diseases need to be popularized and applied through a network laboratory. This laboratory is currently under construction.

5. Early-Warning Mechanism for the Outbreak of Parasitic Diseases. An early-warning response system or mechanism for parasitic diseases refers generally to all factors that influence the occurrence and spread of diseases, including early identification, verification, evaluation, investigation, and control of those factors. Therefore, the complete system of parasitic disease prevention and control includes the management system, early-warning system, detection and processing system, and reserve system. The management system performs timely coordination of professional institutions of parasitic disease prevention and control throughout the country to carry out comprehensive prevention and control in the event of a sudden outbreak. The early-warning system analyzes and verifies all kinds of information, establishing a system of epidemiological analysis and forecasting and issuing a warning or alarm. The detection and processing system includes sample surveys, rapid laboratory tests, and field emergency response for the timely and accurate determination of pathogen and transmission routes and monitoring of epidemic development. Then, prevention and control measures can be rapidly formulated to effectively control the outbreak and spread of infectious diseases.

   The sampling survey quickly clarifies the causes of outbreaks and epidemics, including the determination of the scope of the pathogen and the outbreak and of the vectors and/or routes of transmission as well as follow-up surveys of imported cases. Laboratory rapid testing is implemented for some pathogens, including rare or imported parasitic pathogens, that need further qualitative and quantitative testing or identification through isolation. Emergency response refers to timely control of epidemic outbreaks and strict implementation of emergency treatment in epidemic areas, including patient diagnosis, isolation, and treatment; isolation of latent and suspected patients; disinfection of epidemic areas and harmless management of excrement as well as safe water supplies; the control of vectors; and the carrying out of public health campaigns. An early-warning mechanism for sudden outbreaks of major parasitic diseases and foodborne parasitic diseases in China has been set up, including contingency plans for emergencies, work programs for forecasting, standards for issuing early-warning information, technical standards for emergency disposal and technical operation plans, and early-warning exercise programs for emergency response surveillance.

6. Parasitic Disease Surveillance Team and Talent Cultivating Mechanism. This is an important part of China's surveillance of parasitic diseases and a significant component of the sustainable development of surveillance. In addition to holding seminars and training courses on parasitic disease surveillance in various places, it is necessary to train specialists in different levels of surveillance.

1.2.2.2 Evaluation Index of Surveillance System

1. Sensitivity: refers to the capacity of surveillance systems to identify public health problems. There are two main aspects: the proportion of cases reported by the surveillance system in the actual number of cases and the ability of this system to judge the outbreak or disease epidemic or other health events.

2. Timeliness: refers to the time taken by the surveillance system to identify public health issues and then report to the relevant departments. It reflects the speed of information feedback from the surveillance system.
3. Representativeness: refers to the extent to which public health problems discovered by the surveillance system represent the actual situation of the target population.
4. Positive predictive value: refers to the proportion of real cases reported by the surveillance system.
5. Simplicity: refers to whether the data collection, surveillance methods, and operation of the surveillance system are easily performed.
6. Flexibility: refers to whether the system can be changed or adjusted for new public health issues in a timely manner.
7. Acceptability: refers to staff willingness to participate in the surveillance in all aspects of the system and reflects whether they can provide effective information.

1.3 Classification and Application of Surveillance Techniques

1.3.1 Classification

In recent years, with the deepening of the prevention and control of parasitic diseases and the development of modern information technology, many surveillance techniques have developed rapidly.

According to the technologies utilized, surveillance techniques are classified as case surveillance diagnosis, identification of epidemic event, pathogen tracing, information collection, data statistical analysis, disease prediction and simulation, risk reporting system, or vector information screening technology. The surveillance components include information collection, analysis, reporting, and utilization. The surveillance objects include pathogens, reservoirs, epidemics, health resources, ecological environments, vectors and intermediate hosts, early-warning technology, and response technology. The epidemiological characteristics of diseases under surveillance include geographical distribution, population distribution, time distribution, epidemic factors, and health resources and analysis. Three means used in surveillance, including epidemiological survey, laboratory detection, and surveillance network of infectious diseases.

1.3.2 Application

As mentioned earlier, surveillance techniques are used mainly to identify the prevalence of parasitic diseases, identify new epidemic areas for parasitic diseases, understand prevalence patterns, evaluate the effectiveness of interventions, and predict the prevalence trends of parasitic diseases. The core function is to monitor information processing and provide early warning (Fig. 1.2). The former involves information gathering, collation, analysis, and feedback, which have become routine tasks; the latter involves early warning of disease trends, which is more difficult, and some technologies are still in the research stage.

In information-processing surveillance, information is collected mainly by modern network information transmission technology to combine the medical service information system with the public health surveillance information system, arrange the surveillance sites according to various needs and purposes, and periodically transfer the data of the sites to the Information Analysis Center in a timely manner. China has established a web-based reporting system for the infectious disease information network, which can perform timely transmission of epidemic information covering all towns and villages across the country within 24 h. Time-series data analysis has commonly been used in recent years to achieve the purpose of early warning, including controlling charts, the moving average method, the accumulation method, regression analysis, and spatiotemporal clustering detection. Research and development of disease prediction models has focused mainly on qualitative and quantitative mathematical models, such as the SIS and SIR models in decisive models, gray prediction models in mathematical models, Markov models, neural network models, Bayesian models, and path analysis models, which have all developed rapidly in recent years for the prediction and forecasting of infectious diseases. The risk reporting system is based mainly on infectious disease laws and regulations, the standard of the emergency plan of the Ministry of Health, and risk assessment and investigation as well as verification. The system reports by level and response to all types of emergencies and infectious disease outbreaks. Media information screening, including the classification and analysis of various media and network information and the establishment of special telephone lines for public complaints, whistleblowing, and health counseling (12,320 for public service), is still in its infancy. This system collects information on epidemics and outbreaks for screening and verification in accordance with the procedures prescribed by law to obtain timely information.

Due to the rapid socioeconomic development in China in recent years, people's health needs are constantly increasing, and emergency management needs more support from different early-warning technologies. Therefore, such technologies have developed rapidly. The issues that must be addressed include sharing epidemic information resources through network technology and providing relevant technologies for epidemic surveillance or early warning for emergency disposal sites in the emergency response and early-warning system; the provision of relevant technologies and products for emergency response to outbreaks, including emergency sampling techniques to rapidly determine the cause of the outbreak or epidemic (including the identification of the pathogen and the scope of the epidemic); and laboratory rapid testing to rapidly identify and detect pathogens and vectors (intermediate hosts) that are rare or infective parasites. Additionally, field emergency response technologies are needed for timely control of outbreaks in epidemic areas and strict implementation of emergency treatment, including the diagnosis, observation, and treatment of cases, the disinfection of epidemic areas, the safe disposal of feces, the safe water supply management, and the control of vectors. Under the premise of establishing the epidemic information system, we conducted a model analysis of an epidemic situation combined with various natural and social factors to analyze and integrate the information, comprehensively analyzed the characteristics and trends of the epidemic, and regularly provided the forecast data of the epidemic situation to relevant departments, providing a scientific basis for the formulation and revision of interventions.

1.4 Surveillance Role in Epidemiological Investigation and Control Programs

Surveillance of parasitic diseases has long played a very important role in epidemiological investigation and prevention and control programs. In the field of epidemiological investigation, the surveillance of filariasis in China through longitudinal surveillance revealed that when the population microfilaria rate dropped below 1% and the average density of microfilaria dropped to 5/60 μl, there was no transmission of lymphatic filariasis. This discovery provided the theoretical basis for the formulation of prevention and control strategies in the elimination phase and the development of strategies for eliminating lymphatic filariasis worldwide, thus perfecting the national surveillance system for filariasis and playing a significant role in validating the elimination of this disease. Regarding disease control programs, China has set up fixed surveillance sites and flow surveillance sites since 1992 in combination with World Bank-funded projects for the prevention and control of schistosomiasis. These sites directly well served the control program of schistosomiasis in China, guide prevention and treatment efforts, provide effective information for the midterm assessment and adjustment of the strategies, and offer a scientific basis to confirm the final project assessment indicators. Therefore, surveillance is directly related to the quality and process of parasitic disease prevention and control program in China and provides effective support for epidemiological studies.

Whether in epidemiological investigations or disease control program, the surveillance of parasitic diseases has two core functions. The first is to effectively guide prevention and control according to the information feedback provided by the surveillance data. This allows the prevention and treatment of parasitic disease to be correctly evaluated and guides the effective implementation of interventions through case finding, confirmation, and traceability and the effective use and feedback of epidemiology data. The second is to provide early warning of the potential public health impact of parasitic diseases according to the analysis of surveillance data and to forecast possible future epidemic tendencies of parasitic diseases. The main effects are reflected in emergency preparedness, such as forecasting and emergency response (Fig. 1.1). In its extended role, parasitological surveillance provides scientific data for prevention and control or project evaluation, helping to promote its implementation (Fig. 1.2).

1.4.1 Identification, Confirmation, and Traceability of Surveillance Cases

The discovery, confirmation, and traceability of cases are the basic functions of parasitological surveillance, laying a good foundation for the systematic collection of surveillance data by identifying or confirming each case. Due to variations in clinical symptoms and different influences on the transmission of parasitic diseases, surveillance cases are defined in various programs. For example, surveillance cases of angiostrongyliasis have clinical manifestations of meningoencephalitis and recent cases of transmission by eating mollusks (mainly freshwater snails and terrestrial snails) or catfish. Malaria surveillance cases include epidemics in district housing and chills, fever, sweating, and other clinical symptoms such as cyclical attacks, anemia, and splenomegaly and are diagnosed with etiological examination and serological tests. The schistosomiasis cases under surveillance involve people who were exposed to transmission by water, with symptoms of fever, diarrhea, hepatomegaly, and liver fibrosis-induced portal hypertension, and they are diagnosed by etiological examinations and serological tests.

For some severe or special cases, case surveys should be carried out to offer effective information confirming sources of infection, routes of transmission, or susceptible zones. For instance, a case survey should be conducted for acute schistosomiasis cases, as there is a need for acute schistosomiasis cases to be investigated. Surveillance of malaria requires case studies of patients with imported malaria, particularly when imported from abroad. If necessary, the origin of acute schistosomiasis or imported malaria cases should be traced in the identified infected areas to guide the implementation of control measures and prevent the occurrence of secondary cases or further spread of the disease.

Case surveillance is even more effective when case identification is combined with outbreak reports of notifiable infectious diseases. First, it verifies the data of the epidemic status report; second, it estimates the false-negative (false-positive) rate of the epidemic reports to provide a utilization rate for the epidemic report data; third, it makes further use of group phenomena in epidemic status data and carries out timely proactive surveillance to identify more cases and to prevent and control further occurrences of the disease.

1.4.2 Utilization and Feedback of Epidemiological Data

Making good use of data can achieve multiplier effects in parasitic disease control programs. However, the utilization value of data is not the same at different stages of parasitic diseases control programs. The value of surveillance data depends on the continuity and systematicness of the data at the surveillance sites, which is also the prerequisite and foundation for feeding back the right information to local policy-makers.

For example, disease surveillance was the main means of the schistosomiasis control program in China from 1950 to 1970, as surveillance and investigation of schistosomiasis and snails were conducted with a focus on schistosomiasis occurrences and cures (or deaths) of patients. This provided a reliable basis for treating patients and controlling the schistosomiasis epidemic. The distribution characteristics and transmission patterns of schistosomiasis prevalence in different regions of China and the ecological characteristics of snails were clarified by disease surveillance, providing a scientific basis for the formulation of targeted interventions and greatly accelerating the schistosomiasis control program in China. At the same time, the schistosomiasis control strategies of "put prevention first and comprehensively prevent and cure in accordance with local conditions" was put forward. When some provinces (autonomous regions and municipalities) or regions reached the standards for the blockade of transmission, China formulated a timely "Surveillance Plan in Transmission-block Areas." In view of potential epidemic factors in such areas, disease surveillance, including snail surveillance (local and imported transmission) and illness surveillance (human, livestock, and wildlife), was carried out. Once the epidemic was identified, it would be eradicated in time. It was important to eliminate residual snails and reduce sources of imported infections and transmission risk factors. In the period of 1980–2000, China set up 21 national-level longitudinal surveillance sites in eight provinces (cities) in endemic areas in line with the implementation of the World Bank Loan Project on Schistosomiasis Control. In addition to the disease indicators, the surveillance covered epidemic factors and other information. This was essential for understanding the distribution and prevalence of schistosomiasis in China and to grasp the nature of the epidemic, especially after floods, providing a scientific basis for the formulation of planning and decisions on national schistosomiasis prevention and control program. Additionally, a number of surveillance sites were established in China, including fixed (or longitudinal) and mobile (or lateral) surveillance sites, forming a surveillance network. In particular, a large number of mobile surveillance sites were set up, playing a role in correctly estimating the total number of patients and describing the trend of the schistosomiasis epidemic. This network provided a scientific basis for all endemic areas in China to carry out control measures according to local conditions and to adjust prevention and control strategies and interventions in a timely manner to enhance the prevention, treatment, and surveillance of schistosomiasis.

The Chinese government attached great importance to public health as the country entered the twenty-first century and schistosomiasis surveillance entered a new phase. In particular, a total of 80 surveillance sites in heavy epidemic areas were founded in China in combination with the implementation of the new "Law of the People's Republic of China on Prevention and Control of Infectious Diseases" and "Regulations on the Prevention and Treatment of Schistosomiasis" as well as a comprehensive prevention and control strategy based on the control of sources of infection. In 2014, a state-level surveillance was extended to 458 sites covering all types of endemic areas in China. Nationwide surveillance of schistosomiasis has been comprehensively carried out. It is necessary to adjust the epidemic situation in different areas in a timely manner and to effectively implement control measures according to different environments.

1.4.3 Forecasting and Emergency Preparation

With the development of early-warning technology, the value of surveillance is reflected in the prediction and forecasting of the occurrence and development trends of local parasitic diseases. Additionally, surveillance can directly guide the preparation of emergency technology reserves and emergency resources.

For example, prior to the survey of the national foci of angiostrongyliasis, the prediction model of angiostrongyliasis was established by using surveillance data from each province to select the correct sampling sites. The model predicts the distribution of potential foci, providing a scientific basis for the proper investigation of these foci and the collection of more accurate surveillance data. It also provides basic information for formulating contingency plans for angiostrongyliasis in high-risk areas.

In addition, based on the biologically driven model of schistosomiasis, researchers have predicted that climate warming may lead to a potential transmission area of schistosomiasis from northern Jiangsu Province to southern Shandong Province in the next 50 years. Therefore, a surveillance mechanism for schistosomiasis has been established in these potential transmission regions. Surveillance sites have been set up in high-risk water areas for system surveillance, and technical preparations are being made in advance to provide emergency treatment in new epidemic areas. In another example, according to the historical distribution of malaria, researchers have predicted that the southern border, the central Huang-Huai Plain, and the northeastern border of China will be the major high-risk areas for malaria under the climate warming scenario to guide surveillance together with the prevention and control of imported infection sources of malaria.

1.4.4 Emergency Response

Surveillance is the basis for conducting an effective emergency response to parasitic diseases as well as an effective measure for evaluating the effectiveness of emergency treatment. When the emergency response starts, it is necessary to first understand the nature and scope of the emergency event through the analysis of accumulated case surveillance data to guide the response. Only by comparing the morbidity trends before and after emergency can improve the effectiveness of emergency response.

For example, to identify sources of infection and foci of schistosomiasis in a timely and effective manner, the National Institute of Parasitic Diseases at Chinese Center for Disease Control and Prevention (NIPD) in 2009 cooperated with the Jiangsu Institute of Parasitic Diseases and Jiangxi Provincial Institute of Parasitic Diseases to carry out real-time surveillance in epidemic areas of the Yangtze River in Jiangsu Province and Poyang Lake in Jiangxi Province, respectively, employing surveillance data and special investigation materials. Early warning was issued in high-risk areas to guide local prevention and control efforts and ensure the pertinence and effectiveness of prevention and control measures of schistosomiasis control programs.

In another example, the number of malaria cases in northwestern Anhui Province increased anomalously in the summer of 2005. The data collected by epidemic reporting and surveillance sites showed that the incidence of malaria in this area significantly exceeded the average of the previous 3 years. The accumulated cases within 1 month by village significantly exceeded the stipulation of the Malaria Emergency Response Plan issued by the Ministry of Health in 2003. The transmission and outbreak in the epidemic villages was located in a clustered region. The NIPD promptly issued a timely emergency epidemic notification for the health administrative department of Anhui Province and dispatched expert team to the junction area of Anhui and Henan Provinces. The expert team investigated and verified the epidemic and formulated emergency response plans for the two provinces with prevention and control measures focusing on the management of the source of infection. After two consecutive seasons of effective intervention, the incidence of malaria in Anhui Province dropped significantly, and no outbreak or spread to surrounding areas occurred. This case shows that surveillance provided scientific information for timely early warning, correct intervention measures, and effective incident handling in response and promoted malaria prevention and control in Anhui Province.

1.4.5 Implementation of Control Programs

In another supporting role, surveillance promotes disease control program, thereby promoting the realization of prevention and control targets and reducing morbidity and mortality rates (Fig. 1.2). Parasitic disease control program cannot be separated from repeated checks and cures of diseases to reduce the disease burden of residents in endemic areas. However, a large amount of disease treatment data come into being each year at the same time. The collection, analysis, and utilization of these data lay the foundation for surveillance. Therefore, surveillance, prevention, and control mutually interact with and promote each other. The accumulated data can be used to monitor the discovery of cases and the evaluation of control program.

The pilot project of echinococcosis prevention and control was implemented in China in 2006, and the program has since been implemented in ten provinces. As echinococcosis prevention and control program started not so long time in China, the pilot prevention and control program was designed to carry out disease investigation and determine the scope and extent of the epidemic. The accumulated data provided a scientific basis for the formulation of echinococcosis prevention and treatment plans in 2010–2020 and promoted the implementation of national echinococcosis prevention and control program. In the same way, the surveillance of filariasis, focusing on identifying the epidemic range of Bancroftian and Malayan filariasis, provided a scientific basis for the prevention and control of filariasis in China. On this basis, the establishment of a complete surveillance system, including vertical surveillance, horizontal surveillance sites, and stricken population surveillance, provided important technical support for the evaluation of the effectiveness of filariasis prevention and control throughout China and played an important role in understanding the transmission patterns of filariasis and studying transmission thresholds. Therefore, filariasis surveillance has promoted the elimination of the disease in China.

1.4.6 Evaluation of Control Programs

Continuous and systematic surveillance can support the evaluation of the effectiveness of prevention and control programs or projects for parasitic diseases. On the one hand, surveillance data accumulated through history can be used as baseline data for implementing performance of the evaluation. On the other hand, the surveillance data obtained during the implementation can be directly used for effectiveness evaluation. Among the various evaluation indexes of the control program, the direct index is often the incidence of the disease, while the indirect index is often the influencing factors. Both direct and indirect indicators are inseparable from the data accumulated in the disease surveillance.

For example, during the implementation of the World Bank Loan Project on Schistosomiasis Control (1992–2001), midterm and final evaluations were carried out. In the assessment, the direct indicators were the incidence of schistosomiasis among humans and livestock and the scope and infection density of snails. The indirect indicators were people's correct cognition and behavior rates and the net investment benefit of the project. Data were collected before, during, and after the implementation of the project; then, these indicators were evaluated, and all indicators reached the set targets. However, according to the medium-term assessment, some strategies were changed, and environmental modifications were added in high-risk areas based on the original single chemotherapy interventions so that some indicators could be addressed in a timely manner. Therefore, timely application of surveillance data to the intervention assessment and feedback to decision-makers can directly affect the implementation of a plan and the realization of the expected goals.

1.5 Development Trends and Application Prospects

1.5.1 Development Trends

In recent years, surveillance and early warning of parasitic diseases in China have been increasingly promoted in research and application. On the one hand, public safety standards have been raised in China. The improvement of the standards has promoted surveillance and early warning of parasitic diseases. On the other hand, parasitic diseases vary, and some are prone to be overlooked because of the slow development of epidemics (such as clonorchiasis and echinococcosis); others have developed rapidly (such as malaria and Kala-azar), and emergency response is difficult, which aggravates the difficulty of control and management of infection sources. It is still difficult to achieve the expected targets solely by relying on government investment in disease prevention and control. Therefore, there is an urgent need to develop innovative surveillance and early-warning technology that are compatible with current parasitic disease prevention and control capabilities. For this reason, building sensitive surveillance networks and early-warning systems has become a priority in recent years for the control program of parasitic diseases in China.

The construction of a sensitive parasitic disease surveillance network and early-warning systems includes research on pathogen detection technology, the establishment of a surveillance network, and research on an early-warning index and early-warning models. The surveillance network for parasitic diseases in China includes 81 national schistosomiasis surveillance sites, 64 malaria epidemic surveillance sites, 10 soil-transmitted parasitic disease surveillance sites, 2 foodborne parasitic disease surveillance sites, 3 angiostrongyliasis surveillance provinces, and 10 echinococcosis surveillance provinces. Nearly 200 schistosomiasis surveillance stations are sampled at random throughout China every year. Sampling surveys of the prevalence of schistosomiasis are conducted every 10 years. Disease surveillance is carried out in combination with prevention and control measures and trials. The surveillance of epidemic situations at these sites and prevention and control provide the scientific basis for the calculation of the number of people stricken with parasitic diseases and the analysis of epidemic trends in China. Indicator research on parasitic disease warning is still in its infancy. Although the study of the filariasis transmission threshold has successfully led to the elimination of filariasis, other studies on parasitic disease transmission thresholds are still in the research stage. Research on early-warning models of parasitic diseases has developed a kinetic model of the distribution of insect vectors (intermediate host), a population infection rate and infection degree model, and a transmission dynamics model. In recent years, with the impact of climate change on the spread of tropical diseases, the climate change prediction model based on schistosome biology has undergone new development. It provides a theoretical basis for quantitatively describing the transmission mechanism of parasitic diseases (i.e., the basic reproduction rate) and the rate of transmission (i.e., the transmitted energy) to achieve the goal of controlling spread and ultimately blocking transmission and provides a new tool for evaluating prevention and control strategies and measures (including vaccine strategies, chemotherapy strategies, and infection source control strategies).

1.5.2 Application Prospects

With the development of modern biotechnology engineering, modern information science, and space epidemiology, the overall level of surveillance and early warning for schistosomiasis in China has been improved to some extent. However, compared with the advanced international level and the current demand for prevention and control, there is still a large gap. In particular, with the improvement of national public safety and public health requirements and the national goals of eliminating malaria by 2015 and eradicating schistosomiasis by 2025 in China, it is even more necessary to further improve the surveillance and early-warning system for parasitic diseases. In addition to early-warning response systems or warning response mechanisms for infectious diseases, three aspects need to be strengthened. Firstly, it is the detection and processing system, which includes the detection of genetic variations in pathogens and intermediate hosts, genetic tracing techniques for the spread of genes and vectors (intermediate hosts), the rapid detection of pathogens and early infections in the environment, and surveillance techniques for drug resistance. Secondly, it is the early-warning information release mechanism for parasitic diseases, including the establishment of a special parasitic disease prevention and control system and an early-warning platform for parasitic diseases that integrates surveillance information, early-warning indicators and models, and other technologies to analyze and release epidemic occurrence and development trends in a timely manner and accurately guide field prevention and control. Thirdly, it is the reserve system, including techniques, personnel, and material reserves. In terms of technical reserves, full use of innovative technology should be made of the support of major national infectious disease projects, national science and technology support projects, and major projects of the National Natural Science Foundation of China have to focus on practical and effective surveillance and early-warning techniques. In terms of the talent pool, it is necessary to nurture surveillance and early-warning personnel and carry out multilevel and multidisciplinary training to enable more professionals to conduct research and fieldwork. Regarding material reserves, China needs to reinforce the quality and standardization of serological diagnostic reagents for parasitic diseases, set up reference laboratories at all levels, and explore new anti-insect drugs and combination therapies (Fig. 1.3).

With the development of the prevention and treatment of parasitic diseases in China, the epidemic situation of parasitic diseases in more areas will be reduced and finally eliminated. For example, the action plan for eliminating malaria in China began in 2010. According to the global standard for eliminating schistosomiasis adopted by the latest World Health Assembly, China could achieve its goal of eliminating the disease by 2015 in the control and elimination plan for schistosomiasis. To this end, the Chinese Center for Disease Control and Prevention held the "First Forum on Surveillance and Response System of Elimination of Tropical Diseases" in June 2012. The forum brought together experts from around the world to discuss how to establish an effective surveillance response system and make it the most important intervention in the process of "eliminating" tropical diseases. Only by establishing an effective surveillance and response system can the most basic epidemic situation information be provided and the progress of control program be monitored to provide timely, effective, and reliable public health information and promptly adopt quick and accurate countermeasures for every case. At an early stage, each case will be identified, and every epidemic focus will be dealt with to ensure the achievement of goals and the consolidation of results. Surveillance has become a major intervention in disease elimination. Therefore, the tasks of surveillance and early warning of parasitic diseases in China will become increasingly important. They require a strong support of all parties and cooperation in various fields to establish a more thorough surveillance and response system to eliminate the harm of parasitic diseases.

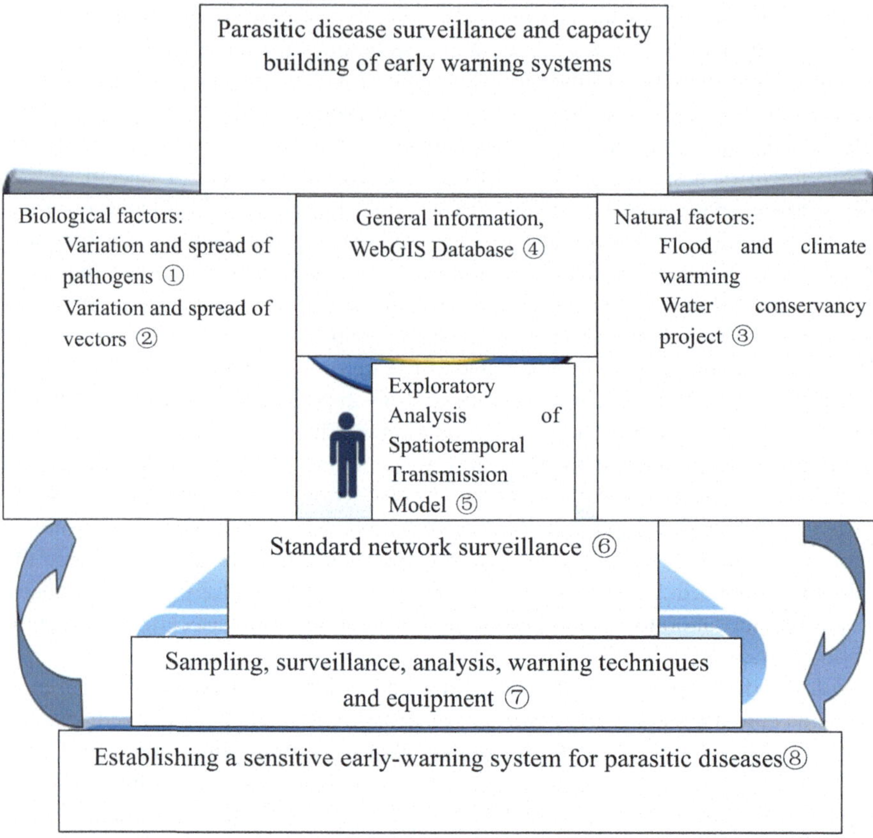

Fig. 1.3 Priority fields of parasitic disease surveillance and early-warning capacity building (*Note:* (1) etiological surveillance techniques; (2) surveillance techniques of vector and intermediate host; (3) surveillance techniques of ecological environmental factors; (4) Internet data publishing system based on large-scale integrated database; (5) analysis technology of disease spatiotemporal model; (6) construction of standardized network laboratory and surveillance techniques; (7) sampling, testing, and analysis techniques and modern technical equipment; and (8) sensitive early-warning system and technical reserve based on multitechnology compounding)

Pathogen Surveillance 2

Jing Xu

Many infectious diseases caused by different pathogens in different parts of the world have been effectively or basically controlled throughout human history, due mainly to effective vaccine application, improvement of sanitation conditions, infection control, food hygiene, and nutritional improvement. When people are overoptimistic about the control of infectious diseases, new and emerging infectious diseases reemerge. For example, schistosomiasis, which was well controlled in the twentieth century, has resurged in the Yangtze River Basin since 2000 due to natural factors such as environmental changes and floods. At the same time, parasitic diseases with slower transmission, such as *Clonorchis sinensis* infections and echinococcosis (*Echinococcus* disease), have risen from year to year, bringing the number of victims and corresponding disease burden to the highest level in the world. The spread of these diseases not only will undermine social and economic development in one area but also may affect the economic stability of the country and even the world. Therefore, to maintain public safety in a region, the surveillance of infectious diseases should be placed on the agenda. Pathogen surveillance is the most important element of the surveillance of parasitic diseases.

2.1 Brief Introduction

2.1.1 Definition

Pathogens of parasitic diseases are the parasites that can cause diseases in humans, including protozoa, helminths, and arthropods. When pathogens invade human or animal bodies, they continue to grow and reproduce and can then be excreted,

J. Xu (✉)
National Institute of Parasitic Diseases at Chinese Center for Disease Control and Prevention (Chinese Center for Tropical Diseases Research), Shanghai, China
e-mail: xujing@nipd.chinacdc.cn

X.-N. Zhou (ed.), *Surveillance and Management of Parasitic Diseases in China*,
https://doi.org/10.1007/978-981-97-1241-0_2

resulting in the transmission and spread of parasitic diseases. Therefore, the main role of pathogen surveillance is to detect and identify the types and distribution of pathogens as soon as possible to identify the route of transmission and provide a basis for effective interventions.

2.1.2 Surveillance Approaches

There are three stages from the invasion to the clearance of pathogens: the incubation, clinical symptoms, and recovery periods. Incubation refers to the period in which pathogens invade the body and the earliest clinical symptoms appear. Different parasitic diseases vary in the length of incubation, with some being as short as hours and some as long as several years, but the same disease generally has a relatively constant incubation period. Changes in incubation may be related to the number of pathogens entering the body, virulence, reproductive capacity, and body resistance. Clinical symptoms are disease-specific symptoms and signs. Because of the large number of pathogens in a patient's body during this period, clinical symptoms help the discharge and spread of pathogens, so transmission is strongest in this phase. Recovery is the phase in which the clinical symptoms have disappeared and the patient's body is gradually recovering. During this period, the patient's immunity begins to appear with the clearance of pathogens. However, for some parasitic diseases, pathogens may still be excreted during the recovery period and may even become a source of lifelong infection.

Pathogen surveillance is achieved mainly through the active search for or reactive discovery of sources of infection. The so-called source of infection refers to humans and animals with pathogens in the growth and reproduction stage that are able to discharge pathogens. Those who can excrete pathogens, regardless of whether their clinical symptoms are obvious or not, can become sources of infection. Patients with overt symptoms are easily diagnosed. They often require bed rest and can be easily treated and managed; therefore, their ability to spread pathogens is limited. Those with mild clinical symptoms, or even no symptoms of infection, can easily be overlooked as a source of infection because mild symptoms do not affect their routine activities. Practically, sources of light or even nonobvious infections cannot be ignored. Mild cases are often easily misdiagnosed or treated with delayed isolation. Insufficient cases generally require serological retrospective diagnosis or pathogen examination. This requires further inspection and clinical diagnosis, using all feasible methods to identify these patients to effectively control the source of infection.

This chapter focuses on surveillance techniques for the distribution of parasite pathogens and their growth, decline, transmission, and mutation.

2.2 Surveillance Techniques for the Distribution of Parasite Species

Patients excrete pathogens in the spread of infectious diseases, which is known as the infection period. The length of the infection period can affect the epidemic characteristics of the disease. For example, diseases with short-term infection often cause many secondary cases to appear in the short term, diseases with long-term infection often cause the succession of cases one after another, and the duration may be longer. In the surveillance of parasitic diseases, parasite distribution surveillance can directly grasp the distribution of parasites in a region and suggest control strategies in different regions. The main surveillance methods are as follows: first, the detection of human-pathogen carriers; second, the surveillance of free-living parasites (often infected larvae) in different environments; and third, monitoring the parasites in the vector or intermediate hosts. The most common method is the surveillance of pathogen carriers in the population because many of them may be free of any clinical symptoms but can nonetheless excrete pathogens. Patients in the incubation period who can excrete pathogens in vitro are known as latent pathogen carriers. Clinical symptoms disappear within a certain period after the discharge of pathogens, and these patients are called recovery pathogen carriers. People who have never suffered from an infectious disease but are able to excrete pathogens are called "healthy" pathogen carriers. Surveillance of pathogen carriers can directly reflect the status of parasitic infections in local populations and enable the disease burden of local parasitic diseases to be calculated.

2.2.1 Investigation of the Distribution of Human Parasites in China

China has carried out three nationwide sampling surveys of human parasitic diseases. This section takes the first survey as an example to explain the distribution of parasite species.

China conducted its first survey of the distribution of parasite species in 30 provinces (municipalities and autonomous regions) (excluding Hong Kong, Macao, and Taiwan) from 1988 to 1992. This was the most important surveillance of the distribution of parasite species in China. Here, we introduce the selection of sampling sites, the investigation of the distribution of intestinal helminths in the population, the statistical analysis of parasite species distribution, the standardized infection rate of intestinal helminths in the Chinese population, and the estimation of the number of people infected with intestinal helminths in China.

2.2.1.1 Selection of Sampling Sites

According to random sampling principles, the survey employed three-tier hierarchical cluster sampling. First, all provinces (municipalities and autonomous regions) were regarded as the main layer with three sublayers. According to landform or geographical orientation, the main tier was divided into several subareas as the first subtier. Then, each subarea was divided into upper, middle, and lower endemicity counties as the second subtier according to economic, cultural, and health indicators. Each county was divided into upper, middle, and lower endemicity townships as the third subtier in accordance with terrain and the above three indicators.

2.2.1.2 Investigation of the Distribution of Soil-Transmitted Helminths

According to the sampling principles, a total of 726 counties (cities and districts) and 2848 investigation sites were randomly selected from 30 provinces (municipalities and autonomous regions). The number of inspected persons was 1,574,219, with the number of actual inspected persons being 1,477,742, and the examination rate was 93.87%. Other eggs in addition to roundworm eggs had to be counted for in each smear through the modified Kato-smear method with an approach of one test, one sample.

2.2.1.3 Statistical Analysis of the Distribution of Parasitic Species

The survey data were entered in the unified database after the site investigation. After collation and verification of the data, the database of parasite species distribution in the whole country was compiled and used for further statistical analysis. The statistical analysis mainly focused on the distribution characteristics of soil-transmitted helminths (roundworms, whipworms, and hookworms) by different regions, sexes, and ages to grasp the key areas and populations of the distribution of parasitic diseases in China. It included infection rates in different regions, sexes, and age groups. In addition, analysis and comparison were performed according to different occupations, educational levels, and ethnic groups. The index of infectivity was the number of stool eggs per gram (EPG), and the average was the geometric mean (GMEPG).

To compare parasitic infection rates among different regions, the sex and age compositions of the national sample census were used as standards to normalize the infection rates and determine the standardization rates of parasitic infections in different regions. The infection rates and infectiousness of roundworms, whipworms, and hookworms in various regions after standardization are shown in Fig. 2.1 and Table 2.1. There was infection rates of STHs without no need to standardize by sex and age (Table 2.2 and Fig. 2.2).

2.2.1.4 Calculation of the Number of Soil-Transmitted Helminth Infections in China

The estimation of the number of infections of soil-transmitted helminths in China is based on the proportion of the population in one province (municipality and autonomous region) among all the provinces (municipalities and autonomous regions) being surveyed to calculate the weighted infection rate, which is then multiplied by the total population of the provinces (municipalities and autonomous regions) being surveyed to calculate the total number of people infected in the country.

The results showed 658,934 roundworm infections, with an average infection rate of 44.59% and a weighted infection rate of 47.00 ± 0.36%. An estimate of 531.31 million (523.34–539.29 million) people in China was found to have roundworm infections. A total of 256,772 people were identified as having whipworm infections, with an average infection rate of 17.38% and a weighted infection rate of 18.80 ± 0.33%. An estimate of 212.49 million (205.22 million to 219.75 million) people were found to have whipworm infections. A total of 213,953 hookworm infections were detected, with an average infection rate of 14.48% and a weighted infection rate of 17.17 ± 0.32%. The estimated number of hookworm infections in China was 194.05 million (186.91 million to 201.20 million).

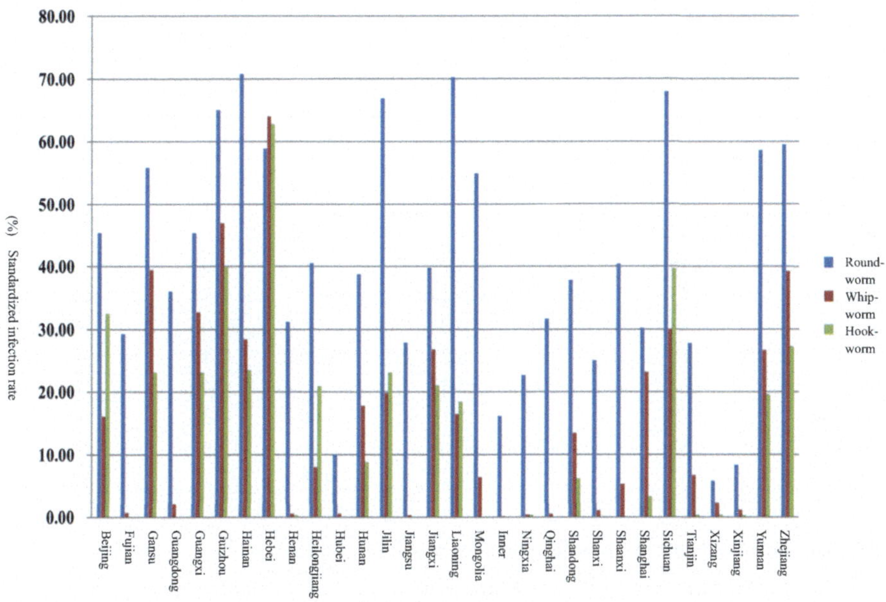

Fig. 2.1 Infection rates of soil-transmitted helminths in different areas

Table 2.1 Infection rates and infectiousness of soil-transmitted helminths in different areas

Province (municipality, autonomous region)	Roundworm		Whipworm		Hookworm	
	Standardized infection rate (%)	GMEPG	Standardized infection rate (%)	GMEPG	Standardized infection rate (%)	GMEPG
Anhui	45.39	1608	16.12	96	32.54	216
Beijing	29.29	408	0.75	48	0.00	0
Fujian	55.84	2064	39.47	192	23.12	168
Gansu	36.08	1224	2.06	72	0.01	192
Guangdong	45.36	720	32.76	144	23.10	192
Guangxi	65.01	3096	46.98	144	39.96	216
Guizhou	70.83	3840	28.46	120	23.49	120
Hainan	58.92	5280	64.02	480	62.79	312
Hebei	31.22	1200	0.58	72	0.31	120
Henan	40.48	1464	8.06	72	20.90	144
Heilongjiang	10.03	408	0.59	120	0.00	0
Hubei	38.72	2184	17.80	144	8.87	144
Hunan	66.90	2016	19.81	96	23.10	144
Jilin	27.88	1128	0.36	144	0.00	0
Jiangsu	39.72	1248	26.77	144	21.04	192
Jiangxi	70.26	2016	16.47	120	18.44	168
Liaoning	54.94	1896	6.44	168	0.03	48
Inner Mongolia	16.18	792	0.20	48	0.07	96
Ningxia	22.66	1152	0.43	72	0.37	72

(continued)

Table 2.1 (continued)

Province (municipality, autonomous region)	Roundworm		Whipworm		Hookworm	
	Standardized infection rate (%)	GMEPG	Standardized infection rate (%)	GMEPG	Standardized infection rate (%)	GMEPG
Qinghai	31.72	1320	0.55	72	0.00	0
Shandong	37.84	1224	13.43	120	6.19	96
Shanxi	24.99	816	1.03	72	0.03	96
Shaanxi	40.36	1056	5.27	96	0.01	432
Shanghai	30.23	1248	23.14	120	3.31	144
Sichuan	67.99	2088	30.04	96	39.65	192
Tianjin	27.78	408	6.71	96	0.40	48
Tibet	5.77	1008	2.37	360	0.40	168
Xinjiang	8.38	984	1.16	96	0.32	144
Yunnan	58.63	2472	26.68	120	19.61	168
Zhejiang	59.52	2136	39.25	168	27.25	144

Table 2.2 Infection rates of soil-transmitted helminths by sex and age

	Roundworm		Whipworm		Hookworm	
Age	Male	Female	Male	Female	Male	Female
0~	40.41	41.64	8.72	8.89	4.49	4.36
5~	50.81	50.48	20.58	20.38	8.17	7.65
10~	50.57	50.90	22.85	22.53	11.40	10.79
15~	47.16	48.53	19.62	20.30	14.54	15.76
20~	43.66	46.40	16.65	18.69	17.05	18.85
25~	41.07	45.04	15.20	16.91	15.61	16.85
30~	40.17	44.85	15.10	17.34	15.99	16.76
35~	39.76	44.66	14.99	17.25	16.87	17.77
40~	39.54	44.63	15.45	17.84	18.38	19.04
45~	39.94	44.85	15.72	18.18	19.44	20.44
50~	38.70	44.20	15.54	18.50	19.99	20.09
55~	38.16	43.78	15.67	18.21	19.66	19.65
60~	36.56	42.86	14.87	17.47	19.18	18.65
65~	34.70	40.97	14.54	16.99	17.83	16.64
70~	33.21	39.47	14.42	16.84	16.10	14.49
75~	32.80	38.83	14.55	18.14	14.83	13.02
80~	30.58	36.41	14.36	16.48	12.78	12.66

Through this survey, we obtained comprehensive knowledge of the prevalence of three major soil-transmitted helminth diseases (roundworms, whipworms, and hookworms) in China in the early 1990s, which provided a basis for formulating corresponding prevention and control strategies across the country and in all regions. The survey showed that three main endemic helminth diseases in China were serious epidemics with high infection rates at that time. Roundworms and whipworms showed a national distribution. Hookworms were distributed in 26 provinces (cities

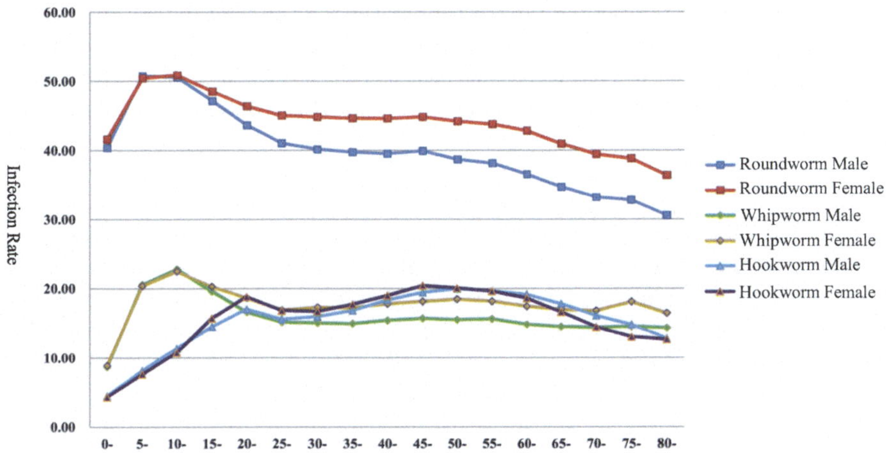

Fig. 2.2 Infection rates of soil-transmitted helminths by sex and age

and autonomous regions) other than Beijing, Jilin, Heilongjiang, and Qinghai Provinces, especially in the southeast, southwest, and central regions. There were differences in the prevalence levels of the three major endemic helminthiases in different regions. For example, the prevalence of roundworms was the highest in most areas, followed by whipworms and hookworms. However, the infection levels of whipworms and hookworms were similar in some areas, while in other areas, all three were basically the same. As the main parasites in different regions differed, prevention and control were targeted in all areas. An analysis of infection by sex and age found that the infection rates of roundworms, whipworms, and hookworms were similar by sex in the lower age group (0–15 years). However, there were differences in infection by sex among those over 15 years old. The infection rates of roundworms and whipworms were higher in females than in males, while that of hookworms also fluctuated between males and females without distinct differences. The roundworm infection rate was highest in those 5–15 years old and then gradually decreased with age. The whipworm infection rate was highest in those 5–15 years old and then decreased and was maintained at a relatively stable level. Hookworm infection levels peaked around the age of 20 and then remained at this level. The results showed that roundworms and whipworms were mainly in the younger group, and females were more affected in the older group. Hookworms dominated the older group, with no obvious sex difference.

2.2.2 Surveillance of the Distribution of Parasites in Special Populations

Monitoring the distribution of parasitic species in some high-risk special populations is conducive to understanding infection pathways and influencing factors in high-risk groups and providing preventive measures to prevent the advance of the disease. In particular, some opportunistic parasites do not develop after infecting the general population. However, for people who are infected with a basic disease that leads to low immunity (such as cancer patients and HIV/AIDS patients), a parasitic infection can cause the deterioration of the original underlying disease and even death.

In some rural areas in central China, there have been more cases of HIV/AIDS due to illegal blood collection and other reasons. Most of these people have poor family economic conditions and live in rural areas with economic backwardness, poor hygiene conditions, weak health service facilities, and a relatively high prevalence of intestinal parasitic diseases. The surveillance of parasitic infections in HIV-infected individuals in these areas can provide a basis for the prevention and of co-infections of HIV/AIDS and parasitic diseases in China.

2.2.2.1 The Choice of a Special Population

An HIV/AIDS treatment site in northern Anhui Province was selected as a surveillance site. The site was mainly for the treatment of HIV/AIDS patients in 21 townships and 126 natural villages. The number of registered patients was 324. HIV-positive cases were detected, and positive parasitic diseases diagnosed by local or provincial health authorities. To correctly grasp the path and influencing factors of HIV/AIDS patients infected with parasites, this surveillance was conducted according to a case–control study. A total of 315 HIV-negative controls were recruited locally using a 1:1 match. The criteria were patients between 6 and 65 years old who were willing to provide written informed consent. Participation in the research was strictly confidential.

2.2.2.2 Surveillance Components and Methods

A face-to-face questionnaire was conducted with the subjects aged 6–65 years. Blood and stool samples were collected at the same time. Ten milliliters of venous blood samples and two stool samples were collected from each study subject. The excrement was coated with three slides by Kato-Katz to be microscopically examined at the on-site laboratory. The remaining fecal samples were stored and brought back to the central laboratory for the detection of other parasitic species. The on-site investigation process is shown in Fig. 2.3.

Laboratory tests included HIV screening and detection of CD4+/CD8+ T lymphocytes, cytokines, hemoglobin, and parasite species. The parasite species detection method is shown in Fig. 2.4.

2.2.2.3 Statistical Analysis of Data

A total of 624 people were investigated, of whom 309 were HIV-positive and 315 were HIV-negative (see Fig. 2.5 for the surveillance results). All survey data were entered into the database using EpiData 3.1. Student's *t* test was used for the two-sample mean. The $\chi 2$ test and logistic regression model were used for univariate analysis and multivariate statistical analysis, respectively. The relationship among

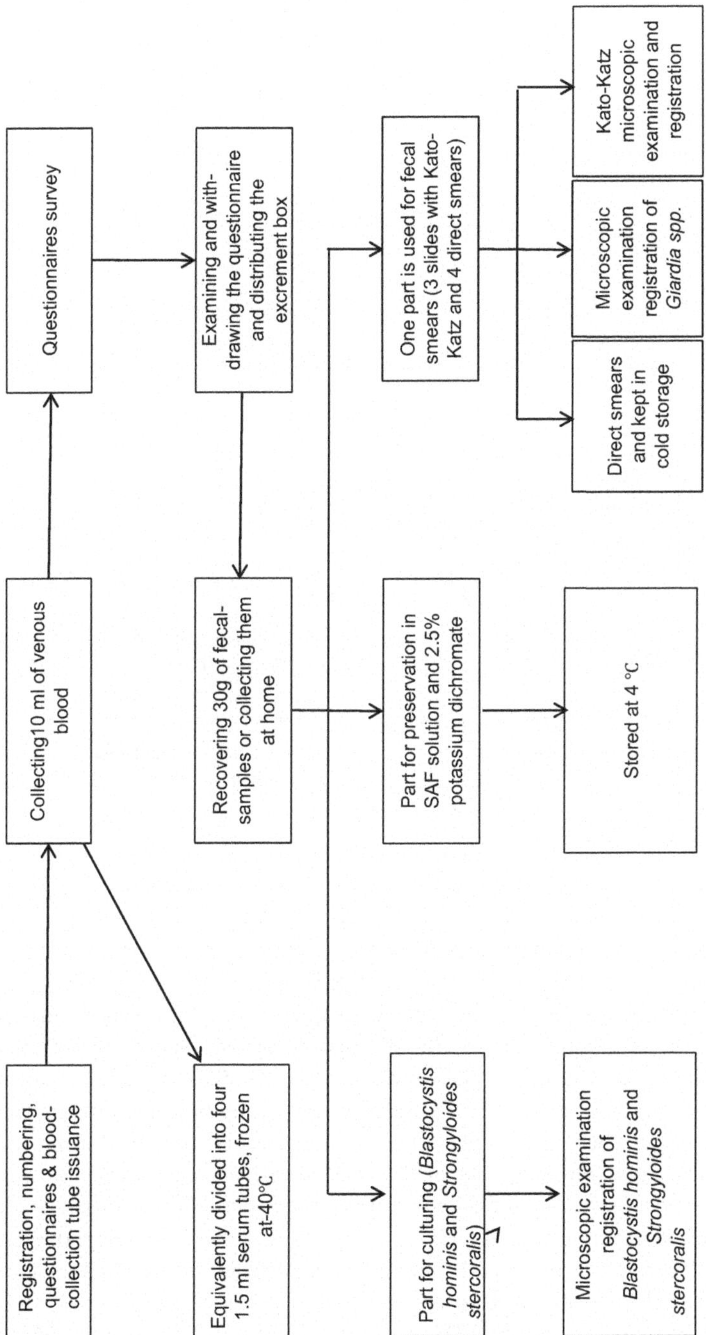

Fig. 2.3 Flow of parasite species in HIV-infected people

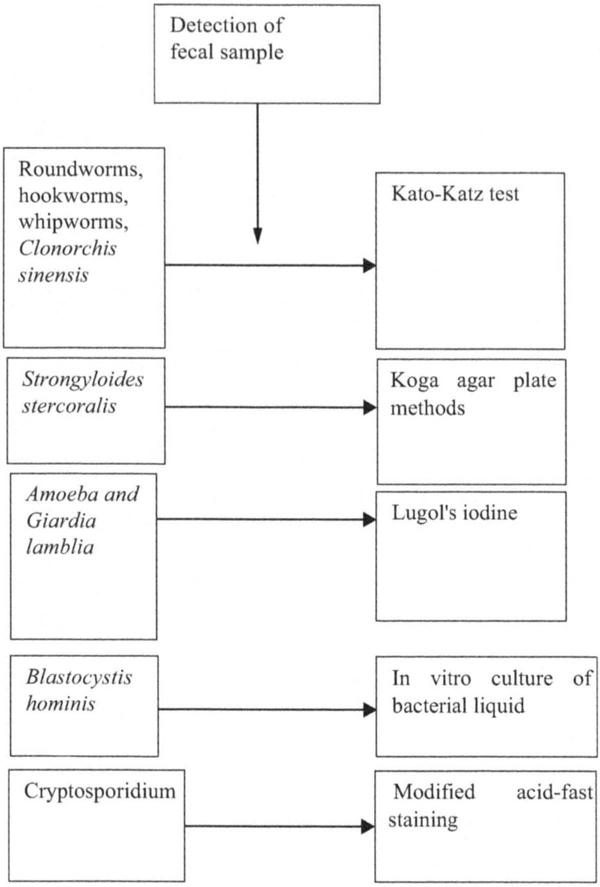

Fig. 2.4 Schematic diagram of parasite species detection in fecal samples

social demography, behavioristics, immune factors, and parasite infection rates was analyzed.

Intestinal parasitic infections were investigated to observe intestinal helminth and protozoa infections. Among the 605 people whose fecal samples were tested, the intestinal helminth infection rate was 4.79% (29/605), and the highest infection was hookworm, at 3.97% (24/605), followed by whipworm and clonorchiasis, both at 0.5% (3/605). The hookworm infection rate in the HIV-positive group was 3.64% (11/302), and the infection rate in the HIV-negative group was 4.29% (13/303). There was no significant difference between the two groups in the $\chi 2$ test. The roundworm infection rate in the HIV-positive group was 0. The negative group was 0.66% (2/303). Fisher's exact test showed no significant difference between the two groups. The whipworm and *Clonorchis sinensis* infection rates in the HIV-positive and HIV-negative groups were 0.66% and 0.33% and 0.33% and 0.66%, respectively. The differences were not significant. *Strongyloides stercoralis* was not

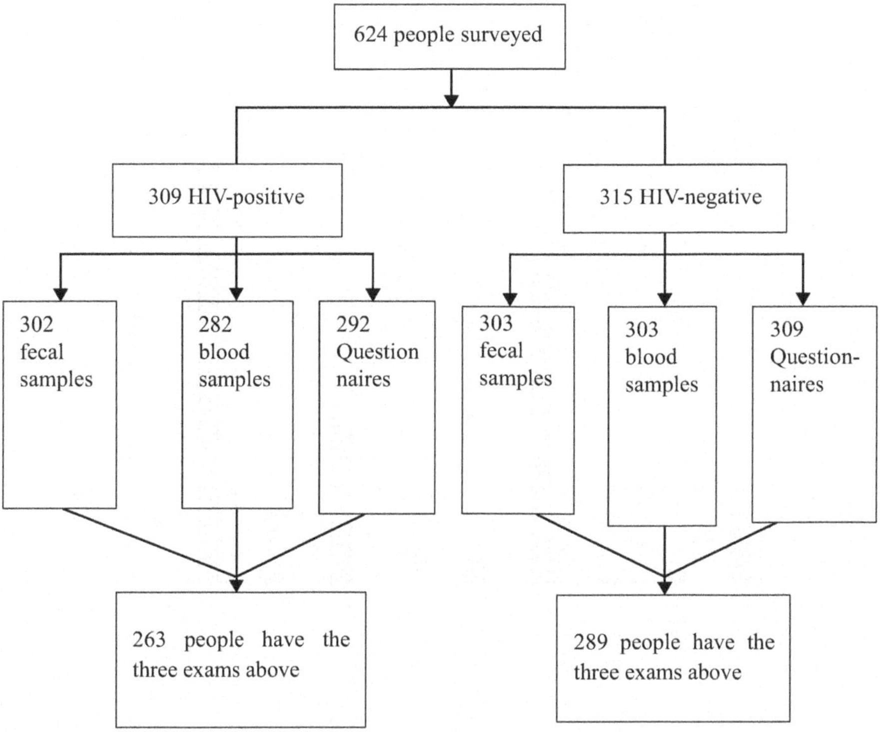

Fig. 2.5 Grouping of case–control surveillance results

detected in either the HIV-positive or HIV-negative group. The intestinal helminth infection rates in the HIV-positive group and the HIV-negative control group were 4.30% (13/302) and 5.61% (17/303), respectively. There was no statistically significant difference between the two tests (Fig. 2.6).

The intestinal protozoa infection rate was 24.13% (146/605). The highest infection was *Blastocystis hominis* (19.17%), followed by *Cryptosporidium* (5.62%); the HIV-positive group had an infection rate of 16.23%, whereas the HIV-negative group had an infection rate of 22.11%. There was no significant difference between the two groups. The *Cryptosporidium* infection rates in the HIV-positive and HIV-negative groups were 8.28% and 2.97%, respectively. There were significant differences between the two groups according to the $\chi 2$ test. The amoebic protozoa infection rates in the HIV-positive and HIV-negative groups were 1.66% and 0.66%, respectively, and the difference was not statistically significant. The *Giardia* infection rate was 1.32% in the HIV-positive group and 0.66% in the HIV-negative group. The difference was not statistically significant (Fig. 2.6).

Among the 605 people surveyed, the *Cryptosporidium* infection rate was significantly higher in men (7.67%) than in women (3.77%), and the difference was statistically significant. In HIV-positive people, the *Cryptosporidium* infection rate was higher in men (12.59%) than in women (4.40%), while the *Blastocystis*

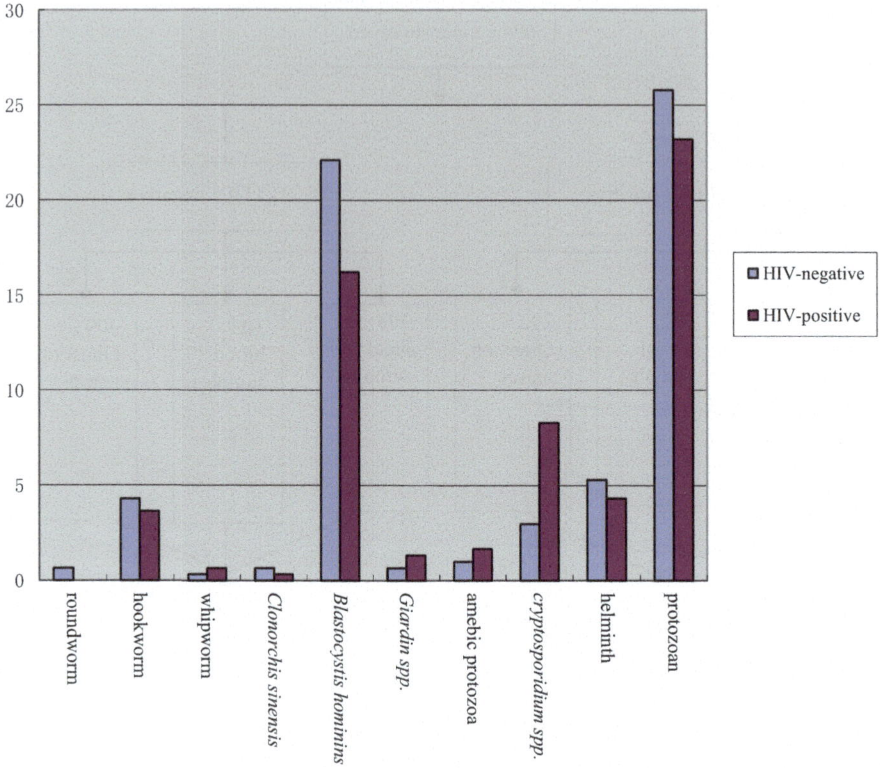

Fig. 2.6 Infection of different parasite species in HIV-positive and HIV-negative populations

Table 2.3 Comparison of parasitic infection rates among HIV-positive and HIV-negative people by different sexes [% (n)]

	HIV-positive				HIV-negative			
Parasitic species	Male	Female	χ^2	*P*	Male	Female	χ^2	*P*
Helminth (N)	2.80 (4)	5.66 (9)	1.50	0.22	5.56 (8)	5.03 (8)	0.04	0.84
Blastocystis hominis (N)	11.89 (17)	20.13 (32)	3.76	0.52	25.00 (36)	19.50 (31)	1.32	0.25
Cryptosporidium (N)	12.59 (18)	4.40 (7)	6.62	0.01	2.78 (4)	3.14 (5)		1.00

hominis infection rate was lower in men (11.89%) than in women (20.13%). In the HIV-negative population, the infection rates of helminths, *Blastocystis hominis*, and *Cryptosporidium* spp. in males and females did not differ significantly (Table 2.3).

The surveillance results revealed that the infection rates of eight kinds of intestinal parasites, roundworms, hookworms, whipworms, *Clonorchis sinensis*, *Blastocystis hominis*, *Giardia lamblia*, amebic protozoa, and *Cryptosporidium spp.*,

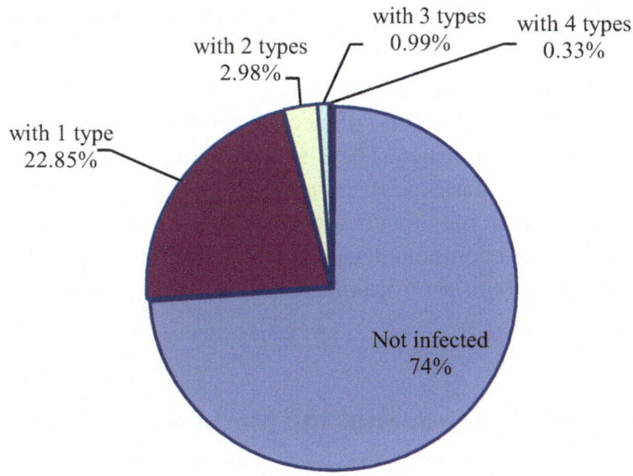

Fig. 2.7 Distribution of parasitic species infections in HIV-positive individuals

among the 302 HIV-positive individuals were 0, 3.64% (11/302), 0.66% (2/302), 0.33% (1/302), 16.23% (49/302), 1.32% (4/302), 1.66% (5/302), and 8.28% (25/302), respectively. The total intestinal parasite infection rate was 26.16% (79/302), of which the intestinal helminths infection rate was 4.3% (13/302) and the intestinal protozoa infection rate was 23.2% (70/302). There were 66 individuals (21.85%) infected with only one type of parasite, 9 (2.98%) infected with two types of parasites, and 3 (0.99%) infected with three types of parasites. One person was infected with four parasites, accounting for 0.33%. No one was infected with five or more parasites at the same time (Fig. 2.7).

2.2.2.4 Surveillance Conclusions and Recommendations

A total of 605 fecal samples were tested. The intestinal helminths infection rate in HIV-positive individuals was 4.30%. There was no significant difference in the intestinal helminths infection rate between the HIV-positive and HIV-negative groups. The intestinal protozoa infection rate among HIV-positive individuals was 23.20%, and the parasite with the highest infection rate was *Blastocystis hominis* (16.23%), followed by *Cryptosporidium* (8.28%). The latter rate was significantly higher than that in the HIV-negative group (2.97%), suggesting that there is an increased chance of *Cryptosporidium* infection among HIV patients. Therefore, surveillance of *Cryptosporidium* infection in HIV patients should be strengthened. Multivariate logistic regression analysis of HIV-associated cryptosporidiosis-infected individuals revealed that males (OR = 6.700), HIV-infected individuals aged 42 years or younger (OR = 4.148), and individuals with poor personal hygiene habits (OR = 0.324) were more likely to be infected with *Cryptosporidium*. It is recommended that these types of people be targeted for the prevention and control of HIV-associated intestinal parasitic infections.

2.3 Surveillance Techniques for Specific Parasite Species

Surveillance of parasite species and their growth is used mainly for vector-borne parasitic diseases, including the surveillance of the distribution of vectors and the infections of vectors with parasites. These surveillance data can be used to predict the spread of vector-borne parasitic diseases and evaluate the effectiveness of control measures. Due to their specific characteristics, different vector-borne diseases have different surveillance components and steps. In this section, the surveillance of the categories and distribution of Kala-azar pathogens in an area is taken as an example.

2.3.1 Surveillance Components and Methods

The surveillance of Kala-azar in an area includes (1) surveillance of species and infection rates of *Leishmania* spp. in the population; (2) surveillance of species and infection rates of *Leishmania donovani* in dogs; (3) surveillance of sandfly species population density, and *Leishmania* infection; (4) investigation of growth and decline factors; and (5) analysis of parasite species and their influencing factors.

2.3.1.1 Investigation of *Leishmania* Infection in Human Beings

1. Extraction of *Leishmania brachial*. Puncture was performed in patients with Kala-azar found at the surveillance site, and bone marrow was cultured in Novy-MacNeal-Nicolle medium.
2. Extraction of protozoal DNA. The culture of *Leishmania* was expanded to collect protozoa and extract DNA.
3. Protozoan sequence amplification. The full-length small-loop DNA sequence of *Leishmania* was amplified using primers (KP1:5′-GGGGTTGGTGTAAAATAGG-3′, KP2:5′-TGAACGGGGTTTCTGTAT-3′) and sequenced.
4. Determination of parasite species. The sequences obtained and *Leishmania* sequences known to identify the species of the obtained *Leishmania* were compared.

2.3.1.2 Investigation of *Leishmania* Infection Rate in the Population

1. Selecting surveillance objects. People of different ages and sexes (residents from May to September) were selected randomly as surveillance targets at the surveillance sites. Younger people could be selected depending on the changes in the population each year.
2. Sampling. Venous blood was collected from the surveillance objects every April, and serum was prepared.
3. Serological tests. One serological test method (enzyme labeling, fluorescent antibody, or strip test) was employed to detect serum antibodies. Once the test method was selected, it would be used every year.

2.3.1.3 Survey of Infection Rates of *Leishmania canis*

In the human-canine communicable Kala-azar epidemic area, all dogs at the surveillance sites were examined by etiological methods (bone marrow puncture smears) every year in April, and venous blood was taken for serum detection to evaluate the infection rate of *Leishmania canis*.

2.3.1.4 Investigation of Species, Growth, and Infection of *Leishmania* in Different Seasons

1. Investigation of Species of Sandflies and Their Growth/Decrease. A survey of sandfly density and species was conducted every half month at the surveillance sites from the beginning of June to the beginning of October every year. For wild or near-wild sandflies, more than five outdoor locations with representative habitats were selected to use human trapping methods to catch sandflies at night. Collection was performed for 1 h per person at a time, and density was calculated and sandfly species identified by "manual hours." Sandflies in homes were collected in human rooms and animal houses in the same way, the density of sandflies was counted, and the species were identified.
2. Investigation of Sandfly Infection. After the captured sandflies were identified, the female sandflies were placed in 70% ethanol and stored at room temperature for later use. After the preserved sandflies were ground and treated with proteinase K, DNA was extracted using a DNA extraction kit, and ITS1 and kDNA minicircle DNA sequences of *Leishmania* were amplified by PCR using specific primers. The amplified product was subjected to agarose electrophoresis, and the positive amplified products were sequenced. The obtained sequences were then aligned to determine the species of *Leishmania* in infected sandflies.
3. Impact Factors of Sandflies Populationg. Data such as temperature, precipitation, air humidity, soil moisture, and wind speed were collected and measured in each month in the monitored area (including the year in which the surveillance was initiated and the last year).

2.3.1.5 Analysis of Parasite Species, and Influencing Factors

Based on the data obtained from continuous surveillance over several years, relevant statistical analysis was conducted. The statistical analysis of the *Leishmania* infection rate in the population and dogs was performed based on mapping (often using the infection rate change curve in different years), which can directly reflect changes in the infection rate in each surveillance year, and annual infection rates were compared with the $\chi 2$ test. At the same time, the surveillance time was plotted with each kind of density to visually reflect the changes of each sandfly population. By comparing the proportion of *Leishmania* species in the population, different species of *Leishmania* infecting sandflies, and their proportion, the dominant vector at the surveillance sites and the major *Leishmania* pathogen could be determined. Regarding the analysis of transmission factors, linear correlation analysis of the relationship between temperature, precipitation, air humidity, soil moisture, wind speed, and infection rates in humans and dogs as well as vector densities was performed with relevant software.

2.3.2 Surveillance Reports and Utilization

The county-level professional institution that undertakes the surveillance is responsible for the collection and management of all data at the surveillance site and the identification of dedicated personnel. The original data must be uploaded to the computer independently for two times, and then a database must be created. The surveillance site database and analysis report are reported to provincial CDCs before the end of January of the following year. The provincial CDCs submit the database and surveillance reports to the NIPD before the end of February of the following year. The NIPD gathers national data and writes an annual epidemic surveillance report, reporting to the China CDC. The China CDC deliver or disseminate the annual surveillance results to the provinces (districts) in a timely manner.

Based on the surveillance and analysis of the species and distribution of pathogens of Kala-azar in a certain area, the report can provide timely information on the prevalence, epidemic dynamics, epidemiology, and trends of the monitored diseases and their effects, thus providing a scientific basis for formulating and evaluating prevention strategies and measures for monitored diseases.

2.4 Surveillance Techniques for Parasite Transmission

This section describes the role of farm cattle in a selected region in the transmission of schistosomiasis as an example.

Schistosoma japonica is a zoonotic parasitic disease, and cattle are the main source of animal infection. Studies have shown that the number of schistosomiasis eggs is much larger in infected cattle than in humans, and schistosomiasis survives much longer in cow dung than in human excrement, up to 140 days. The role of cattle in the transmission of schistosomiasis is determined by the population, the number of schistosomiasis eggs excreted by the body, the degree of contamination of the snail, and the degree of density associated with human activities. To effectively control the infection of livestock, especially cattle, it is necessary to systematically carry out epidemiological investigation and surveillance in the epidemic area to evaluate its role in the transmission of schistosomiasis. The main components of the surveillance are the investigation of the intermediate host, the infection of the cattle, and the pollution of the snail environment.

2.4.1 Surveillance Components and Measures

2.4.1.1 Selection of Research Area

The infection rate of schistosomiasis in farm cattle is directly related to the type of endemic area, the distribution and density of snails, and the habits of grazing. Therefore, the surveillance of farm cattle should be carried out according to the historical epidemic data of schistosomiasis in endemic areas. Surveillance should be carried out in the snail environment or susceptible zone accessible to humans and animals.

2.4.1.2 Surveillance of the Number of Intermediate and Definitive Hosts

The snail survey was carried out in spring and autumn. The snails and any suspicious snail environments were checked by systematic and environmental sampling methods. The density of live snails and infectious snails and the snail infection rate were calculated from the results of the sample survey, and the snail area size was calculated. An administrative village was used as a unit to register the snail distribution and density change from year to year in the form of maps and tables, which was useful for monitoring. For farm cattle, it is possible to coordinate with the livestock department, collect basic information on the prevention and control of cattle in the survey area, conduct on-the-spot investigations and visits, establish a livestock check register, and monitor dynamic changes in the numbers of local cattle and sick cattle.

2.4.1.3 Surveillance of Fecal Contamination and Potential Pollution Index at Snail Breeding Sites

The investigation of wild animal manure was conducted in snail areas accessible to humans and animals. It was carried out in one-way cross-sectional or periodic continuous surveys. During the investigation, all wild animal manure in the investigation environment was collected, and the type of investigation environment, the area of investigation, the amount of manure, and the type of animal source were recorded. The weight of each field manure sample was recorded, and the density of field manure was calculated (number of field manure per unit area) as well as the total weight. When periodic continuous surveys were conducted, all wild animal feces in the survey environment were removed after each investigation to avoid duplication. All manure samples were brought back to the laboratory and tested by capping hatching (one fecal sample for three tests). The weight of manure from different animal sources differed (50 g feces of bovine and equine animals and 30 g, 10 g, and 10 g of cattle, sheep, and dogs, respectively) in calculations of the positive rate of wild animal feces:

$$\text{Positive rate in feces of certain livestock} = \frac{\text{No.of positive feces of certain livestock}}{\text{Total feces under surveillance of certain livestock}} \times 100\%$$

To carry out the surveillance of the potential pollution index, it was necessary to monitor the livestock manure by collecting eggs through sediment collection or a nylon sieve pocket. The sediment was directly smeared and microscopically examined, and the number of eggs per gram of feces (EPG) was calculated. The following formula was used to calculate the potential pollution index (IPC) and the relative potential pollution index (RIPC):

$$\text{IPC} = \frac{\begin{array}{c}\text{Positive rate in feces of certain livestock} *\\ \text{EPG of a certain livestock} * \text{percentage of the number}\\ \text{of feces of a certain livestock to the total number of feces}\end{array}}{100}$$

$$\text{RIPC} = \frac{\text{IPC of feces of certain livestock}}{\text{Total IPC}} \times 100\%$$

2.4.1.4 Surveillance of Infection and Reinfection in Farm Cattle

A total of 100 farm cattle were randomly selected in the village as a unit for investigations of schistosomiasis infection. Villages with fewer than 100 cattle were investigated by the hatching method (one fecal sample for three tests). After the livestock were inspected, registration cards were made with unified components and format so that one card was created for each animal. The owner, livestock, sex, age, inspection date, method, and result were investigated in detail. After the inspection and treatment were completed every year, the data were collated and aggregated in a timely manner to calculate the infection rate of cattle of different ages. Reinfection surveillance of the cattle was carried out 3 months after the end of the previous transmission season to carry out the efficacy assessment. Pathogen testing was carried out to determine whether a diseased cow had been cured, and the investigation was carried out again after the end of the new transmission season to calculate the new infection rate and reinfection rate of cattle:

$$\text{Infection rate of farm cattle} = \frac{\text{No.of cattle with positive results in feces test}}{\text{Total no.of cattle under test}} \times 100\%$$

$$\text{New infection rate of farm cattle} = \frac{\begin{array}{c}\text{No.of cattle with positive result in feces test} -\\ \text{No.of unhealed cattle among cattle with positive result in the last test}\end{array}}{\begin{array}{c}\text{No.of cattle under feces test} -\\ \text{No.of unhealed cattle among cattle with positive result in the last test}\end{array}} \times 100\%$$

$$\text{Reinfection rate of farm cattle} = \frac{\text{No.of reinfected cattle that were cured after the first infection}}{\text{No.of cattle with positive result in the last feces test}} \times 100\%$$

2.4.2 Surveillance Report and Its Utilization

Cattle surveillance should be included in the routine surveillance system in areas with endemic schistosomiasis. A written report should be issued after the end of the annual communication season. Alternatively, surveillance should be carried out in the risk assessment of schistosomiasis transmission to issue a special report. After the report is written, it should be promptly fed back to the relevant units and individuals to respond quickly to the epidemic.

To evaluate the role of cattle schistosomiasis in the spread of local schistosomiasis through the survey of snails, the pollution of beaches with snails and the investigation of cattle infection can provide a basis for the formulation and evaluation of local schistosomiasis control strategies.

2.5 Surveillance Techniques for Parasite Mutation

This section describes the surveillance of variation in *Schistosoma japonicum* in different regions as an example.

Owing to the natural isolation barrier, *S. japonicum* in China is distributed in different geographical locations and ecological environments, showing different degrees of differentiation in terms of genetic characteristics, infectivity, and pathogenicity. Chinese and foreign scholars have suggested that *Schistosoma japonicum* in Mainland China is not a single strain but a strain complex composed of a number of differentiated geographical strains. In addition, large-scale praziquantel chemotherapy can exert strong selection pressure on schistosomiasis, which may cause genetic variation. Methods for performing genetic variation analysis of schistosomiasis populations include biology, morphology, protein- and enzyme-based assays, and DNA-based molecular biology. DNA-based molecular biology can directly reflect the gene itself; thus, it can obtain more abundant information and can be used for genetic variation analysis within and between populations. This section mainly describes the general surveillance methods and steps of molecular biology for genetic variation among schistosomiasis populations.

2.5.1 Surveillance Components and Methods

2.5.1.1 Sample Acquisition

Schistosoma adults are the best source of DNA-based genetic analysis. Therefore, it is necessary to collect infectious snails from different areas with schistosomiasis. Infectious larva should be collected from the infected snails to infect the host animals, and schistosomiasis adults can be obtained after 6 weeks by dissection. They should be washed with sterile physiological saline and then stored for later use.

2.5.1.2 Genomic Extraction

Appropriate amounts of adult schistosomiasis specimens are used to lyse and extract the DNA genome by the phenol chloroform method or genomic extraction kit, and the integrity of the genome is detected by electrophoresis. The purity of DNA is determined in a cuvette, and the OD260/OD280 ratio is usually between 1.80 and 1.90, meeting the requirements. After the DNA concentration is measured, it is stored at −20 °C until use.

2.5.1.3 Nucleic Acid Amplification and Results Analysis

1. Selection of Nucleic Acid Amplification Methods. Genetic variation analysis of *Schistosoma* based on molecular biology is related to the selected method and the analyzed DNA sequence. Nucleic acid sequence determination, specific PCR, PCR-RFLP, PCR-SSCP, and other methods can be used for direct detection of genetic markers that can more accurately reflect the nature of genetic variation. The candidate markers for the determined DNA fragments are mainly rDNA, mtDNA, and GST glutathione transferase. mtDNA has the characteristics of no recombination, independent evolution, and serogrouping between genomes, so it can reflect the evolutionary history of maternal lines. It is especially suitable for the analysis and surveillance of genetic variation in the study of schistosomiasis classification and phylogeny. Using these methods requires knowing the relevant sequence of the target gene in advance, and there are high standards for the DNA requirement. Random amplified polymorphism DNA (RAPD) is an excellent tool for interpopulation polymorphism analysis, but it is a dominant marker in population genetic analysis, which limits the clarity of results within the population due to the lack of allele expression. Microsatellite (SSR) markers, which are simple repetitive sequences in the genome, have collinear expression and are stable throughout life history. The polymorphism is revealed by designing complementary oligonucleotides as single primers to amplify the repeats. The polymorphism can be used for individual identification, subspecies identification, population outcome prediction, and historical analysis of evolution.
2. Nucleic Acid Amplification. The DNA genome extracted from adults is used as a template and amplified by a predesigned specific primer, random primer, or microsatellite primer according to the optimal system and reaction conditions. The RAPD amplification products can be determined by electrophoresis, and the SSR amplification products can be determined by Genescan to determine the length of the DNA fragments and analyzed by specific software.
3. Results Analysis. The nucleic acid amplification products based on specific gene sequences or SSR markers can be purified and sequenced for analysis. The composition and variation of nucleic acid, statistics of allele numbers, observation of heterozygosity, and expected heterozygosity should be analyzed. The sequence should be analyzed by the neighbor-joining method to compare the genetic differences between populations.

   Regarding RAPD technology, according to the results of electrophoresis of amplification products, the bands between different populations or geographical strains can be compared to calculate the similarity coefficient and genetic distance of different populations or geographical strains of schistosomiasis. The specific calculations are as follows:

$$S = \frac{2Nab}{(Na + Nb)}$$

$$D = 1 - S$$

where *Na* is the number of all amplified bands of the population or geographical strain a, *Nb* is the number of all amplified bands of the population or geographic strain b, and *Nab* is the number of shared bands of the population or geographic strains a and b.

According to the genetic distance D, the phylogenetic tree or clustering map is drawn by the unweighted pairing method using SAS or professional technical software.

4. Transmission Index. The genetic variation of schistosomiasis is significantly correlated with spatial distribution. Therefore, the evolutionary history of schistosomiasis can be analyzed by analyzing the correlation coefficient between the genetic distance and geographic distance of schistosomiasis in different regions.

2.5.2 Surveillance Report and Utilization

The genetic variation of schistosomes in different regions can be analyzed in the form of a genetic distance matrix and phylogenetic tree.

Monitoring the genetic variation of *S. japonicum* in different regions not only helps understand the biological characteristics of *Schistosoma japonicum* but also clarifies its evolutionary dynamics and variation under different habitat conditions. It is significant for guiding the epidemiology and control of schistosomiasis and the development of vaccines.

2.6 Case Analysis

2.6.1 Surveillance of Infection Status of Human *Clonorchis sinensis* in China

2.6.1.1 Proposing the Issue

The first national survey of human parasitic diseases in 1988–1992 showed that the total infection rate of *Clonorchis sinensis* in China (excluding Hong Kong, Macao, and Taiwan) was 0.37%. The second national survey on the prevalence of human parasitic diseases in 2001–2004 showed that the total infection rate of *C. sinensis* had increased to 0.58%. At the same time, according to the survey, the total *Clonorchis sinensis* infection rate in 27 prevalent provinces (municipalities and autonomous regions) was 2.40%, and it was estimated that the national infected population reached 12.49 million. *Clonorchis sinensis* had become the main food-borne parasitic disease in China. Therefore, in 2006, the Ministry of Health established eight demonstration areas for the comprehensive prevention and control of soil-transmitted parasitic diseases in China and established two demonstration zones in Zhaoyuan County of Heilongjiang Province and Yangshan County of Guangdong Province to explore comprehensive integrated prevention and control measures for *Clonorchis sinensis*. After 3 years of intervention, a survey in 2009 showed significant control effects. To grasp changes in the *Clonorchis sinensis*

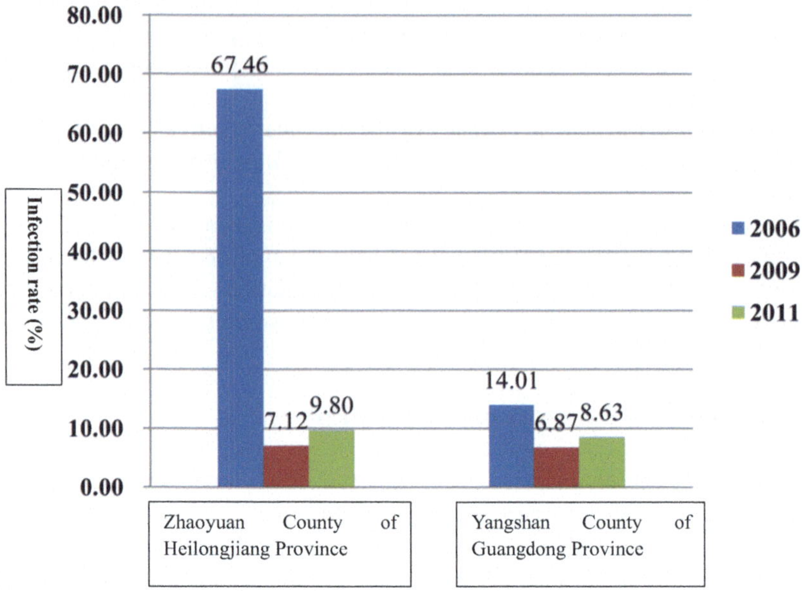

Fig. 2.8 Changes in the *Clonorchis sinensis* infection rates at the observation sites

epidemic after the completion of comprehensive prevention and treatment measures and to provide guidance for further prevention and treatment, the Ministry of Health in 2010 turned the above two *Clonorchis sinensis* demonstration areas into observation sites similar to those for soil-transmitted helminthiasis.

2.6.1.2 Surveillance Methods and Results

1. Surveillance Methods. The observation of *Clonorchis sinensis* was conducted every year in the east, west, south, north, and central regions of China. One township was selected from each direction, one village was selected from each township, and 200 people were selected from each site for a total of 1000 people. The modified Kato thick smear (a fecal sample for three tests) was used to check for *Clonorchis sinensis* infection.
2. Surveillance Results. Figure 2.8 shows changes in the prevalence of *Clonorchis sinensis* in the 2006 baseline survey (before intervention), in 2009 (after intervention), and in 2011 (2 years after intervention). Before the intervention, the *Clonorchis sinensis* infection rates at the two sites were 67.46% and 14.01%, respectively; after the intervention, they decreased to 7.12% and 6.87% with decline rates of 89.45% and 50.96%, respectively. Two years after the end of the intervention, the infection rates at the two observation sites showed a certain degree of recovery, with increases of 37.64% and 25.62%, reaching 9.80% and 8.63%, respectively.

2.6.1.3 Case Illustration

At present, the demonstration areas for *Clonorchis sinensis* in Zhaoyuan County of Heilongjiang Province and Yangshan County of Guangdong Province have become observation sites, similar to the surveillance sites. Observation sites will help improve our understanding of changes in the prevalence of *Clonorchis sinensis* in these areas after the intervention. Obviously, comprehensive prevention and control has led to a significant decline in the prevalence of *Clonorchis sinensis* in the two demonstration areas. However, 2 years after the end of the intervention, the epidemic level in the population had rebounded to some degree. Although the prevalence level was still far below the baseline level before the intervention in 2006, the infection rates would certainly return to the baseline level after a certain period with no interventions. At present, the observation has been carried out for only 1 year. As the observation continues for up to 5 years or more, changes in prevalence will provide more information. This information will provide guidance for the prevention and treatment of *Clonorchis sinensis*; for example, it will help determine when it is necessary to intervene again. Of course, current observations have shown that even if the prevalence of the *Clonorchis sinensis* epidemic is reduced to a lower level after intervention, it is necessary for such work to be continuous.

2.6.2 Analysis of Surveillance Results for Intestinal Parasitic Diseases at Surveillance Sites in China

2.6.2.1 Proposing the Issue

Soil-transmitted helminths are widely distributed in China, and the large number of infections seriously endangers public health and hinders economic development and social progress. The first national survey of human parasitic diseases conducted in China from 1988 to 1992 showed that the standardized rate of soil-transmitted helminthiasis in China was 53.21%, with 43.77% for roundworms, 17.27% for whipworms, and 15.47% for hookworms. The estimated total number of infected people was 536 million. The second national survey of parasitic diseases conducted in 2001–2004 showed that the standardized infection rate of soil-transmitted helminths was 19.34%, with 12.57% for roundworms, 4.56% for whipworms, and 6.08% for hookworms. The estimated total number of infected people was 129 million. Although the second survey showed that the general infection level was significantly lower than in the first national survey, it was still high, equivalent to the level of Japan in the 1960s and Korea in the 1980s, and there was still a large gap between China and developed countries, such as Europe and the United States, which was not compatible with the current requirements for building a harmonious society. Therefore, to grasp the epidemic dynamics and patterns of soil-transmitted helminths in China in a timely manner, provide a scientific basis for formulating strategies and measures for the prevention and control of soil-transmitted helminths, evaluate the effects, and provide relevant information for the creation of national and local health cities (towns), it is necessary to carry out the surveillance of

Table 2.4 Changes in population infection rates of soil-transmitted helminths at surveillance sites in 2006–2009 (%)

Year	Roundworm	Whipworm	Hookworm	Soil-transmitted helminths
2006	10.10	5.90	8.90	20.90
2007	8.90	5.70	9.00	18.90
2008	7.40	6.60	6.80	16.60
2009	6.40	5.20	5.80	13.30

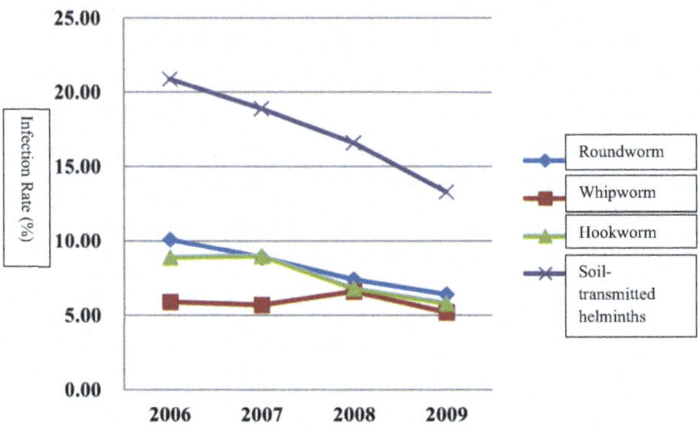

Fig. 2.9 Changes in population infection levels of soil-transmitted helminths at surveillance sites in 2006–2009

soil-transmitted helminthiasis in a planned, continuous, and systematic manner nationwide.

2.6.2.2 Surveillance Methods and Results

1. Surveillance Methods. In 2006, the Ministry of Health established surveillance sites for soil-transmitted helminths in 22 provinces (municipalities and autonomous regions) in China. Among them were ten sites with infection rates >20% (Hainan, Guizhou, Sichuan, Guangxi, Hunan, Chongqing, Fujian, Hubei, Jiangxi, and Yunnan Provinces), seven sites with infection rates of 5% to 20% (Anhui, Guangdong, Shandong, Gansu, Qinghai, Henan, and Jiangsu Provinces); and five sites with infection rates <5% (Heilongjiang, Shanghai, Hebei, Shanxi, and Beijing). Each surveillance site was based on a village, and the level of soil-transmitted helminth infection was close to the average prevalence level for the province (city, district). Using the modified Kato thick smear method (one fecal sample for three tests), soil-transmitted helminths were tested every year for 1000 permanent residents aged 3 years and older.
2. Surveillance Results. The surveillance results of soil-transmitted helminths are shown in Table 2.4 and Fig. 2.9. In 2006, the total roundworm, whipworm, hookworm, and soil-transmitted helminth infection rates were 10.10%, 5.90%, 8.90%, and 20.90%, respectively, and in 2009, they were 6.40%, 5.20%, 5.80%, and 13.30%, respectively. The overall trend was downward.

2.6.2.3 Case Illustration

Two nationwide surveys of parasitic diseases were carried out in China in the early 1990s and early twenty-first century. The results clearly revealed the prevalence of soil-transmitted helminths at that time in China and provided a basis for prevention and control nationwide. However, large-scale nationwide surveys can be carried out only once every 10 years or so. To continuously and systematically grasp changes in the prevalence of soil-transmitted helminths, it is necessary to continue to carry out relevant surveillance. In this context, surveillance sites for soil-transmitted helminths have been established nationwide. The surveillance sites for soil-transmitted helminths were set up in villages with a prevalence level close to the average prevalence level for each province (city, district) to enable the provinces (cities, districts) to master the situation of soil-transmitted helminths in the region. Once the prevailing changes have taken place, corresponding measures will be taken accordingly. The establishment of surveillance sites for different prevalence levels in 22 provinces (cities, districts) across China will enable the relevant departments to grasp the overall epidemic dynamics of soil-transmitted helminths in China, which will be conducive to the formulation of national-level control measures. The 4-year surveillance showed that the overall epidemic level of soil-transmitted helminths in China presented a downward trend, which was consistent with the changes in social and economic development. However, compared with the higher decline in the infection levels of roundworms and hookworms, whipworms have declined less and remain at their original levels. Therefore, it is necessary to strengthen research on the prevention and control of whipworms, identify relevant factors, and provide guidance for the development of targeted prevention and control measures.

3 Surveillance of Vectors

Qing-Biao Hong

Vectors are organisms that transmit diseases; the term generally refers to organisms that transmit human diseases. Broad-based vectors include vertebrates and invertebrates. Vertebrate vectors are mainly rodents of the mammalian rodent group. Invertebrate vectors are mainly insects, such as mosquitoes, cockroaches, and crickets; arachnids, such as ticks and mites; and mollusk snails. Vector organisms can not only affect or harm the normal life of human beings directly through bites and contaminated food but also spread a series of important infectious diseases through various channels. This chapter introduces surveillance techniques for the distribution, growth, migration, and propagation energy of vectors.

3.1 Overview

3.1.1 Definition

Biomonitoring of parasitic vectors refers to the long-term, continuous, and systematic collection of vector organisms, such as mosquitoes and snails, by scientific methods and the analysis of types, quantities, and distributions of data as well as seasonal changes in data. Interpretation and feedback of the results for health administrative agencies and CDCs are good for developing, implementing, evaluating, and adjusting strategies and measures for vector control. Systematic development of parasitic vector surveillance can provide not only a basis for the development of programs of parasitic disease control but also predictive and warning information for tracking epidemic trends of parasitic diseases.

Q.-B. Hong (✉)
Jiangsu Institute of Parasitic Diseases, Wuxi, Jiangsu, China

X.-N. Zhou (ed.), *Surveillance and Management of Parasitic Diseases in China*,
https://doi.org/10.1007/978-981-97-1241-0_3

3.1.2 Species

Many infectious diseases reported in China are vector-borne diseases, such as plague, epidemic encephalitis, dengue fever, and typhus. The parasitic diseases transmitted by vector organisms are mainly malaria, leishmaniasis, filariasis, and hymenolepiasis, and those transmitted by snails as the intermediate host include schistosomiasis, clonorchiasis, angiostrongyliasis, and fascioliasis (Table 3.1).

Table 3.1 Main vectors for transmission of parasitic diseases

Main vectors	Main parasitic diseases	Role
Mosquito	Malaria, filariasis	Vector
Sandfly	Leishmaniasis	Vector
Fly	Thelaziasis, myiasis	Vector
Cockroach	Filariasis, roundworm infection, hookworms disease, whipworm infection, pinworm infection, tapeworm infection, amoebiasis, etc.	Vector
Beetle	Filariasis, hymenolepiasis, *Pseudanoplocephala crawfordi* infection	Vector
Flea	Hymenolepiasis	Vector
Tick	Babesiasis, tick paralysis, etc.	Vector
Acarid	Acne, demodicosis, etc.	Vector
Snail		
Oncomelania	Schistosomiasis	Intermediate host
Bithynia	Clonorchiasis	First intermediate host
Parafossarulus	Clonorchiasis, metorchiasis, echinostomiasis	First intermediate host
Alocinma	Clonorchiasis	First intermediate host
Bellamya	Echinostomiasis	First intermediate host
Potamididae	Paragonimiasis	First intermediate host
Planorbidae	Fasciolopsiasis	First intermediate host
Radix plicatula	Fascioliasis	First intermediate host
Galba pervia	Fascioliasis	First intermediate host
Ampullaria gigas	Angiostrongyliasis	First intermediate host
Achatina fulica	Angiostrongyliasis	First intermediate host
Limax	Angiostrongyliasis	First intermediate host

3.2 Surveillance Techniques for Vector Distribution and Growth

3.2.1 Surveillance of Mosquito Distribution

Mosquitoes are mainly vectors for parasites such as malaria and filarial larvae. The main vectors of transmission of malaria are *Anopheles sinensis*, *Anopheles anthropophagus*, *Anopheles minimus*, and *Anopheles dirus*. The main vectors for transmitting filariasis are *Anopheles sinensis*, *Anopheles anthropophagus*, *Culex pipiens* pallens, and *Culex fatigans*. Mosquitoes are insects that metamorphose, and their life history has four stages: spawning, larval development, adulteration, and emergence of adults. The ecological habits of various mosquito species differ. Each mosquito has a certain habitat. Larval breeding sites can be divided into hydrostatic, slow-flow, jungle, sewage, container, etc. The activities of adult mosquitoes (referring mainly to sucking the blood of the host) are related to temperature, humidity, light, wind, and other environmental conditions. Most mosquito species are active in the morning, at dusk, or at night. After the female mosquito sucks blood, it seeks dark, damp, and windless places, such as nets, under beds, on ceilings, behind doors, on walls, and on sundry other indoor surfaces or outdoor locations such as grass, cavities, under trees, and in crops. Mosquito species can be divided into family type, half-family type, and wild type according to their habitats.

In areas with endemic mosquito-borne parasitic diseases or related potential endemic areas, different regions and environments are selected for surveillance sites by setting up sites or sampling procedures according to the ecological habits of mosquitoes. Then, surveillance of mosquitoes related to the spread of parasitic diseases such as *Anopheles* and *Culex* is needed to understand their distribution, ecological habits, and disease-transmitting effects. This understanding provides a scientific basis for the development of control measures for mosquito-borne parasitic diseases and supports a correct evaluation of the effects of these measures. According to different mosquito species and their ecological characteristics, different terrains and environmental settings are selected. During the active mosquito season, they are regularly or irregularly caught during the day or night in human houses, barns, cave dwellings, empty houses, basements, and spaces in stacked materials as well as caves, bridges, culverts, and plant plexuses around residential areas and other places that mosquitoes may inhabit. Mosquitoes are caught to identify their species, or larvae can be caught for identification after they have emerged. Searching and trapping methods can be used for investigation.

3.2.1.1 Mosquito Trapping in Nets

All mosquitoes in the mosquito nets were caught in the morning in a representative residential area, and the captured mosquitoes were first identified and classified. This measure can be used for the surveillance of mosquito density.

3.2.1.2 Host Trapping

This measure includes trapping on human and animal hosts. The trapping timing and length depend on the density of the mosquitoes in sight, and trapping is generally conducted at the peak of mosquito blood sucking, such as starting half an hour after sunset to catch mosquitoes overnight or at a certain time. This measure can be used for mosquito density surveillance:

1. Human Trapping. There are many forms of human Trapping. One is to hang an open mosquito net between a residential area and the breeding ground. The bottom of the net is 35 cm above the ground. One person sits in the net and catches mosquitoes as they fly into the net. Another method is to hang an unopened mosquito net with the bottom touching the ground. One person sleeps inside this net. A large opened mosquito net is placed outside the first net. Another person enters the large net to catch the mosquitoes landing in it. A third method involves one person sitting with bare legs at the observation site to capture the blood-sucking mosquitoes that land on his or her legs and count the numbers (this individual must pay close attention to self-protection).
2. Animal Trapping. Animals (bovines, pigs, etc.) are tied in a location between a mosquito habitat and a residential area. The investigators constantly capture mosquitoes on the bodies of the animals, surrounding walls, or other objects. Another approach is to hang a special open mosquito net with an animal tied inside it and to catch the mosquitoes that fly into the net.

3.2.1.3 Light Trapping

Using mosquito phototaxis, mosquito traps are placed in a certain place to collect mosquitoes. Trapping starts at sunset, and the mosquito box is removed the next day to identify the species and sex as well as the count and estimated density of each species.

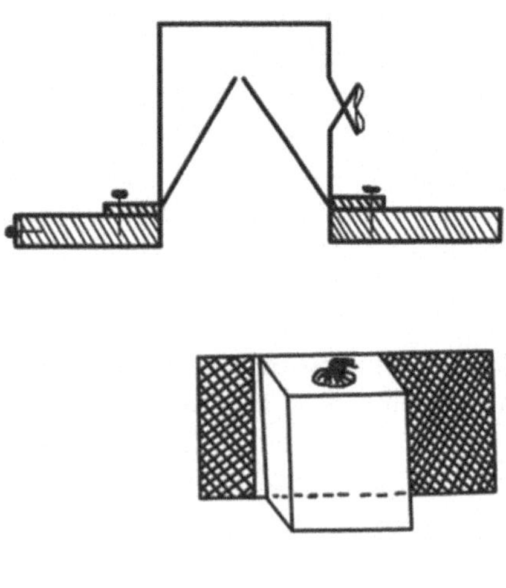

Fig. 3.1 Window trap

3.2.1.4 Window Trapping

A window trap is placed on a wall or window facing the east side of a house or barn (Fig. 3.1) to collect mosquitoes for identification and counting. This measure can be used for mosquito-density surveillance.

3.2.1.5 Mosquito and Egg Traps

This method is commonly used to monitor *Aedes* mosquitoes and other species. *Aedes* residential areas, such as human residential areas, parks, construction sites, hospitals, and waste collection stations, are selected, and mosquito traps and oviposition traps are deployed. After a certain period of time, adult mosquitoes and mosquito eggs are checked and collected.

3.2.2 Surveillance of Mosquito Ecological Habits

3.2.2.1 Seasonal Population Fluctuations

One or several density survey methods are selected to investigate a fixed site during the mosquito activity season every 10 days. Years of observation using the same method at the same location reveal seasonal fluctuations on mosquito density based on 10-day or monthly density. During the density survey, local temperature, humidity, rainfall, wind power and direction, moon phase, and crop planting should be recorded or collected for reference when analyzing the results.

3.2.2.2 Inhabitant Habits

Inhabitant habits refer to the characteristics of the female mosquito during the time from blood sucking to ovary development until the time of spawning. Mastering the inhabitant habits of mosquitoes is the basis for developing measures to control adult mosquitoes. The inhabitant habits of mosquitoes can be roughly divided into three types, namely, family, semifamily, and wild type. The survey methods are as follows:

1. Mosquito Searching. Mosquitoes are regularly caught during the day in human houses, barns, cave dwellings, empty houses, basements, spaces in stacked materials, etc. in human residential areas as well as caves, bridges, culverts, plant plexuses, and other places around residential areas to learn about the habitat selection and habitat characteristics of various *Anopheles* mosquitoes.
2. Window Trap. Indoor mosquitoes in houses where window traps are placed must all be caught or killed by insecticide fumigation during the day. Mosquitoes in the window traps and the house must then be collected and caught the next morning to understand the proportion of mosquitoes entering the room from outside after blood sucking.
3. Anatomical Observation. Blood-sucking *Anopheles* mosquitoes are captured from human houses and barns during the day, dissected and observed for ovarian development. If the ovarian development of the mosquito is in the fourth to fifth phase and the number of such mosquitoes is equal to or close to that of the sec-

ond to third phase of the mosquito, it indicates that the *Anopheles* mosquito is a domestic mosquito species; if the former is significantly less than the latter, the *Anopheles* mosquito is a semidomestic mosquito species; and if the former is not caught, it is a wild mosquito species.

3.2.2.3 Blood-Sucking Habits

Understanding the blood-sucking habits of mosquitoes is an important basis for determining their role as a vector. Regarding blood-sucking habits, the focus of the investigation is as follows:

1. Blood-sucking Tropism. This refers to the tendency of mosquitoes to choose animals for blood sucking. The investigation methods are as follows: mosquitoes are directly trapped on people and livestock (30–50 m apart) at night in the same place, the number of mosquitoes trapped by one person and one animal is compared, and the blood-sucking tropism of a certain mosquito can then be generally understood. At the same time, the number of mosquitoes trapped by humans and livestock is multiplied by the number of people in the residential area and the number of heads of various animals to roughly estimate the proportion of human blood-sucking mosquitoes in the mosquito population.
2. Human Blood Inhalation Rate. This refers to the proportion of human blood-sucking mosquitoes in the mosquito population. For family-habitat and close-to-family-habitat *Anopheles*, the survey method commonly used is the full-catch method, which involves selecting a representative small natural village. In the peak season of *Anopheles* density, *Anopheles* mosquitoes are caught after blood sucking in all houses, barns, and other places in the morning. Stomach blood samples are immediately prepared for identification by serological methods, and the human blood-sucking rate of *Anopheles* mosquitoes collected from various places is calculated. It is difficult to use a wild habitat to capture wild or close-to-wild blood-sucking *Anopheles* mosquitoes and obtain stomach blood samples to perform gastric blood identification. The formula for inhaling human blood is as follows:

   $$\text{Blood-sucking rate} = \frac{\text{No.of examined stomach blood samples with human blood}}{\text{Total no.of stomach blood samples examined}} \times 100\%$$

   If the formula is not multiplied by 100%, the result will be the blood absorption index.
3. Blood-sucking Places. Understanding the blood-sucking places of *Anopheles* mosquitoes is significant for the development of control measures. The survey involves simultaneous trapping using the same number of humans as bait indoors and outdoors and calculating the ratio of indoor to outdoor blood-sucking mosquitoes (O/I ratio):

$$\text{O / I ratio} = \frac{\text{No.of outdoor blood-sucking mosquitoes}}{\text{No.of indoor blood-sucking mosquitoes}}$$

4. The Pattern of Blood Sucking at Night. Mastering the pattern of night-time blood sucking of *Anopheles* mosquitoes is greatly significant for strengthening personal protection measures and reducing human exposure to mosquito bite. The method is to survey once every 10 days. Human bait is used to trap mosquitoes indoors and outdoors half an hour after sunset. The number of mosquitoes can be recorded in hours. After 1 or 2 years of observation, the blood-sucking activity of local *Anopheles* mosquitoes in different seasons can be understood.

3.2.2.4 Life Span

Understanding the life span of a female *Anopheles* mosquito is directly related to determining whether the *Anopheles* species transmits malaria and its transmission energy:

1. Individual Life Span. There are two commonly used survey methods:
   (a) Direct observation: Observe the number of living days of a certain number of *Anopheles* mosquitoes, and determine the average life expectancy (days). However, *Anopheles* mosquitoes under experimental conditions are not affected by the wind, rain, and natural enemies that are part of their natural conditions, so their average life expectancy will be longer than that in nature.
   (b) Indirect observation: A nodule forms on the follicular duct of the *Anopheles* ovary when it produces one egg. Anatomizing and observing the number of follicular nodules in *Anopheles sinensis* indicate a mosquito's physiological age and the number of times it has laid eggs. According to observation under certain temperature conditions, knowing the number of days it takes for the *Anopheles* mosquito to complete one reproductive cycle enables the number of days of survival of each *Anopheles* mosquito to be calculated.
2. Life Expectancy of the Mosquito Population: Understanding the life expectancy of the population enables scientists to measure the role of certain *Anopheles* mosquitoes in transmitting malaria. The life expectancy of the mosquito population is calculated from the daily survival rate, and the daily survival rate is obtained from the rate of ovipositing mosquitoes in the population.

   Rate of ovipositing mosquitoes (M): Anatomical observation of the shape of the tracheal branches of the *Anopheles* ovarian surface in various places or the presence or absence of nodules on the follicular tube is conducted to determine whether a mosquito is oviparous (egg-producing); if so, then the rate of oviposited mosquitoes (M) can be calculated. The formula is as follows:

$$M = \frac{\text{No.of oviparous mosquitoes}}{\text{No.of mosquitoes under observation}}$$

   The daily survival rate (P) is calculated as:

$$P = M^{1/x}$$

where x indicates the number of days from mosquito emergence to the first spawning.

According to the above two results, the life expectancy (1/−lnP) can be calculated. For example, if 1000 *Anopheles*, including 729 ovipositing mosquitoes, are under anatomical observation, assuming that the first spawning is 3 days, $M = 729/1000 = 0.729$, $P = 0.7291/3 = 0.9$ (90%), and 1/lnP = 1/−ln0.9 = 9.5 (days).

3. Life Expectancy of *Anopheles*. This refers to the expected number of days of survival of *Anopheles* mosquitoes that already have the ability to transmit malaria. It can be calculated based on the expected duration of life. First, the ratio of infectious mosquitoes (P^n) in the mosquito population should be calculated. With n being the number of days required for *Plasmodium* to proliferate in mosquitoes, the life expectancy of *Anopheles* (P^n/−lnP) is calculated. Based on the above description, the formula to calculate the life expectancy of infectious mosquito is as follows:

   Rate of infectious *Anopheles*: $P^n = 0.9^{10} = 0.349$

   Life expectancy of infectious *Anopheles*: $P^n/-\ln p = 0.9^{10}/\ln 0.9 = 3.31$ (days)

When investigating the rate of oviposited mosquitoes in a mosquito population, attention must be paid to the representativeness of the sampling. *Anopheles* should be captured as anatomical specimens from different locations during the day and night. In addition, the rate of oviposited mosquitoes is often affected by a variety of factors, such as the start and end of the mosquito breeding season and sudden increases or decreases in the breeding ground. These factors may bias calculations of the ratio of oviposited mosquitoes and the daily survival rate derived therefrom. Therefore, a certain number of specimens should be regularly dissected throughout the mosquito breeding season to obtain the average ratio of oviposited mosquitoes.

3.2.2.5 Breeding Habits

Through the investigation of mosquito breeding habits, we can understand how various mosquitoes select breeding grounds and thus develop targeted measures to control larvae. The investigation combines a larval density survey, a census of various water bodies, and a detailed record of the types and characteristics of larval water bodies, such as size, running or still water, water turbidity, water temperature, pH value, water flow rate, water plants, presence or absence of shade, surrounding environment, and distance from residential areas.

3.2.3 Surveillance of Sandflies

Sandflies are the vector of *Leishmania spp.*, which needs to pass through four stages: egg, larva, pupa, and adult. Its lifetime is 30 days in the tropics, while in the temperate zone, it can last for more than 10 months because of the diapause period. There are approximately 20 species of sandflies that spread *Leishmania spp.*, of which four species are found in China: such as *Phlebotomus sinensis*, *Phlebotomus longiductus*, *Phlebotomus wui*, and *Phlebotomus alexandri*. Different sandflies have different landscape distributions and ecological habits. When surveilling sandflies, the local surveillance time should be determined according to the specific conditions of the ecological environment, seasonal climate, and topography. In general, the surveillance time in Jiangsu, Shandong, Anhui, eastern Henan, southern Shanxi, the Central Shaanxi Plain, Hubei Province, and other regions is mainly from May to June, while in northwestern Henan, northern Shanxi, Miyun in Beijing, northwestern Shaanxi, northwestern Sichuan, Gansu, Ningxia, Qinghai, Xinjiang, and similar regions, it is mainly from June to July. The surveillance includes the distribution of sandflies, the composition of the species, and changes in population density. When sandflies are surveilled in caves, it is necessary to check and identify whether their digestive tract is infected with *Leishmania* promastigotes; when surveillance is in a plain area where family-based sandflies are dominant, it is necessary to identify family-habitat sandflies and the near-nature habitat to perform a correct evaluation and develop the correct prevention strategy.

Surveillance methods for sandflies are mainly manual collection, lure lamps, viscous oil paper, and human/light trapping. The relevant data obtained through these measures can be used to calculate sandfly density, type, composition ratio, and curve of monthly growth and decline for statistical analysis.

3.2.3.1 Manual Collection

Appropriate locations are selected from mid-June to the end of July. The trapping tube/bottle is used to collect sandflies in wild caves during the day and in residential areas (including houses, barns, toilets, and temples) at night. The density can be calculated according to the number of sandflies collected per unit of time; that is, each person uses the trapping tube/bottle for 1 h, and sandfly density is calculated according to the number trapped per "manual hour."

3.2.3.2 Lure Lamps

This method can also be used for the investigation of sandfly density. At the time of surveillance, one mosquito (sandfly) lamp is used beginning 30 min after sunset. After 12 h of trapping, the sandflies are dissected to determine the species and sex and count the number of captured individuals. The formula for calculating the density index of the lure lamp is as follows:

$$\text{Blood-sucking rate} = \frac{\text{No.of examined stomach blood samples with human blood}}{\text{Total no.of stomach blood samples examined}} \times 100\%$$

$$\text{Density}(\text{no./hour}) = \frac{\text{No.of captured sandflies}}{(\text{Capture time} * \text{no.of lamps})}$$

3.2.3.3 Sticky Paper

Book strips/kraft paper (28 cm × 39 cm) are coated with castor oil/mineral oil, and the sticky paper is then hung indoors 5–8 cm from a wall or close to a bed and pillow or artificial light source. The sticky paper is hung outdoors close to the source of blood (such as a bird's nest, rat hole, or livestock bar); sticky paper hung at night must be taken down before sunrise the next morning. Sticky paper inserted into a hole is oiled on one side, and the oiled surface is the inner side of the paper tube. After trapping is conducted for six consecutive hours, the number of sandflies collected on the paper is summarized and then averaged according to the number of sheets used. The average number of sandflies captured on each sticky paper on that day can be calculated. Then, the number of sandflies captured by each sticky paper is calculated by month to calculate sandfly density. The advantage of this method is that it saves labor and can be used in places where manual methods cannot be used; the disadvantage is that the captured sandflies are dead and cannot be used for further specimen preparation.

3.2.3.4 Human/Light Trapping

Humans can choose an appropriate place 30 min after sunset in a wild desert or mountain areas to set up and illuminate screens or mosquito nets. They can lure sandflies with light and then collect them on the screen or net with live traps/bottles. If each person captures for 1 h at a time, the sandfly density can be counted as the captured number by "manual hour."

3.2.4 Surveillance of Flies

The life cycle of flies is a full metamorphosis. Except for a few flies that directly produce larvae, the life history of most flies includes four stages: such as egg, larva, pupa, and adult. The life span is generally 1–2 months. According to the main fly species and their ecological characteristics and occurrence rules, researchers can choose a targeted farmers market, the environment outside a catering area, or a greenbelt or residential area at an appropriate time and under appropriate meteorological conditions for surveillance. When conducting surveillance, fly traps and trapping methods can be used.

3.2.4.1 Fly Traps

A fly trap can be made as follows. A cotton screen or thin cotton cloth is cut into four large triangles of the same size, with each piece having a base length of 21 cm. Each piece is then fashioned into a conical mesh bag. Then, a lead wire with a diameter of 3 mm is rolled into a round frame with a diameter of 26 cm.

3.2.4.2 Temptation

This method includes cage trapping and food trapping. For example, when trapping fruit flies, suspended fruit bags and mosquito nets with chopped fruit are commonly used:

1. Cage Trapping. A certain number of fly traps are placed at the selected site between 9 and 10 o'clock in the morning and recovered from 3 to 4 o'clock in

the afternoon. After the cage is closed, the flies are killed by diethyl ether or chloroform and then classified, and the number of each fly species is counted. The formula for the density of adult flies is as follows:

$$\text{Density of adult flies}(\text{no.per cage}) = \text{No.of flies} / \text{no.of cages}$$

Blood-sucking rate =

$$\frac{\text{No.of examined stomach blood samples with human blood}}{\text{Total no.of stomach blood samples examined}} \times 100\%$$

2. Suspended Fruit Bag. This method is often used to trap fruit flies by placing their favorite ripe apples or pears at 25–30 °C for 2 to 3 days of moderate fermentation. Then, they are wrapped in gauze and plastic window screens and tied with white ropes. The "fruit package" is used to trap flies near rural residents' homes. The package is hung from a tree approximately 1 m above the ground. The investigator can put fly traps around the fruit bag at approximately 0.8 m to catch the flies.

3. Mosquito Net with Chopped Fruit. Light blue double-nylon nets supported by bamboo poles should be placed near residential areas. The area around the bottom of the net is rolled up to approximately 25 cm above the ground. The fermented chopped fruits are packed in pots and then placed in the mosquito nets. Fruit flies have a certain phototaxis and will gather at the top of the net. Investigators can catch them in the morning and afternoon.

3.2.5 Surveillance of Ticks

Ticks that have a strong-shelled shield at the back of their body are commonly known as hard ticks and belong to the Ixodidae family. Ticks without a hard shield are called soft ticks and belong to Argasidae. Approximately 800 species of ticks have been found around the world. Approximately 100 species of hard ticks and 10 species of soft ticks have been recorded in China. Hard ticks are temporary parasites that feed on the surface of many species of vertebrate animals. They are vectors and hosts for some zoonotic diseases. Hard ticks live in forests, bushes, open pastures, grasslands, mountain soil, etc. Soft ticks inhabit livestock sheds, the caves of wild animals, bird nests, and the wall chinks of human houses. When not sucking blood, they are as small as a mung bean or even a rice grain; after the blood is sucked, some are the size of a soybean, and some can reach the size of a fingernail. Therefore, a survey of tick distribution can be divided into two categories: ticks on livestock and free ticks.

3.2.5.1 Survey of Ticks on the Livestock

Surveying ticks on livestock mainly involves investigating ticks on dogs, cattle, and sheep. For dogs, investigators should focus on the ears, neck, hind legs, and tail root.

For cattle and sheep, they should focus on the inside of the hind legs and body surface. Long-haired animals must be searched by hand. When ticks are found sucking blood, a small forceps is used to clamp the head; gently pull it, shake it from left to right, and then remove it. When catching ticks, the animals should be properly fixed in place.

3.2.5.2 Survey of Free Ticks

According to the distribution characteristics of natural ticks, surveys of indoor free ticks and environmental free ticks are often conducted:

1. Surveys of Indoor Free Ticks. Generally, in places where the population is concentrated, the number of ticks captured per hour is used to investigate free ticks in animal housing. Wall surfaces should be checked, especially corners, in barns, laps, dog rooms, etc. If a nest is encountered that is not easy to enter, investigators can use a flannel probe to conduct adhesion investigation. If the investigation site is a kennel, adobes or bricks can be turned over to look for ticks.
2. Surveys of Environmental Free Ticks. This type of survey is conducted with a cloth flag and divided into the investigation of the free environment around residences and in the wild environment. In the environment surrounding a concentrated residential area, a suitable grassy environment should be chosen, and in a wild environment, a hilly slope with grass and vegetation should be chosen.

The cloth flag is a white cotton cloth of 90 cm × 60 cm with the narrow ends fixed by ropes. The flag is laid on the ground and dragged forward by the ropes. The investigator should stop and check the number of attached ticks every 10 m. Different methods can be chosen according to the vegetation conditions in the lot for the timing or distance of the drag. To investigate density, the investigator can evenly drag or wave the flag over the selected sites and calculate density index statistics based on the number of ticks captured per person per hour or per 100 m per person. The investigation distance should not be less than 200 m. During the investigation, the captured ticks should be placed in collection bottles, the location marked on the outside of the bottle, and the questionnaire completed. The investigation time should be morning or evening.

3.2.6 Surveillance of Beetles

Methods of general insect investigation can be used for beetle monitoring. The habits of the beetles as well as their completeness should be considered during the investigation. When collecting beetles, the investigator must first identify their hosts and habitats. Many insects have characteristics of suspended animation. When collecting beetles, host plants can be shocked to make them fall for easy collection.

Net catching is the most commonly used method for collecting insects such as beetles, and various types of traps can also be utilized according to insect tropism.

3.2.7 Surveillance of Cockroaches

According to different geographical orientations and regions, a certain number of surveillance sites are selected. Different environments, such as farmers markets, restaurants, hotels, hospitals, and residential areas, can be randomly selected for monitoring at each surveillance site. However, sites that have been treated with insecticides within a week cannot be selected as surveillance sites. Surveillance can be carried out throughout the year or fixed on a certain day of each month.

Sticky traps are often employed in surveillance. The adhesive paper is generally 17 cm × 10 cm, sweet bread is used as bait (2 g/piece), and ten sheets of sticky paper are placed at each location. The selected sites are, for example, food processing and sales counters of a general market, the operation room and restaurant of a catering service and hotel, the wards of a hospital, and the kitchen of each household in a residential area. The traps are placed in the evening and collected in the morning. One paper is placed in each standard room (approximately 15 m^2), and one is placed in each kitchen per household. The sticky paper must be updated for each monitoring effort.

The total number of captured cockroaches is the total number of cockroaches stuck to the sticky paper. The types of cockroaches caught, the numbers of female and male cockroaches, and the total number of recycled sticky papers are registered to calculate the cockroach density and invasive rate. The formulas are as follows:

$$\text{Cockroach density}\left(\text{no.}/\text{sticky paper}\right)=\frac{\text{No.of captured cockroaches}}{\text{Total no.of recycled sticky papers}}$$

$$\text{Invasive rate}=\frac{\text{No.of positive sticky papers}}{\text{Total no.of recycled sticky papers}}\times 100\%$$

3.2.8 Surveillance of Snails

In endemic areas of schistosomiasis or related potential transmission areas, monitoring the distribution and populations of *Oncomelania* snails, or *Oncomelania hupensis*, the intermediate snail host of *Schistosoma japonicum*, in selected different regions, environments, etc., through point-by-point or sampling methods can provide a scientific basis for the development of schistosomiasis prevention strategies and measures, that enable the accurate evaluation of snail-control measures and effects.

3.2.8.1 Surveillance Scale

According to the historical records of snail distribution, specific snail locations of surveillance should be determined, including the current snail environment, the historical snail environment, and places connected to snail water habitats or the existing snail environment as well as any environment with common plants and aquatic resources that attract snails and environments where snails are suspected, such as submerged flooded areas.

3.2.8.2 Investigation Methods

1. Systematic Sampling. This method is suitable for environments where snails are both scattered and dense. In general, the water area is framed at 20 to 50 m × 20 to 50 m, and inland and hilly areas are 5 to 10 m × 5 to 10 m. This method enables researchers to count the density of snail distribution in the environment. The snails seized during the investigation are packaged according to the environment, line number, and frame number to identify living and dead snails and infectious snails in the environment. This helps the investigators count the density of snails and infectious snails.
2. Environmental Sampling. This method is applicable to places with an environment where snails are suspected or where there is low-density snail distribution. It is used to determine whether there are snails in an environment where it would be easy for them to live. According to the environmental characteristics of vegetation, low-lying land, and snail habits, investigators can detect a suspected environment by a framed survey. The number of survey frames depends on the size of the environment. After snails are found by environmental sampling, systematic sampling is used to determine the snail density in the environment.
3. Comprehensive scrutiny. This method is applicable to snail monitoring surveys in endemic areas or complex environments where snails are close to elimination. During the investigation, frames should be placed every step to check the entire environment gradually. Once snails are found, the environmental characteristics and actual size of the area should be recorded for further investigation.
4. Attracting snails. This method is suitable for snail investigations in underwater environments or those with high banks, steep slopes, and other factors that make it difficult to directly observe them with the naked eye. To attract snails, straws are first made into a curtain of 0.1 m^2 and placed in a suspected environment to be monitored. After 1–3 days, the straw curtains are repeatedly washed in clear water to determine whether they received snails. The snail density in the environment can be calculated based on the numbers of snails seized and curtains placed.
5. Remote sensing technology. This method employs remote sensing technology to detect the relevant characteristics of snail habitats, thereby indirectly grasping the potential distribution areas of snails.
6. Infectious snail investigation. Snails seized by systematic sampling, environmental sampling, comprehensive scrutiny, or attraction are subjected to crushing microscopic examination to enable researchers to understand and grasp the distribution and population of infectious snails. The specific operation is as follows: the captured snails are placed on thick glass slides and kept at a certain distance from one another to prevent cross-contamination. Another thick glass slide is used to gently crush the snal shell, and a drop of water is then added to each snail. To place the snails under the anatomical mirror, the anatomical needle (or ophthalmological forceps) is used to open the outer casing, tear the soft tissue,

and observe each one. An infectious snail is confirmed when the schistosomiasis larvae or metacercaria or cercaria are found. This method is used to count the infection rate of snails in the environment, and system sampling can be used to count the density of infectious snails in the environment.

3.2.8.3 Calculation of the Size of the Snail-Infested Area

1. Hills and Water Network Areas. To identify snails, rivers, gullies, ponds, and ridges are all framed at equal distances every 10 m (every 15 steps in the field). If there are snails in adjacent frames, then those frames are snail segments. When the area with no snails between two snail segments is within 30 m, the area is regarded as one snail segment; if it is farther than 30 m, it is calculated as two snail segments. Snail area (m^2) = length (m) × width (m). The length of the snail segment extends 15 m from the farthest point of the snail segment, and the length of an isolated snail segment is calculated as 30 m. If a river ditch has water and the water level is stable year-round, the actual slope height of the river ditch is the width; if the river ditch has water in summer and no water in winter, the width is the actual height of the ditch plus the bottom width; and if the snails are found only on one side, the height of the side is the width, and the width of the ridge is calculated as 1 m. If special terrains such as cold-soil fields, mountains, graves, bamboo forests, or wood forests are found to have snails, the area is calculated as the actual area of the habitat.
2. Marshland and Lake Areas. If the snails are found in marshlands of not more than 200 mu (1 mu = 0.0667 ha), the total area is calculated as a snail area. For a marshland with a total area of more than 200 mu, when two snail frames are within 300 m, they are regarded as one snail area. When the distance between the frames is greater than 300 m, the two frames are calculated separately. First, the length and width of the snail area are calculated; then, the length and width of each end are extended by 50 m to calculate the snail area. In general, equidistant line and frame surveys are set up, with a line spacing of 20 m and a frame distance of 20 m. The length (or width) of the snail area (m) = $(A + B) \div 2 + 50 \times 2$, where A is the longest spacing and B is the shortest (narrowest) spacing; snail area (m^2) = length (m) × width (m). The calculation of the snail area in an embankment epidemic area is similar to that in a water network area.
3. Calculation of Snail-infected Area. Calculation of the area of infectious snails in hills and water network areas is based on the calculation method of snail area. That is, the infected snail area is first determined, and if there are infectious snails in adjacent frames, then those frames are infectious snail segments. When a non-infectious snail area between two infectious snail segments is within 30 m, it is regarded as one infectious snail segment; if it is farther than 30 m, it is calculated as two infectious snail segments. The length of the infectious snail segment extends 15 m from the farthest point of the infectious snail segment, and the length of the isolated infectious snail segment is calculated as 30 m. To calculate the width of the infected snail segment, if a river ditch has water and the water level is stable year-round, the actual slope height of the river ditch is the width; if the river ditch has water in summer and no water in winter, the width is

the actual height of the ditch plus the bottom width; and if infectious snails are only on one side, the height of the side is the width, and the width of the ridge is calculated as 1 m. If an isolated infectious snail area is found in a marshland, the area can be extended 50 m to the surrounding area, that is, 10,000 m^2; if two infectious snail sites are within 50 m, the distance between the two points is added and spread 50 m to each side to calculate the infected snail area; and if the two infected snail sites are more than 50 m apart, they are calculated as isolated snail sites. The calculation of the snail area in an embankment epidemic area is similar to that in a water network area.

3.2.9 Surveillance of Other Types of Snails

In addition to *Oncomelania* snails, there are many other types of freshwater or terrestrial snails that can serve as intermediate hosts or vectors of parasites, such as *Bithynia*, *Bithynia manchourica*, *Alocinma*, *Bellamya*, Planorbidae, apple snails, *Achatina fulica*, *Radix plicatula*, and *Galba pervia* (Table 3.1).

According to the epidemic characteristics of local snail-transmitted parasitic diseases, surveillance and investigation of relevant snails can be carried out in a focused and planned manner. In surveillance, qualitative or quantitative collection should be carried out in accordance with the ecological and living habits, and habitat characteristics of different snails in different distribution areas.

In humid areas, ditches, ponds, streams, lakes, and springs, snails are mostly attached to stones and dead leaves in the water and can usually be collected by tweezers. If snails that live on the bottom of the water or on grass that is slightly farther from the shore are collected, a long-handled dip net can be used. The collection of snails in lakes and rivers can be carried out by triangular trawling. The collection of terrestrial snails can be carried out in suspected environments, such as grass, tree trunks, roots, rocks, and wet corners. Apple snails, *Achatina fulica*, and other snails that are often used as foods can be sampled and monitored in markets and catering establishments. The collected snails should be registered and preserved in a timely manner to identify the snail species and parasitic infections.

3.3 Surveillance Techniques for Vector Development, Migration, and Transmissivity

Global economic integration and climate change can directly affect the development and migration of a vector; that is, the density and regional distribution of a vector population may show significant changes. These changes directly affect changes in the capacity of vector transmission in a certain region and ultimately affect the spread of diseases. Therefore, surveillance of vector development, migration, and transmissivity is extremely important for the control of vector-borne diseases. This section uses mosquitoes, snails, etc., as examples to introduce surveillance techniques for vector development, migration, and transmissivity.

3.3.1 Surveillance of Vector Development

Through regular and continuous density surveillance of a certain vector, its development and decline can be understood. According to the habitats and ecological characteristics of different vectors, one or more density survey methods can be used to regularly carry out vector population investigations at fixed sites during the vector abundance season. Years of observation using the same method at the same location make it possible to grasp the local population and decline of the vector population in different time periods. During the density survey, local temperature, humidity, rainfall, wind power, wind direction, moon phase, and crop planting should be recorded or collected for reference when analyzing the results. The mosquito is taken as an example to introduce relevant surveillance methods for vector density.

3.3.1.1 Density of Adult Mosquitoes

1. Manual Hour. A certain number of residential houses in a representative residential area are selected, and one person catches mosquitoes in each room for 15 min and counts them in the morning. The captured mosquitoes are first identified by species and then counted. The total number of a mosquito type caught is the indoor density of this type of mosquito (no./manual hours). This method can also be used to investigate adult mosquito density in barns, vacant buildings, and other locations. When selecting or setting up surveillance sites, it is necessary to fully consider residential neighborhoods in different areas, such as urban central areas, suburban junctions, and rural areas, as well as livestock sheds, such as cowsheds and pigpens, and related wild environments. Surveillance can generally be performed twice a month at intervals of 15 days. The surveillance time is 1 h after sunset, and manual or electric mosquito traps in indoor locations of each surveillance site (corridors can also be selected in residential areas) are used to catch mosquitoes for 15 min. The captured mosquitoes are counted by species. The surveillance starts 1 month before the mosquito wintering activity. If mosquitoes are not caught during two consecutive surveillance efforts, then the surveillance can stop.
2. Mosquito Capture in Nets. All mosquitoes in a representative residential area are caught in the morning in 50 to 100 mosquito nets where people sleep. The captured mosquitoes are first identified by species and then classified and counted. The average number of mosquitoes caught in each mosquito net is used as the mosquito density (no./trap).
3. Host Trapping. People or livestock are used as bait to trap mosquitoes at a fixed location in the evening or overnight. The number of certain types of mosquitoes trapped within a certain period of time is the mosquito density (no./night or hour). Human or animal trapping can be used. The time of trapping depends on the visual density and the use of the survey. Trapping is generally carried out at

the peak of mosquito blood-sucking activity, such as from half an hour after sunset to overnight, or is performed for a certain amount of time. However, the timing should be the same for each survey site for comparison:

(a) Human trapping. There are many forms of human trapping. One is to hang an open mosquito net between a residential area and a breeding ground with the bottom of the net 35 cm above the ground. One person sits in the net and continually catches the mosquitoes that fly into the net. The second method is to hang an unopened mosquito net with the bottom touching the ground. One person sleeps inside the net. A large open mosquito net is placed outside the net, and another person regularly enters the large net to catch the mosquitoes landing on it. The third method is one person sitting with bare legs at the observation site to capture the blood-sucking mosquitoes that land on the legs in a timely manner and count them.

(b) Animal trapping. The animals (bovines, pigs, etc.) can be tied in a location between a habitat and a residential area. The investigators then constantly capture mosquitoes on the animal's body, surrounding walls, or other objects. Alternatively, a special open mosquito net can be hung with an animal tied inside it, and the mosquitoes that fly into the net can then be caught.

4. Light Trapping. Using the phototaxis of mosquitoes, mosquito traps are placed in certain spots to collect mosquitoes. The number of mosquitoes caught in a certain period of time is used as the mosquito density (no./hour or night). At the time of trapping, one mosquito lamp can be used at each location. The surveillance starts 20 min after sunset, and trapping is conducted continuously for 6 h. The next day, the mosquito collection box is removed; the mosquito type, sex, and number are identified; and the density of each mosquito species is counted.
5. Window Trapping. A window trap is placed on a wall or window facing the east side of a house or barn (Fig. 3.1), and the trapped mosquitoes are collected to count mosquito density. The density of blood-sucking and non-blood-sucking mosquitoes can be calculated separately as needed.
6. Mosquito and Mosquito Egg Traps. This method is commonly used to monitor *Aedes* mosquitoes. *Aedes* areas, such as human residential areas, parks, construction sites, hospitals, and waste collection stations, are chosen, and a certain number of traps and oviposition traps are deployed and surveilled every mid-month. When surveillance is carried out, those traps and ovipositors should have been in place for four consecutive days, and adult mosquitoes and mosquito eggs should be collected for checking on the fourth day.
7. Container Index. A certain number of residential areas, parks, construction sites, waste collection stations, or other suspected environments can be selected according to different geographical locations to check the positive rates of *Aedes spp.* distributions mosquito larvae and eggs in various indoor and outdoor water containers. Various types of water containers are classified as permanent (such as water tanks and pools) and temporary (such as vases, tires, and empty bottles). The density index of positive containers can be calculated separately. The formula is as follows:

$$\text{Container index (CI)} = \frac{\text{No. of positive containers with } \textit{Aedes} \text{ mosquito larvae and eggs}}{\text{Total no. of containers under surveillance}} \times 100\%$$

3.3.1.2 Density of Mosquito Larvae

A water body with a relatively stable water source should be selected as a survey site. For larger water bodies, a larvae spoon should be used regularly (every week or 10 days) to scoop 10–20 spoons at the edge of the breeding ground or the slow flow with water plants, with a 1-m-long water surface per scoop, and the number of larvae and eggs per scoop should be taken as the density (no./scoop). For small water bodies such as footprints and caves, a small spoon or a large-diameter pipette should be used to capture all larvae, and the number of mosquito larvae captured from each water body is the density.

3.3.1.3 Relative Density Calculation

Since mosquito density is affected by many factors, the antimosquito effects depend not only on the density change in the mosquito-killing area but also on the density change in the control area where mosquitoes are not killed. A commonly used correction indicator is the relative density ratio (RPI), which is calculated as follows:

$$\text{RPI} = \frac{\text{Density of control area before control measures}}{\text{Density of mosquito-killing area before control measures}} \times \frac{\text{Density at a certain time of mosquito-killing area after control measures}}{\text{Density at a certain time of control area after control measures}}$$

RPI is also an indicator of the antimosquito effects. The smaller the RPI value, the better the control effect is. For example, if the RPI is 0.4, the density is reduced by 60%. If the RPI ≥ 1, it is invalid.

3.3.2 Surveillance of Vector Diffusion and Migration

The vector population is generally located in a certain suitable ecological environment. However, changes in natural and social factors such as the ecological environment and human activities often cause the spread and migration of vector populations. This often leads to the spread of diseases and epidemics or to the emergence of new endemic areas. Therefore, surveillance of vector diffusion and migration provides predictions and early warning of the prevalence of parasitic diseases. In the following section, snails are mainly used as an example to introduce surveillance methods for diffusion and migration.

Snail migration and diffusion mainly occur through active crawling or swimming, drifting while attached to objects, and being carried by people and animals. Active crawling or swimming of snails is weak. Therefore, the migration and diffusion caused by these activities occur mainly around the original breeding environment, and surveillance can be achieved by expanding the original snail range. Through drifting with water and other objects or being carried by humans and animals, snails may migrate to distant areas. Once they are brought to a snail-free area that is suitable for them to live or an area where they have previously been eliminated, snail populations in these areas will emerge or rebound. The rebound or nonsnail area becomes a snail area, which causes the schistosomiasis epidemic to rise or spread. Therefore, strengthening the monitoring of snail diffusion is of great significance.

3.3.2.1 Surveillance of Floating Objects

This method is used mainly for water systems that communicate with endemic areas of schistosomiasis. The situation of snails carried by floating objects drifting from the water system in the epidemic area should be monitored. For example, in the Yangtze River Basin, surveillance is conducted mainly in the flood season and periods of water withdrawal and should be carried out once a month or semimonthly in the flood season. The surveillance sites are mainly beaches that have been flooded (including inland areas that have been flooded after breakwaters have been damaged) as well as river courses and their tributaries. There are two aspects of surveillance: the first is to monitor floating objects, that is, to classify, weigh, and register floating objects and check whether there are snails on each object, and the second is to check the areas surrounding the floating objects for snails. Once snails are found on a floating object or in the surrounding environment, a comprehensive investigation of the entire beach or river course should be carried out to identify the location and area of snail diffusion, deal with the snail environment in a timely manner, and prevent snails from further breeding and spreading. The slopes of hilly areas often have snails. After heavy rains or mountain flooding, snails can be washed down from these slopes with the loosened plants and then flow downstream along streams or deep valleys.

3.3.2.2 Surveillance of Aquatic Plants

Snails climb along aquatic plants. Snails often migrate and spread to other places by attaching to plants such as reeds, water lotus, and wormwood. When monitoring, the situation of local aquatic plants, especially the distribution and sources of aquatic plants with economic or use value, must first be understood. Through comprehensive or sampling inspection of aquatic plants introduced from endemic areas, plant placement sites, and planting sites, snail surveillance is carried out year-round.

3.3.2.3 Surveillance of Carriers

This method mainly involves surveillance of the toe gaps of mammals such as cattle, sheep, and pigs introduced from epidemic areas to check for snails; in addition, vessels from endemic areas, aquatic animals such as fish and shrimp, fishing tools, etc. are checked for snails. Tree species introduced from endemic areas should also be monitored.

3.3.3 Surveillance of Vector Transmissivity

Vector transmissivity refers to the number of people to whom a disease may be transmitted by a vector in a certain period from a primary case. It is important to measure the capacity of a certain vector to transmit diseases, evaluate the effects of vector control measures, and predict the spread of vector-borne diseases. Such measurement is currently used in the surveillance of insect-borne diseases such as malaria. In this section, *Anopheles sinensis* is used as the main example.

3.3.3.1 Parameters and Calculation Methods for Vector Transmissivity

1. Biting Rate. This parameter is the number of bites by a vector *Anopheles* per person per night, expressed in *ma*. The biting rate should be investigated based on the specific circumstances of each place.

   For outdoor blood-sucking *Anopheles* mosquitoes in areas where residents generally sleep outside, the number of mosquitoes trapped by one person outdoors in a night is the biting rate. For indoor blood-sucking *Anopheles* mosquitoes in places where few residents sleep outside, the biting rate can be calculated in different ways in different situations. In areas where people have few outdoor activities during the night and mostly do not use mosquito nets, the number of *Anopheles* trapped by one person indoors can be used as the biting rate; in areas where residents stay outdoors to cool off and then go indoors to sleep or sleep in mosquito nets, the biting rate can be calculated by counting the average number of mosquitoes trapped by one person in one place.
2. Biting Habit. This parameter is the ratio of blood drawn per mosquito per day, expressed in *a* with two factors: the daily blood-sucking frequency and the human blood index. Daily blood-sucking frequency × human blood index is the formula for calculating the habit of biting human beings.
3. Daily Survival Rate. This parameter is the daily survival number divided by the mosquito population, expressed as *P*.
4. Life Expectancy of Infectious Mosquitoes. This parameter is the time period that infectious mosquitoes can live, expressed as pn/−lnp. Being infective means that *Plasmodium* has completed spore proliferation in the mosquito. The life expectancy of infectious mosquitoes has three factors: daily survival rate (*P*), spore growth period (*n*), and life expectancy (1/−ln*p*).

3.3.3.2 Calculation of Vector Capacity

Vector capacity (*C*) is the product of the rate of biting humans (*ma*), the habit of biting humans (*a*), and the life expectancy (pn/lnp) of the infectious mosquito. It is calculated as follows:

$$C = \mathrm{ma} \times a \times p^n \times \frac{1}{-\ln p} = \frac{\mathrm{ma}^2 p^n}{-\ln p}$$

In fact, *Anopheles* mosquitoes may not be infected after they have sucked *Plasmodium* gametophytes, and various *Anopheles* mosquitoes have different infectivity susceptibility to different species of *Plasmodium*. Therefore, the above formula should include the infectivity susceptibility of mosquitoes to different species of *Plasmodium* (*b*). The *b* value is usually expressed as the manual infection rate. After *b* is added, the formula for *C* is as follows:

$$\mathrm{C} = \frac{\mathrm{ma}^2 b p^n}{-\ln p}$$

3.4 Case Analysis

3.4.1 Surveillance of Snails in the Baoying and Gaoyou Sections of the Li Canal Along the East Route of the South-to-North Water Transfer Project

3.4.1.1 Proposing the Issue

The Li Canal is a section of the Beijing-Hangzhou Grand Canal between the Yangtze River and the Huaihe River. It has been planned as an important water conveyance section of the East Route of the South-to-North Water Transfer Project. Both the Baoying section and the Gaoyou section of the Li Canal pass through historically endemic areas of schistosomiasis. Therefore, surveillance of the distribution of snails in the Baoying and Gaoyou sections was needed to provide a basis for the scientific evaluation of whether the East Route Project of the South-to-North Water Transfer Project could cause the spread of schistosomiasis epidemic areas north of 33°15′ north latitude. To this end, special investigation and monitoring of the distribution of snails in the Baoying and Gaoyou sections were carried out.

3.4.1.2 Surveillance Components and Methods

1. Investigation of the Historical Distribution of Snails. A retrospective investigation was used to check local schistosomiasis epidemic and prevention data, and the distribution of snails in the Baoying and Gaoyou sections of the canal was recorded.
2. Dynamic Surveillance of Snails in the Li Canal. The snail investigators were organized every spring from 2001 to 2010 to investigate the snails in the Gaoyou section through systematic sampling and environmental sampling. The distribution of the snails was determined by GPS, the area and density of the snails were counted, and the snails were dissected to observe whether they were infected with schistosomiasis.
3. Snail Survey of the Water body of the Li Canal. Dragging in the river was used to investigate snails carried in the canal water flotage in 2006–2010. The investigation was conducted once a month from June to September every year. Floating objects on the 1000 m^2 surface of the river near the shore were surveyed and weighed. The seized snails were counted and identified. Snails in the canal were investigated by monthly trapping from June to September. Approximately 10–20 straw curtains were placed on the water surface near the shore of the monitoring section. The distance between the curtains was 10–20 m. After

7 days, the straw curtains were removed and then washed and sieved, and the trapped snails were counted and identified.

3.4.1.3 Surveillance Results

1. Historical Distribution of Snails. According to the survey, snails were discovered in 1965 on the eastern slope of the canal at the Qingkan dyke in the Baoying section of the Li Canal. They were distributed from the opposite side of the Ziying sluice gate to the Qilizha crossing, covering an area of 4800 m^2. In 1984, snails were found in the Longdukou part of the Baoying section, the eastern slope of the canal, covering 4935 m^2 (this was also the northernmost boundary of snail distribution in the inner canal), and snails were not found in the Baoying section of the canal afterward. In 1955, snails were first discovered in the Gaoyou section of the Li Canal from the Cheluo sluice gate to the Ershilipu area, with an area of 1.4 hm^2. Snail surveys were carried out every year afterward. In 1971, the snail area was the largest, reaching 8.3795 hm^2. After the stone revetment cement was grouted in 1977–1980, snails could not be found, but after 1993, snails were found several times in damaged areas of the stone revetment.
2. Dynamic Surveillance of Snails in the Li Canal. In 2001–2010, dynamic surveillance of snails in the Gaoyou section of the Li Canal showed that the snail density was low, with fluctuations in small areas. In 2010, the active snail density increased. The snails were distributed in three areas of Gaoyou city in a range from 119°24′36″ to 119°25′34″ and from 32°48′30″ to 32°54′'05″". No northward spread of the snail was found.
3. Snail Survey in Water. In 2006–2010, a total of 1506 kg of floating materials were dragged in the Baoying and Gaoyou sections of the Li Canal for snail diffusion monitoring inside water. A total of 2148 other kinds of snails were found, but no *Oncomelania hupensis* was found. A total of 520 straw curtains were placed. A total of 14,509 other snails and no *Oncomelania hupensis* were trapped.

3.4.1.4 Recommendations

According to historical data, snails were found in the Gaoyou section of the Li Canal in 1955. The northernmost point of the canal snail distribution was the Longdukou part of Dongpo-Baoyingdi, the west dike of the canal. From 2001 to 2010, dynamic monitoring of snails in the Gaoyou section of the Li Canal showed that the snail density was low with fluctuations in small areas, and the snails had not diffused northward. The surveillance data showed that since the discovery of snails in the Li Canal 55 years ago, including the construction and operation of the Jiangsu North Transfer Project in Jiangsu Province over nearly 50 years, there was no evidence that the snails had spread northward and no report of snails north of 33°15′ north latitude. However, the source of the East Route Project of the South-to-North Water Diversion Project still provided suitable habits for snails living and spreading, and with its long water transmission line and many influencing factors, the scale of the project is much larger than that of the Jiangsu North Transfer Project in Jiangsu Province. In particular, the observation and evaluation of the impact of the project on the ecological environment are quite long processes. Therefore, it is necessary to carry out long-term and continuous surveillance of snails and schistosomiasis along the eastern line of the South-to-North Water Transfer Project.

3.4.2 Surveillance of *Anopheles sinensis* in Jiangsu Province from 2008 to 2009

3.4.2.1 Proposing the Issue

Jiangsu Province is located in the lower reaches of the Huaihe River and the Yangtze River. The plain area accounts for 68% of the total area of the province. The water area of rivers, lakes, and reservoirs accounts for approximately 17% of the total area of the province, and the climate and environment are very suitable for breeding *Anopheles* and malaria transmission. *Anopheles sinensis* and *A. lesteri* are the most important malaria vectors in the province. To understand the development trends of the population and density of *Anopheles* mosquitoes and determine prevention effects, surveillance sites for *A. sinensis* were established in Sihong, Suining, and Qixian Counties, and the surveillance results of 2008–2009 were analyzed to provide a scientific basis for the development of future malaria control measures and to correctly evaluate the effects of vector control measures.

3.4.2.2 Surveillance Components and Measures

Density Surveillance of *A. sinensis*

1. Selection of surveillance sites. One residential area each was selected from Sihong, Suining, and Qixian as surveillance sites for *Anopheles* spp. In the period from late June to early October 2008–2009, manual trapping of *A. sinensis* was carried out semimonthly:
   (a) Surveillance ways
      - Bait trapping. A mosquito net was placed between the residential area at the surveillance site and the breeding site from 1900 to 2400. The bottom of the net was 35 cm above the ground. One person sat in the net and continually caught mosquitoes that flew into the net. Another person stayed outside the mosquito net and caught *Anopheles* spp. mosquitoes. After all *Anopheles* spp. captured were identified, density was calculated by species.
      - Trapping in pig shed. The trapping started half an hour after sunset. One person stayed in a pig shed in the village to catch mosquitoes for 15 min. After all captured mosquitoes were identified, density was calculated by species.

Surveillance to *A. sinensis* Sensitivity to Insecticides

1. First-generation mosquito adults. In 2008, a total of 498 *A. sinensis* were captured at field vector surveillance sites in Sihong, Suining, and Qixian Counties.

In 2009, 279 were captured, and the mosquito breeding room was used for single-tube feeding. The eggs produced were identified by mosquito species together with the egg masses of *A. sinensis* at the monitoring points. The mosquitoes were then raised to the first generation of adult mosquitoes for testing. Mosquito room conditions: room temperature (25 ± 1) °C, relative humidity (75 ± 5) %, light 10 h/day, and 5% glucose water per day.

2. Insecticide and concentration test: 0.05% deltamethrin, 0.15% cyfluthrin, and 5% malathion.
3. Sensitivity test by experimental method: A filter paper contact tube for adult mosquitoes was used to distinguish doses of 0.05% deltamethrin (0.0178 g/m^2), 0.15% cyfluthrin (0.0534 g/m^2), and 5% malathion (1.78 g/m^2) to test *A. sinensis* sensitivity. Approximately 20 to 25 female *Anopheles* spp. mosquitoes were used per test tube, and each sample was tested at least twice with a blank control group. The first knockdown time and the number of mosquito knockdowns at 10, 15, 20, 30, 40, 50, and 60 min were recorded, and the knockdown rate was calculated. After 60 min of contact, all mosquitoes were placed in a recovery cylinder and kept in a mosquito room. After 24 h, the number of dead mosquitoes and the number of surviving mosquitoes were checked, and mortality was calculated. According to the WHO drug resistance evaluation index, the mortality rate is 98%–100% for the sensitive group (S) and 80%–97% for the preliminary resistance group (M) (resistance needs further verification); those with mortality <80% are the resistant group (R).

3.4.2.3 Surveillance Results

1. Mosquito capture density and biting rate in the net for half a night. A total of 322 *Anopheles* spp. mosquitoes were collected at midnight at three surveillance sites in 2008. The mosquitoes were identified as *A. sinensis*, and their biting rates were 0.80, 4.24, 1.56, 2.28, 2.04, 1.20, 0.64, and 0.12/(per labor hour) every 10 days from late June to early October. The average biting rate was 1.61/(per labor hour). A total of 349 mosquitoes were captured at the three surveillance sites and identified as *A. sinensis* in 2009. The biting rates in late June to early October were 0.74, 3.51, 2.23, 1.09, 0.94, 0.89, 0.57, and 0/(per labor hour), with an average biting rate of 1.25 (per labor hour).
2. *Anopheles* spp. density in the pig shed. A total of 886 *Anopheles* spp. mosquitoes were captured at the three pig shed surveillance sites in 2008. All were identified as *A. sinensis*. The densities were 152, 271, 164, 187, 72, 123, 120, and 93 (per labor hour) every 10 days from late June to early October, and the average density was 147.66/labor hour. In 2009, a total of 652 *Anopheles* app. mosquitoes were captured at the three pig shed surveillance sites, and all were identified as *A. sinensis*. From late June to early October, the mosquito densities were 164, 296, 75, 61, 43, 112, 89, and 29/labor hour every 10 days, and the average density was 108.67/labor hour.
3. Sensitivity of *A. sinensis* to insecticides. In 2008, a total of 286 first-generation *A. sinensis* were tested for deltamethrin sensitivity. The first *A. sinensis* was

knocked down by deltamethrin after 12 min. The knockdown rate was 67.48% after 60 min and 69.58% after 24 h, indicating that the resistance of *A. sinensis* to deltamethrin had reached the level of R. The test of 197 first-generation *A. sinensis* sensitivity to cyfluthrin indicated that the first knockdown by cyfluthrin occurred after 7 min, the knockdown rate was 88.83% after 60 min, and the mortality rate was 90.36% after 24 h, indicating that *A. sinensis* resistance to cyfluthrin had reached the level of M. In 2009, a total of 44 first-generation *A. sinensis* were tested for sensitivity to deltamethrin. The first *A. sinensis* was knocked down by deltamethrin after 28 min, and the knockdown rate after 60 min was 34.09%. The mortality rate was 29.55% after 24 h, indicating that the resistance of *A. sinensis* to deltamethrin had reached the level of R. There was no knockdown at 60 min when the sensitivity of 41 *A. sinensis* to malathion was tested, and the mortality rate after 24 h was only 2.44%, indicating that *A. sinensis* has high resistance to malathion.

3.4.2.4 Recommendations

The biting rate of *A. sinensis* in Jiangsu Province began to rise in late June and decreased significantly in early October, with the peak season in July, which differed from seasonal variations in *A. sinensis* populations in other provinces. Temperature, rainfall, and farming systems were related to the biting rate in different places. The density of *A. sinensis* in the pig shed remained high during the whole malaria transmission season, which may be related to the blood-sucking habit. However, in the future, if the external environment changes, for example, if the number of livestock decreases, the blood-sucking habits and habitats of *A. sinensis* may also change. Then, the biting rate may increase significantly. Therefore, to control the incidence of malaria, it is necessary to strengthen health education for local residents, raise awareness of mosquito prevention in communities, and reduce the chances of mosquito contact with people. The monitoring results also showed that in 2008, *A. sinensis* had resistance to deltamethrin in some areas and had initial resistance to cyfluthrin. In 2009, the resistance of *A. sinensis* to deltamethrin increased further. Phosphorus has almost failed. However, due to the small number of mosquitoes tested in 2009, the sensitivity of *A. sinensis* to insecticides needs to be further and keeping monitoring. When using insecticide-treated mosquito nets or indoor residual spraying to treat outbreak sites or unstable areas, the appropriate insecticide should be selected to achieve optimal control effects.

Surveillance of Reservoir Hosts

4

Yao Deng

In the host-parasite relationship, hosts provide the nutrients needed by the parasites and even a temporary or permanent habitat for the parasite. Reservoir hosts have an important biological significance for the transmission and prevalence of parasitic diseases. This chapter elaborates surveillance techniques for species, transmission roles, and demographic characteristics of reservoir hosts.

4.1 Overview

4.1.1 Definition

Some animals serve as hosts of human parasites, similar to the parasitic phase in humans, and those animals are called reservoir hosts of human parasites. Dogs and cats are examples of reservoir hosts of *Paragonimus spp.*, as adult parasites of *Paragonimus spp.* that can live and reproduce on them to complete their life cycle.

4.1.2 Varieties

The reservoir hosts of parasites are mainly domestic animals (such as dogs, cats, pigs, cattle, and sheep) and wild animals (such as rodents, wild boars, wildcats, and foxes). Their types and distributions are of great significance in controlling sources of infection and cutting off transmission routes. This section considers several reservoir hosts for major human parasites (Table 4.1).

Y. Deng (✉)
Jiangsu Institute of Parasitic Diseases, Wuxi, Jiangsu, China

X.-N. Zhou (ed.), *Surveillance and Management of Parasitic Diseases in China*,
https://doi.org/10.1007/978-981-97-1241-0_4

Table 4.1 Common human parasites and their reservoir hosts

Parasite classification			Distribution	Pathopoiesia	Main reservoir hosts
Protozoan	Lobosea (amoebae)	*Entamoeba histolytica*	It distributes worldwide but is common in tropical and subtropical regions, such as India, Indonesia, the Sahara Desert, tropical Africa, and South America. It is distributed mainly in northwest, southwest, and north China, and the infection rate in Yunnan, Guizhou, Xinjiang, and Gansu Provinces exceeds 2%	Amebiasis	Cats, dogs, monkeys, and mice
	Flagellate (trypanosomes)	*Giardia lamblia*	It has a global distribution and occurs frequently in both developing and developed countries With nationwide distribution in China, rural infection rates are higher than those in cities	Giardiasis (lambliasis)	Cattle, sheep, pigs, rabbits, cats, dogs, beavers, and muskrats
		Leishmania donovani	It is widely distributed in Asia, Europe, Africa, and Latin America. The main areas are India and Mediterranean coastal countries. In China, nazar is prevalent in the vast rural areas north of the Yangtze River in 17 provinces (municipalities, autonomous regions), including Shandong, Tianjin, Henan, Jiangsu, Anhui, Shaanxi, Gansu, Xinjiang, Ningxia, Qinghai, Sichuan, Shanxi, Hubei, Liaoning, Inner Mongolia, and Beijing	Kala-azar	Dogs and raccoons
		Trypanosoma brucei gambiense	It is found in West and Central Africa	Sleeping sickness (West Africa, Equatorial Africa, and the Gambia)	Cattle, pigs, goats, sheep, antelopes, dogs, and other animals
		Trypanosoma brucei rhodesiense	It is found in East Africa and South Africa	Sleeping sickness (East Africa and Rhodesia)	African antelopes, cattle, lions, and hyenas
		Trypanosoma cruzi	It is distributed mainly in South America and Central America	American trypanosomiasis, Chagas disease	Dogs, cats, bats, rats, and other mammals
	Sporozoan	*Blastocystis hominis*	It is distributed worldwide, especially in Southeast Asia, South America, and other developing countries The insect infection has been found in 22 provinces and cities in China	Blastocystis	Monkeys, orangutans, dogs, pigs, cats, mice, and other animals
		Toxoplasma gondii	It has worldwide distribution and has been found in 30 provinces, municipalities, and autonomous regions in China	Toxoplasmosis	Cats
		Cryptosporidium spp.	It has worldwide distribution and has been reported in more than 90 countries thus far; the infection rate is lower in developed areas than in developing areas	Cryptosporidiosis	Sheep, cats, dogs, rabbits, and cattle
	Infusorian (paramecia)	*Balantidium coli*	Worldwide distribution, including tropical and subtropical areas	Balantidiasis	Pigs and trench rats

Helminth	Trematoda	*Clonorchis sinensis*	Japan, North Korea, South Korea, North Vietnam, and small parts of Russia and China	Clonorchiasis	Cats, dogs, pigs, rats, minks, foxes, badgers, wildcats, otters, and 33 other kinds of animals
		Fasciolopsis buski	Temperate and subtropical regions of Asia	Fasciolopsiasis	Domestic pigs, but natural infections have also been reported in wild boars, macaques, and dogs
		Fasciola hepatica	Sporadic cases around the world	Fascioliasis	Cattle, sheep, and other mammals
		Paragonimus westermani	Japan, North Korea, Russia, the Philippines, Malaysia, India, Thailand, China, Africa, and South America	Paragonimiasis	Tigers, leopards, wolves, foxes, big cats, and civet cats
		Paragonimus skrjabini	Gansu, Shanxi, Shaanxi, Henan, Sichuan, Yunnan, Guizhou, Hunan, Hubei, Zhejiang, Jiangxi, Fujian, Guangdong, and Guangxi Provinces in China	Paragonimiasis	A variety of domestic or wild animals from the cat family and dog family, such as civet cats, domestic cats, dogs, and ocelots
		Schistosoma japonicum	China, Japan, the Philippines, and Indonesia	Schistosomiasis japonica	Domestic animals, including cattle, sheep, dogs, cats, pigs, rabbits, and horses; wild animals, such as brown rats, hares, wildcats, wild boars, foxes, and ocelots
	Tapeworm	*Echinococcus granulosus*	Worldwide distribution, mainly in developed areas with animal husbandry	Cystic echinococcosis	Dogs, jackals, wolves, and other canid carnivores
		Echinococcus multilocularis	High latitudes and permafrost zones in the Northern Hemisphere, from northern Canada, Alaga, Canada, and the United States to Hokkaido, Japan, and Siberia, Russia, throughout North America, Europe, and Asia. It has been reported in Ningxia, Xinjiang, Qinghai, Gansu, and ten other provinces and cities in China	Alveolar echinococcosis	Foxes, dogs, cats, etc.
		Hymenolepis nana	The Americas, Europe, Asia, Oceania, and Africa	Hymenolepiasis	Rodents, including various house mice, voles, etc.
		Raillietina celebensis	It is widely distributed in tropical and subtropical regions, with human infections found in Thailand, Myanmar, Vietnam, the Philippines, Japan, Australia, and Madagascar; in China, it has been reported in Taiwan, Fujian, Guangdong, Jiangsu, Zhejiang, and Guangxi	Raillietiniasis	Rodents, including *nigicus*, *norvegicus*, and small mice
		Bertiella studeri	Mauritius, the Philippines, East Africa, Indonesia, India, Vietnam, and Singapore	Bertielliasis	Monkeys and other primates
	Nematode	*Trichinella spiralis*	It is distributed worldwide and still occurs at serious rates in Russia and Eastern European countries, Mexico, Chile, Argentina, and Thailand. Epidemic areas of human trichinellosis in China are mainly Yunnan, Xizang, Guangxi, Sichuan, Hubei, Henan, and three northeastern provinces	Trichinosis	Pigs, rats, cats, bears, and many other vertebrates
		Strongyloides stercoralis	In tropical, subtropical, and temperate regions, with sporadic cases appearing in cold areas; infected people were found in 26 provinces (municipalities and autonomous regions) in China, mainly in the southern regions	Strongyloidiasis	Dogs, cats, and foxes
		Trichostrongylus orientalis	It is widely found in Japan, China, and North Korea and is prone to epidemic in rural areas where manure is widely applied. Infected people have been found in 18 provinces (municipalities) in China, with high infection rates in Hainan, Jiangxi, Zhejiang, Yunnan, Qinghai, Fujian, Guizhou, and others	Trichostrongyliasis	Sheep, camels, horses, cattle, donkeys, and other animals
		Gongylonema pulchrum	International distribution, with cases occurring in Italy, the former Soviet Union, Bulgaria, Morocco, New Zealand, Sri Lanka, and China. The three reported cases in China were in regions north and south of the Yangtze River	Gongylonemiasis	Sheep, cattle, horses, mules, camels, pigs, bears, monkeys, dogs, cats, rats, etc.
		Angiostrongylus cantonensis	Distribution in tropical and subtropical regions. It is prevalent mainly in Southeast Asia, the Pacific islands, Japan, and the United States, while in China, it is located mainly in Taiwan, Hong Kong, Guangdong, Zhejiang, Fujian, Hainan, Tianjin, Heilongjiang, Liaoning, Shanghai, Hunan, and other places	Angiostrongyliasis cantonensis	Mainly rodents, including *Rattus norvegicus* and *Rattus*. In addition, it occurs in *Rattus tanezumi*, *Rattus edwardsi*, and *Rattus rattoides*
		Gnathostoma spinigerum (G. siamense)	It is distributed mainly in Southeast Asia, with the most human infections in Thailand and Japan. In China, it has been found in 15 provinces (cities), including Zhejiang, Jiangsu, Anhui, Hunan, Hubei, and Shanghai	Gnathostomiasis	Dogs, cats, etc. It can also be parasitic in tigers, lions, leopards, etc.
		Thelazia callipaeda	Distributed mainly in Asia. Cases have been reported in India, Myanmar, the Philippines, Thailand, Japan, Korea, and Russia's Far East. Cases have been reported in 25 provinces (cities, autonomous regions) in China, with Jiangsu, Hubei, Anhui, Henan, Shandong, and other places having more cases	Thelaziasis	Dogs, cats, etc.

4.2 Surveillance of Disease Transmission by Reservoir Hosts

Reservoir hosts are critical sources of infection in the transmission of human parasitosis. The role of reservoir hosts in the transmission of parasitic diseases includes not only preserving the parasites as a reservoir but also complexifying routes of transmission with various potential sources of infection. The goal of reservoir host surveillance is to investigate the transmission role of reservoir hosts.

In endemic areas, it is recommended that reservoir host surveillance be conducted on livestock and wild animals based on the level of local transmission. This chapter presents the reservoir host surveillance of human echinococcosis (hydatidosis or hydatid disease) transmission as an example.

4.2.1 Surveillance Components

4.2.1.1 Choose the Investigation Area

In endemic areas, according to the needs of echinococcosis surveillance, surveillance sites are established in village units. If a village has a population of less than 2500, a neighboring village with a population over this threshold will be chosen instead. Additionally, abattoirs and large-scale slaughterhouses in the surveillance area will be investigated.

4.2.1.2 Investigation of Definitive Hosts

At every surveillance site of the targeted counties, 100 dog fecal samples from each household with *Canis familiaris* or 20 fecal samples from unclaimed dogs in villages with homeless canines (if the village being investigated has fewer than 20 unclaimed dogs, fecal samples should be collected from all homeless canines) should be collected. Each collected sample is tested for echinococcosis-Ag with ELISA (enzyme-linked immunosorbent assay). The test results are documented in the infection tracking log for echinococcosis.

After antigen tests are performed at county-level CDCs, ten positive samples (if this threshold is not met, all positive samples should be kept) and ten negative samples are randomly selected for double-checking with ELISA at provincial CDCs.

County-level CDCs should send Ag-positive samples with the infection tracking log to provincial CDCs for confirmation with PCR tests (if the provincial CDCs lack the resources to perform PCR tests, the Ag-positive samples should be sent to the national CDC as a reference laboratory for confirmation). The results of the PCR tests should be recorded.

4.2.1.3 Investigation of Intermediate Hosts

In abattoirs and large-scale slaughterhouses of the investigation area, the entrails of 1000 sheep or 300 cattle (it is preferable for them to be from different counties) should be tested annually. The examination includes documenting key information about the farm animals (such as their age by dentition and sources), inspecting the surface of each lung and liver for cysts or lumps, performing autopsy on detected cysts or lumps to identify echinococcosis-infected regions, keeping entrails infected with echinococcosis, recording the details of echinococcosis infection (such as the locations, number, and sizes of infections), filing the investigation form for animal echinococcosis infection, and sending each infected sample with the investigation form to provincial CDCs for confirmation with biopsies.

At least 500 small rodents and rabbits should be caught in surveilled counties in mixed echinococcosis endemic areas from April to June every year around human settlements, and the livers and lungs should then be examined by autopsy and diseased organs kept. The questionnaire of echinococcosis of rodents and rabbits should be filled out, and the diseased organs and questionnaires sent to provincial CDCs for pathological section identification.

4.2.2 Surveillance Reports and Their Usage

County-level CDCs are responsible for allocating specialists to collect and manage data from surveillance sites, conduct data analysis, draft annual surveillance reports for the surveillance sites, and submit the annual surveillance data and reports to prefecture-level CDCs at the end of each December. Prefecture-level CDCs are responsible for summarizing, reviewing, and analyzing the surveillance data submitted by county-level CDCs and submitting annual surveillance data from surveillance sites and reports from the corresponding prefectures to provincial-level CDCs by the end of the following January. Provincial CDCs are responsible for managing and analyzing the surveillance data submitted by prefecture-level CDCs, drafting annual surveillance reports for the provinces, and submitting the annual surveillance data and reports to the NIPD China CDC by the end of the following February. The China CDC is responsible for summarizing national surveillance data and reports, drafting annual national surveillance reports, and providing annual surveillance outcomes to all provinces in a timely manner.

By continuously monitoring the infectious level of reservoir hosts in endemic areas of human echinococcosis, data can be provided to evaluate public health interventions for human echinococcosis and to support evidence-based decision-making.

4.3 Surveillance of Natural Population Changes in Reservoir Hosts

4.3.1 Natural Population Changes in Reservoir Hosts

Natural population changes in reservoir hosts are natural decreases or increases in the quantity of species, population density, and abundance. In a specific region, clarifying the density, species, and distributions of reservoir hosts and identifying the main species are of significant value for parasitic disease control and prevention.

Various species can act as reservoir hosts, including domestic animals (such as pigs, dogs, cats, cattle, and sheep) and wild animals (such as tigers, leopards, and mice). Mice, as the largest group in terms of the number of species, distribution, and number of individuals, are reservoir hosts for many critical parasitic diseases because of their wide distribution, high productivity, and proactive characteristics. With increasingly interconnected global travel, imported mouse-transmitted diseases and mice as potential reservoir hosts are increasing. Therefore, it is critical to conduct surveillance of their natural ebb and flow. Compared to other wild animals, mice are relatively easy to surveil. Therefore, most literature on reservoir host surveillance is on mice.

4.3.2 Investigation of Species and Abundance of Mice

Rat traps, rat sticking boards, etc. are arranged in indoor, farmland, outdoor, or other environments according to specific conditions such as area size and terrain characteristics and the corresponding parasitic disease monitoring scheme. The captured rodents are packed, brought to the laboratory, and anesthetized with ether until deep coma or death, and the species and sex are then identified. By catching rodents regularly every month, monthly fluctuations in local rodent density can be obtained.

4.3.3 Population Estimation of Wild Animals

Capture and recapture techniques can be measured to estimate the population of wild animals. This technique was invented in the eighteenth century and developed rapidly in the twentieth century. The main idea is to "capture" a batch of randomized samples and then release them with markings; when "recapturing" a second batch of randomized samples, the number of animals marked in the first capture attempt should be recorded. By assuming that the percentage of marked animals in the second batch of randomized samples is similar to the percentage of animals captured in the first attempt to the whole animal population, the animal population can be calculated (the number of animals captured in the first attempt/the number of the whole animal population × 100% = the number of marked animals in the second batch of randomized samples/the number of animals in the second batch of randomized samples × 100%. That is, if 200 animals are captured, marked, and released in the first attempt, and 10 marked animals are found among 100 samples recaptured samples, then it is estimated that the headage of this animal is 2000):

$$\text{Formula}: N = \{(M+1)(n+1)\} / (m+1) - 1$$

$$\text{Var}(N) = \{(M+1)(n-1)(M-m)(\text{n}-m)\} / \{(m+1)(m+2)\}$$

where *N is the animal population, M is the sample size of the first capture, n is the sample size of the second capture, and m is the number of marked animals.*

4.4 Sentinel Surveillance

4.4.1 Surveillance Purposes

Sentinel surveillance refers to surveillance activities conducted in a selected population or environment with standardized components and methods for specific surveillance purposes. Sentinel surveillance has the advantages of low cost and high efficiency. Using mice to conduct sentinel surveillance in schistosomiasis-endemic areas, which is called sentinel mouse surveillance, is the most common sentinel surveillance method. It is commonly conducted within zones where schistosomiasis is suspected by placing mice in the area of a to-be-tested water body for a period of time and then keeping the mice indoors for 30 days. Then, the schistosomiasis pollution level of the water body is identified by dissecting the mice to determine their infection status by counting the number of parasites found. Through planned, continued, and systematic surveillance of water bodies in endemic areas with this method, surveillance, parasite monitoring, and early warning can be achieved. Sentinel mouse surveillance is critical in China's national monitoring and early-warning system for schistosomiasis. With the progress achieved in schistosomiasis control and prevention in China, the human and animal schistosomiasis infection rates are decreasing, and an increasing number of schistosomiasis-endemic areas are coming under control. To protect this achievement, the need for monitoring and early warning for schistosomiasis is growing.

4.4.2 Surveillance Techniques

This section considers the example of surveillance of sentinel mice in endemic areas of schistosomiasis japonica. After the larvae of *Schistosoma japonicum* escape from infected snails, 98% of the larvae rise and float statically on the surface of the water body. Therefore, good detection effects can be obtained by using sentinel mice to contact the water surface.

The common technique of sentinel mouse surveillance includes placing approximately ten schistosomiasis-free mice in a metal cage with empty bottles or plastic floats tied to both sides. The cage has enough space for the mice to move yet not to leave the bottom of the cage. After the cage is placed in the to-be-tested water body, all legs, abdomens, and tails of the mice should contact the water body. There is no standard time and frequency for the water contact. Commonly, the water contact for mice could continue for 5–7 h or 1–2 h daily for a few successive days. Based on the distance to the shore of the water body, the cage can be tied with a long rope to a tree or piling on the shore to avoid losing it. To test the infection status of a specific water body, it is best to conduct this field experiment on windless days to avoid interfering with the natural distribution of the schistosomiasis larvae. Since the mice might die in cold weather, it is best to conduct this experiment before noon and to avoid days with low air or water temperatures. The mice should be guarded from attacks by water snakes, other animals, or humans during the period in the water. To investigate the distribution of schistosomiasis larvae across the surface of the water body, numerous cages with 5–10 mice per cage can be placed various distances from the shore at 10–20 m intervals. To investigate the temporal distribution of the larvae, a new cage of mice can be placed in the same location every 4 h. During the experiments, air temperature, water temperature, flow velocity, wind direction, etc. should be recorded. After being exposed to the water body, the treated mice should be taken back to the laboratory, kept under regular conditions for 28–35 days, and then dissected to check for schistosomiasis infection in the hepatic portal vein and mesenteric vein.

The infection of the tested water body is represented by the infection rate and infection level of the sentinel mice:

$$\text{Infection rate of sentinel mice}\left(\%\right)=\frac{\text{No.of infected mice}}{\text{No.of dissected mice}}\times 100\%$$

$$\text{Infection level of sentinel mice}\left(\%\right)=\frac{\text{Total no.of adult schistosomes identified}}{\text{No.of dissected mice}}$$

4.5 Case Study

4.5.1 Investigation of Hosts of *Paragonimus* spp. in Zhejiang Province

4.5.1.1 Proposing the Issue

Paragonimus spp. were first found in the lungs of wild animals such as otters, tigers, and civet cats, and adults or eggs of the parasite were then found in the lungs or sputum of patients who come from Taiwan and Xiamen of China, and Japan. Later, it was proven that domestic animals such as cats, dogs, and pigs are also served as the definitive hosts. Understanding the current situation of *Paragonimus* hosts is of great significance in studying *Paragonimus* etiology, epidemiology, and control measures. From 1960 to 1980, some studies investigated the natural infection of *Paragonimus* hosts caught in mountainous areas or animals kept in zoos in Zhejiang Province.

4.5.1.2 Surveillance Components and Methods

A total of 626 animals of 53 different species were investigated. The reservoir hosts were determined from three different sources: (1) dissection performed on dogs and cats purchased from schistosomiasis-endemic areas in the southern, western, and eastern areas of Zhejiang Province (the endemic areas were identified based on the addresses of patients recorded by Wenzhou Medical College); (2) dissection or fecal examination performed on wild animals purchased from hunters living in endemic areas; and (3) dissection or fecal examination of zoo animals, including tigers, lions, jackals, wolves, leopards, civets, bears, monkeys, boars, minks, cranes, and peacocks, in Wenzhou, Ningbo, and Hangzhou cities.

4.5.1.3 Surveillance Results

A total of 62 animals tested positive for paragonimiasis infection. There were nine feline species, five canine species, four civet species, four weasel species, two Ursidae species, one panda species, two pig species, four primate species, and four bird species. The positive rate of *Paragonimus* was the highest in cats, which accounted for 76.1% (35/46). *Paragonimus* had parasitized three tigers caught in Lishui, southern Zhejiang. The pathogens were found in clouded leopards, leopards, and red spring leopards. Among 20 cats, including leopards, wildcats, and domestic cats, 18 (90%) were found to have worms. The results showed that these animals were important hosts of *Paragonimus* in the mountainous area of Zhejiang Province.

There were five canine species, among which the positive rate of *Paragonimus* was 47.76% (51/107). The infection rate of dogs purchased from sick families accounted for 57.65% (49/85), and wolves and foxes were also infected, indicating that these animals were also important hosts in highly epidemic areas. However, the infection rate was lower in canines than in cats. A total of nine raccoons and jackals in this inspection were all negative.

There were 410 felines of 4 species, among which 20 were found to be parasitic, for a positive rate of 5%. The fecal smear method was used to seek *Paragonimus* eggs, which may be missed in mild infection cases. A total of 51 civet cats with negative fecal smears were dissected, and 3 worms were found in one lung. Therefore, it is advisable to use section examination or the washing precipitation method to look for eggs.

There were four mustelid species, and only one hog badger was found to have *Paragonimus* in the endemic area of Qingtian. Two weasels were free of infection. These findings indicated that mustelids are not as important as felines, canines, and civets in the epidemiology of *Paragonimus*.

4.5.1.4 Recommendations

Paragonimiasis, as a zoonotic disease, was distributed in over 50 counties (towns and districts) of Zhejiang Province. In this case, 62 infected animals from over 30 species were reported, with the highest infection rate in the Felidae family, followed by the Canidae, Viverridae, and Mustelidae families. Those animal families are the main source of infection for human paragonimiasis. Therefore, it was recommended that food safety have to be improved and health education promoted to change dietary habits can lead to high infection risks. Furthermore, the investigation of reservoir hosts can also assist in the identification of natural epidemic foci to provide evidence for developing future health security measures in the region.

4.5.2 Sentinel Mouse Surveillance Conducted in Key Water Areas

4.5.2.1 Proposing the Issue

As the human infection rate of schistosomiasis in most endemic provinces in China has dropped to less than 3%, it has become difficult to reflect the prevalence and transmission of schistosomiasis in endemic areas by using the traditional investigation method of focusing on the infected snail distribution. Therefore, the focus of the current schistosomiasis control program is deploying proper techniques for schistosomiasis monitoring and early warning and enhancing the sensitivity level of the current schistosomiasis surveillance system. To this end, in 2010, the National Institute of Parasitic Diseases at China CDC conducted sentinel mouse surveillance in 72 surveillance sites in 34 counties (towns and districts) of 7 provinces such as Hunan, Hubei, Jiangxi, Anhui, Jiangsu, Sichuan, and Yunnan Provinces, for schistosomiasis monitoring and early warning in key national water areas.

4.5.2.2 Surveillance Components and Methods

1. Surveillance scope. The following regions were selected as key water areas for schistosomiasis monitoring and early warning: counties (towns and districts) of Hunan, Hubei, Jiangxi, Anhui, and Jiangsu Provinces (the five provinces with regions along the Yangtze River, Dongting Lake, and Poyang Lake) that did not reach the schistosomiasis control level and counties (towns and districts) of Yunnan and Sichuan Provinces (the two provinces with mountain regions) along the river system that had not reached the schistosomiasis control level in the previous 5 years.
2. Surveillance Methods
   (a) In each county, 1–3 surveillance sites were selected at intervals of 20–50 km. GPS was used for positioning and on-site photography at each surveillance site to establish the geographic information database. The selection principles of the surveillance sites were as follows:
      - A center for fishing boats and people where snail distributed
      - Environment with snail distribution that is near riverside industries, large construction sites, and wharves
      - Environment with snail distribution that is close to residential areas (including river channels)
      - Environment with snail distribution that is near an administrative area with a history of schistosomiasis outbreaks
      - Water system connected with the environment with snail distribution and with residential areas
      - Agricultural land using water from reservoirs for the irrigation purpose where snail distributed

   (b) Surveillance methods. During 2 days of each flood season, including mid-June (the first period) and mid-September (the second period), sentinel mouse surveillance was conducted to determine the infectious level of the tested water bodies. At each site, 20 mice were placed in contact with the water body from 10 a.m. to 2 p.m. (4 h each day) for 2 days. After the exposure, the mice were taken back to the laboratory, kept for 35 days, and then dissected to check for schistosomiasis infection.

   (c) Data collection and analysis. A database was established with geographic information, basic information, and sentinel mouse infection status of each surveillance site. Based on the infection rate of the surveillance sites, snail infection rate, and infection level of sentinel mice, the distribution map of surveillance sites and infected sites was drawn. The spatial and temporal distribution, environmental characteristics, etc. were analyzed at the surveillance sites.

4.5.2.3 Surveillance Results

1. The distribution of surveillance sites. In 2010, the surveillance sites for schistosomiasis in key water bodies were distributed across 34 counties (towns and districts) of 5 provinces with lake regions and 2 provinces with mountain regions, for a total of 72 sites. There were 10 sites each in Hunan, Hubei, Jiangxi, and Jiangsu Provinces, 9 sites each in Yunnan and Sichuan Provinces, 14 sites in Anhui Province, representing 54 sites in the lake regions, and 18 sites in mountain regions. Among them, the ten surveillance sites in Jiangxi Province were active only in September, while all the other sites were active in both June and September.
2. The deployment and recovery of the sentinel mice. In the 72 surveillance sites, 2667 mice were deployed, and 2613 of them were recovered for a recovery rate of 97.98% (87.78%–100%). A total of 177 mice died during laboratory rearing for a total mortality rate of 6.77% (0–18.05%). A total of 2436 survived until dissection.
3. Detection and distribution of mouse positive sites. Among the 72 surveillance sites, 17 infected sites (excluding duplicates) were detected for an infection rate of 23.61%. The 17 infected sites were distributed in Hunan, Hubei, Anhui, Jiangsu, and Sichuan Provinces with infection rates of 40% (4/10), 90% (9/10), 14.29% (2/14), 10% (1/10), and 11.11% (1/9), respectively. Among these, the infection rate of Hubei Province was the highest, accounting for 52.94% of the total number of infected sites. In the first period, 58 sites were under surveillance, and 10 sites were infected for an infection rate of 17.24%; in the second period, 68 sites were under surveillance, and 10 sites were infected for an infection rate of 14.71%. No statistically significant difference was detected in the infection rates of the surveillance sites between the two periods ($\chi^2 = 0.151$, $P = 0.698$).
4. Sentinel mouse infection rate and infection level. Among the 2436 dissected mice, 90 sentinel mice were found to be infected, with 459 schistosomes detected. The total sentinel mouse infection rate was 3.69%, and the average worm load was 5.10 per infected mouse. Hunan Province had both the highest sentinel mouse infection rate, at 14.12%, and the highest average worm load, at 6.37 per infected mouse. Among the mice deployed during the first period, 1099 were dissected and 31 were infected; 76 schistosome worm loads were detected for a total infection rate of 2.82% and an average worm load of 2.45 per infected mouse. Among the mice deployed during the second period, 1337 were dissected, 59 infected mice were found, and 383 schistosome worm loads were detected, for a total infection rate of 4.41% and an average worm load of 6.49 per infected mouse. The sentinel mouse infection rate was significantly higher in the second period than that in the first period ($\chi^2 = 14.681$, $P < 0.01$).

5. Relationship between the infected mouse detection rate and the time period when the infected snails were detected. The 72 surveillance sites were divided into 3 types. If the infected snails were detected in 2010, then the sites were grouped as type I; if the infected snails were detected in 2009, then the sites were grouped as type II; and if no infected snails had been detected during 2008–2010, then the sites were grouped as type III. The infected mice were detected in eight type I sites, five type II sites, and four type III sites, with infected mouse detection rates of 29.63%, 41.67%, and 12.12%, respectively. The difference in the infected mouse detection rate among the three different surveillance groups was not statistically significant ($\chi^2 = 5.227$, $P = 0.071$). The sentinel mouse infection rates of the type I, type II, and type III sites were 9.38%, 3.98%, and 0.59%, respectively, and the sentinel mouse infection levels of the type I, type II, and type III sites were worm loads of 5.99, 2.8, and 2 per infected mouse, respectively. The difference in the mouse infection rates among the three different surveillance groups was statistically significant ($\chi^2 = 20.489$, $P < 0.001$); the sentinel mouse infection levels of the type II and type I sites were 1.40 and 2.99 times higher than that of the type III sites, respectively.

4.5.2.4 Recommendations

The national schistosomiasis surveillance system is a systematic project with a wide range, large workload, and time sensitivity. It should not only establish a working mechanism linked at the national, provincial, city, county, and township levels but also strictly implement a unified plan, testing methods, and time and work norms.

In this study, sentinel mouse monitoring was applied to the infection detection of high-risk schistosomiasis-susceptible water bodies at the national level. Monitoring and early warning were carried out in key water bodies in 2010. The results showed that the infection rate of sentinel mice was 3.69%, the occurrence rate of positive sites was 23.61%, and the positive infection rate, positive rodent load, and environmental positive sites were significantly higher in Hunan and Hubei Provinces than in other provinces, which was basically consistent with previous schistosomiasis epidemics. The analysis also found that about 12.12% to 41.67% of undetected infected snails in 2010, especially in snails where infection was undetected for more than 3 years, the environmental positive rate reached 12.12%, indicating that monitoring a susceptible environment will significantly improve the sensitivity of schistosomiasis monitoring and early-warning systems.

Epidemic Surveillance 5

Qing Fu

Parasitic disease epidemics refer to the current status of parasitic disease transmission involving infection sources, transmission routes, and susceptible populations, and various factors affecting the transmission of parasitic diseases, that reflecting the intensity and scope of parasitic disease transmission. Surveillance of parasitic disease outbreaks refers to continuous, and systematic collection of parasitic disease data and influencing factors through analysis and exploitation that can objectively reflect the occurrence and development trends and changes so that relevant personnel can gain a timely understanding of certain time and space distribution characteristics, risk factors, occurrence patterns, and progress. Then, they can make early-control plans or preparation to take targeted preventive measures, respond rapidly to possible parasitic disease outbreaks, and evaluate the effects of interventions.

The role of parasitic disease epidemic surveillance has three aspects: first, to grasp the trend of a parasitic disease epidemic through the accumulation of long-term monitoring data in order to predict and forecast the occurrence and development of the parasitic disease; second, to determine the focus group of prevention and control in order to formulate interventions and strategies; and third, to adjust the intervention measures in order to ensure that the intervention is reasonable and effective.

5.1 Outbreaks and Epidemic Surveillance

Parasitic diseases have different epidemic intensities in transmission, with three grades of parasitic disease outbreaks. First, when a parasitic disease occurs in a certain area for years with cases scattered throughout the population and no linkage

Q. Fu (✉)
National Institute of Parasitic Diseases, Chinese Center for Disease Control and Prevention (Chinese Center for Tropical Diseases Research), Shanghai, China

X.-N. Zhou (ed.), *Surveillance and Management of Parasitic Diseases in China*,
https://doi.org/10.1007/978-981-97-1241-0_5

between the times and places of the cases, it is called sporadic parasitic disease. Second, when a parasitic disease occurs in a local area or collective unit for a certain period (usually in a short time), it is called an outbreak of parasitic disease. Third, when a parasitic disease occurs in a certain area and the incidence at a certain time significantly exceeds the sporadic incidence level of the disease over the years, it is called an epidemic of parasitic disease. Outbreaks and epidemics of parasitic diseases are the most direct causes of the transmission of parasitic diseases, and they are also the most noteworthy situation for parasitic diseases.

5.1.1 Cause Investigation

When there is an outbreak or epidemic of a parasitic disease, professionals in CDCs should first determine the causes. Although many factors contribute to outbreaks or epidemics of parasitic diseases, transmission is inseparable from the number of infection sources and pathogenicity, the accessibility of transmission routes, and the immune status of susceptible people. Communication factors are inseparable from the influences and restrictions of natural, biological, and social factors. Therefore, cause surveillance of an outbreak or epidemic of parasitic disease is based mainly on the epidemic situation, aiming to identify the source of infection, the route of transmission, and each link of influencing factors, tracing back to the source of infection and finally determining the causes of the occurrence of the index cases and the main and secondary factors affecting transmission to provide a scientific basis for further effective control measures. The following steps should be taken to investigate the causes of outbreaks and epidemics:

1. Combined with symptom monitoring, timely spatial and temporal aggregation analysis of all suspected cases and relevant data should be conducted, and hypotheses for further investigation should be proposed.
2. To determine the causes of an outbreak or epidemic, as many pathogen samples as possible should be collected, including pathogen samples from people, animal hosts, and vectors (intermediate hosts), especially from the first cases or index case, and comprehensive analysis and judgment of these pathogen samples should be conducted from clinical, laboratory, and epidemiological perspectives.
3. Epidemiological case investigation should be conducted for every clinically diagnosed or suspected case. Case survey components generally include basic personal data (name, age, sex, nationality, address, residence, occupation, etc.), clinical data (onset, hospitalization, recovery or death date, disease symptoms and signs, laboratory results, current outcome, etc.), epidemiological data (past history, vaccination history, contact history, possible exposure date, possible source or route of transmission, etc.), and treatment measures (clinical treatment, prevention measures, etc.).
4. Combined with the investigation results of pathogen detection of clinically diagnosed or suspected cases, the hypotheses should be verified by case control, cohort study, and epidemiological experimental study or appropriately revised

and reverified. Finally, the outbreak causes, epidemic characteristics, and trends should be clarified and effective measures taken to deal with the causes.

5.1.2 Investigation of the Distribution

The epidemic characteristics of parasitic diseases are manifested through the distributions of population, time, and region. For parasitic diseases with known causes, epidemic characteristics are the basis for judging and explaining the causes. For parasitic diseases with unknown etiology, the epidemic characteristics are the external manifestation of the etiology and thus are an important source for forming a hypothesis about the etiology. Therefore, it is key to master the epidemic characteristics of parasitic diseases transmission by monitoring them in humans, time, and space.

5.1.2.1 Investigation of the Infected Population

Employing the disease surveillance network to carry out active surveillance and/or reactive surveillance of parasitic diseases is an effective means to detect and monitor possible parasitic disease outbreaks in a timely manner and prevent the spread of the epidemic. To carry out population monitoring, the monitoring objects should first be determined, and the principles of territorial management should then be followed. Usually, the health administration department organizes health personnel and other relevant personnel to investigate and collect data on a certain parasitic disease, such as symptoms, signs, and treatment of side reactions. For the collected monitoring data, a unified format report should be designed, and the data should be reported and summarized by each monitoring point within the specified time to realize the establishment of a database and conduct statistical analysis.

Population surveillance can be used to understand the occurrence and prevalence of parasitic diseases, but it will lead to a certain level of overreaction, and the cost is often relatively high. An efficient, economically feasible, and suitable approach to population monitoring for normal management must be established.

5.1.2.2 Investigation of Temporal Distribution

The temporal distribution of parasitic diseases is an important part of the epidemiological study of such diseases. Time aggregation is the main part of time distribution monitoring, as it can provide not only clues for etiological research but also support for the study of exposure time or incubation period. Additionally, it can be used to evaluate the prevention and control effects of infectious diseases in different periods. It can also help provide a scientific basis for formulating effective prevention and control measures and health decisions. Three aspects should be considered in monitoring the onset time of parasitic diseases: first, the time range of concern, such as whether the observation is for many consecutive years or a certain year, month, or epidemic period that may cross years; second, the time level or time unit, for example, analyzing the onset time data of the same year at different levels such as a month, 10 days, a week, and a day; and third, the monitoring method, as different analysis methods of the same dataset may lead to different conclusions, showing the presence or absence of temporal aggregation and the degree of aggregation. In monitoring and analyzing the onset time, commonly used analysis techniques include cluster analysis, run-length test, concentration ratio, and circle distribution:

1. Cluster analysis is performed when the appearance of the classification of objects is not clear; in this case, those with similar properties are classified as one category, and those with different properties are classified as different categories. Euclidean distance is commonly used as a distance index to measure the similarity of objects in sample clustering. In case analysis, the object is the month, the index is different years, and the observed value is the number of cases.
2. The run-length test, also known as the coherence test, is a method for determining the number of runs formed according to the representative arrangement of sample marks, which can be used to test the randomness of samples. The number of runs is used to test the randomness of the samples to observe whether the occurrence of a parasitic disease is randomly distributed or aggregated. For example, the incidence of a parasitic disease is listed in time sequence in different time units (month, week, day, etc.) within the time range of concern, and the incidence data are divided into different situations with the median or average value. The run-length numbers formed by upward and downward interlacing are used to test whether the samples are random.
3. Concentration ratio refers to the seasonal strength of the onset time, which is calculated by the ratio of the number of monthly onsets to the total number of annual onsets, where M refers to the concentration ratio, $M = 1$ indicates that the cases are concentrated within 1 month, M between 0.9 and <1 indicates that the disease has strict seasonality, M in a range from 0.7 to <0.9 indicates that the disease has strong seasonality, M in a range from 0.5 to <0.7 indicates that the disease has a certain seasonality, a score below 0.3 indicates that the distribution of onset time is relatively uniform, and $M = 0$ indicates that the cases are evenly distributed over 12 months.
4. The circular distribution method is a statistical method that transforms data with periodic changes into linear data by trigonometric function transformation. If a set of circular distribution data has the tendency of centralized distribution, this tendency can be expressed by the average angle. The circle distribution method can calculate the peak time and peak period of parasitic diseases.

5.1.2.3 Investigation of Spatial Distribution

Carrying out spatial distribution monitoring of parasitic diseases and determining their spatial distribution characteristics will help guide the formulation and implementation of disease control plans and the adjustment of surveillance. To collect the incidence data of a certain parasitic disease in a certain period, the media monitoring data in the same period, and the digital map of the corresponding area at a certain scale, spatial local interpolation analysis should be carried out with the support of Arc GIS spatial analysis software. Additionally, the best distribution map of parasitic disease incidence and vector population should be selected with the help of the cross-check index of the distribution map and the principle of distribution map selection to explore the prevalence of a certain parasitic disease and the spatial distribution characteristics of vectors in a certain period.

Spatial exploratory analysis of data is the premise of applying spatial local interpolation in order to test whether the data are spatially autocorrelated, to test the characteristics of the intrinsic hypothesis, and to preprocess the spatial local interpolation. The spatial distribution characteristics described by spatial local interpolation provide a scientific basis for macroscopic spatial monitoring, prediction of parasitic diseases, and health administrative decision-making. The spatial distribution pattern of parasitic diseases identified by spatial local interpolation technology is of great significance to the control of parasitic diseases. More comprehensive and profound results can be obtained by studying the spatial distribution pattern of the population and combining the distribution pattern of population statistics with geostatistics to monitor the spatial distribution of diseases.

5.1.3 Outbreak Response

On the basis of mastering the causes and scopes of outbreaks and epidemics, it is necessary to respond in a timely manner and implement effective disposal measures according to the transmission characteristics and intensity of different parasitic diseases.

5.1.3.1 Epidemiological Investigation

Parasitic disease cases in outbreak and endemic areas should be investigated individually. The epidemic sites and their scope should be determined according to the case investigation clues. The hosts and exposure factors should be investigated. Additionally, related factors such as water level, rainfall, temperature, natural disasters, human and livestock flow, and residents' production and lifestyle should be investigated.

5.1.3.2 Outbreak Reports

Medical institutions and CDCs at all levels, when investigating outbreaks and epidemics of parasitic diseases, should report to the health administrative department of the local government within the time specified by the relevant laws, regulations, or provisions. The health administrative department that receives the report should report to the government at the corresponding level within the specified time and step by step through the parasitic disease epidemic reporting management information system.

5.1.3.3 Organizing the Outbreak Response

After an outbreak of a parasitic disease or epidemic, the relevant departments of health, finance, agriculture, water conservancy, publicity, and education should, according to hierarchical response procedures, organize, command, and coordinate the parasitic disease outbreak response or epidemic emergency treatment. The local administrative department should establish a technical steering group for the medical treatment, epidemic control, investigation, and evaluation of parasitic diseases within the administrative region.

5.1.3.4 On-Site Intervention

In case of a parasitic disease or epidemic outbreak, the administrative department of health at or above the county level should immediately organize medical teams to travel to the epidemic sites to provide treatment and should promptly treat all identified parasitic disease patients. If necessary, early preventive treatment should be provided to people with an exposure history based on the principles of early detection, early diagnosis, and early treatment.

5.1.3.5 Health Education

At the same time as on-site disposal, health education should be vigorously carried out using various forms of publicity, parasitic disease control information should be quickly disseminated, the self-protection ability of the people should be promoted, and all departments should actively cooperate and participate in control measures. Emergency response staff must pay attention to personal protection when carrying out prevention and control on site.

5.1.3.6 Supports

Governments at all levels should strengthen their leadership in the emergency handling of parasitic disease outbreaks and epidemics and coordinate with relevant departments to arrange the necessary personnel, funds, materials, and technologies in a timely manner according to their specific responsibilities in order to guarantee the appropriate epidemic response.

5.2 Surveillance of Transmission Routes and Risk Factors

The occurrence, transmission, spread, and termination of parasitic disease in the population are the epidemic process of parasitic disease. The prevalence of parasitic diseases, like that of other infectious diseases, has three basic conditions, namely, the source of infection, the route of transmission, and the susceptible population. When these three links are present simultaneously in a certain area, they will trigger an outbreak of parasitic disease.

The route of transmission refers to the process experienced by the pathogen after it is discharged from the infected body and then invades another susceptible body with the help of certain transmission factors. The transmission route is composed of various factors of the external environment. Each parasitic disease has certain transmission factors that affect its epidemic spread. Some parasitic diseases follow a single transmission route, whereas others have transmission modes that combine multiple factors. Recognizing the transmission routes of parasitic diseases and strengthening the monitoring of those routes should play an important role in the prevention and control of parasitic disease epidemics.

5.2.1 Transmission Routes and Their Characteristics

The main routes of transmission of parasitic diseases are by means of water, food, soil, airborne (droplets), arthropods, direct human transmission, and transplacental vertical transmission (Table 5.1).

5.2.1.1 Transmission by Water

Some parasites infect people by water transmission. People are infected with parasites when they drink or make contact with water contaminated by parasitic pathogens (infested water). The pollution of drinking water sources can be caused by the infiltration of infected sewage from tap water or by the pollution of water sources by feces and dirt. The epidemic intensity depends on the type of water source, the scope of the water supply, the intensity and frequency of water pollution, the survival time of pathogens in the water, drinking water hygiene management, and residents' hygiene habits. Common parasitic diseases transmitted by drinking water are cryptosporidiosis, microsporidiosis, cyclosporiasis, and *Entamoeba histolytica* infections. Parasitic diseases are also caused by human contact with infested water because parasitic pathogens invade the body through the skin and mucosa. Their pathogenicity depends on the area and number of times and duration of human contact with infested water. Schistosomiasis and other parasitic diseases are commonly transmitted by infested water.

The transmission of parasitic diseases by drinking water often leads to outbreaks or epidemics. The distribution of cases is consistent with the distribution of the water supply among people with a history of drinking from the same water source.

Table 5.1 Main routes of transmission of parasitic diseases

Routes of transmission	Representative parasitic diseases	Common means of infection
Waterborne transmission	Cryptosporidiosis and schistosomiasis	Infection with *Cryptosporidium* spp. caused by drinking untreated water contaminated with oocysts; schistosomiasis caused by contacting water with schistosomes cercaria
Foodborne transmission	Clonorchiasis	Infection with *Clonorchis sinensis* by eating raw or undercooked freshwater fish or shrimp
Soil-transmitted infection	Ancylostomiasis	Invasion by larvae in the soil through the damaged skin
Airborne transmission	Enterobiasis	Infection caused by breathing in floating pinworm ova from the air
Transmission via arthropods	Helopyra	Blood-transmitted *Plasmodium* spp. through *Anopheles* spp. mosquito bite
Spread through direct human contact	Trichomoniasis	Infection with *Trichomonas vaginalis* through sexual activity
Vertical transplacental transmission	Toxoplasmosis	The parasite is transmitted to a fetus through the maternal placenta

The disease can occur in people of all ages, sexes, and occupations except for nursing infants. The incidence rate does not differ among people who drink untreated water at similar rates. Epidemics or outbreaks can be quelled by stopping the use of contaminated water or using water after purification treatment. For example, when a water source is often polluted, the number of cases can be maintained throughout the year, and the disease then becomes endemic. Commonly used monitoring indicators are pathogen distribution in water, exposed population, and effects after emergency treatment.

The epidemic characteristics of parasitic diseases transmitted by contact with infested water are that patients have a history of contact with infested water, such as through swimming, bathing, fishing, harvesting, rescue, and disaster relief in epidemic areas that are exposed to infested water and infection. The incidence is local, seasonal, and occupational, and it is common in water network areas. When a large number of susceptible people enter an epidemic area and come into contact with infested water, outbreaks or epidemics can occur. Strengthening infested water treatment and personal protection can control the occurrence of diseases. The commonly used monitoring indicators are monitoring of infested water infections, monitoring people exposed to infested water, and monitoring the effects of treatment of infested water.

5.2.1.2 Transmission by Food

Foodborne parasitic diseases are transmitted by eating food infected by pathogens. In many areas of China, human feces are used as fertilizer for crops. Parasite eggs in feces easily contaminate food, and contaminated (parasitic) food can cause infection if eaten unwashed, raw, or undercooked. Foods vulnerable to parasitic pathogen infection are mainly uncooked food such as shrimp and crab, fish, shellfish meat, vegetables, and fruits. Common foodborne parasitic diseases are clonorchiasis, Angiostrongyliasis, trichinosis, taeniasis, cysticercosis, and echinococcosis.

The epidemic characteristics of foodborne parasitic diseases are that patients have a history of eating certain infected (parasitic) foods, and those who do not eat the foods do not become sick. An outbreak can easily occur, and the number of cases is determined by the number of people who eat infected food. After the supply of infected food is stopped, the outbreak of the disease can be suppressed. Commonly used monitoring indicators are suspected food parasite infection, the infection situation of people who eat the infected foods, the response to outbreaks, and treatment effects.

5.2.1.3 Transmission through Soil

Some directly developing nematodes, such as *Ascaris lumbricoides* and hookworms, lay eggs with strong resistance to the external environment and need to develop into infected eggs or larvae in the soil. Human infection is related to contact with the soil. Examples include human infection by ingestion of soil contaminated with infectious eggs or contact with hookworm metacercariae. Soil-transmitted parasitic diseases are often associated with the duration and viability of pathogens in the soil, the opportunity and frequency of individual contact with the soil, and personal hygiene conditions and practices. Patients often have a history of contact with contaminated soil. The prevalence of the disease can be controlled by improving personal hygiene, providing information about personal protection, strengthening feces management, and carrying out treatment of feces to render them harmless. Common monitoring indicators are pathogen distribution in soil, people's hygiene habits, and fecal pollution.

5.2.1.4 Air Transmission

The eggs of some parasites can be spread by air or droplets during the infection period. For example, pinworm eggs can float in the air and enter the human body through breathing. Pneumocystis disease can be transmitted by droplets. Improving self-protection awareness, changing unhealthy living habits and personal hygiene, and strengthening regular disinfection of public health facilities and public places can reduce the occurrence of parasitic diseases transmitted through airborne routes. Common monitoring indicators are exposure to pollutants, people's hygiene habits, and disinfection of sanitary facilities.

5.2.1.5 Transmission by Arthropods

Some parasitic diseases can develop and proliferate in arthropods; then enter susceptible organisms through the saliva, vomit, or feces of arthropods; and cause parasitic diseases. Biological blood-sucking transmission is the main way that parasitic diseases are transmitted through arthropods. Common parasitic diseases transmitted by arthropods are malaria, kala-azar, filariasis, and trypanosomiasis.

The prevalence of parasitic diseases transmitted by arthropods is often regional, and the distribution of cases is consistent with that of vector insects. These diseases have obvious seasonality, and the seasonal increase in cases is consistent with or slightly lags behind the season of vector insect reproduction. People generally do not spread these diseases to each other directly. Strengthening environmental remediation and paying attention to personal protection can reduce the occurrence of parasitic diseases transmitted by arthropods. Commonly used monitoring indicators are vector insect population distribution, vector pathogen infection, and people's exposure to vectors and pathogen infection.

5.2.1.6 Transmission by Direct Contact

Some parasitic diseases are transmitted by direct contact when parasitic pathogens infect susceptible people through direct contact between people or contact with daily necessities contaminated by the effluents and secretions of infection sources. Common parasitic diseases transmitted by contact are *Trichomonas vaginalis* infections and sarcoptic mites skin disease.

Parasitic diseases transmitted through direct contact mostly cause the occurrence of individual cases, which are generally sporadic and rarely lead to epidemics. There is no obvious seasonality, and more diseases occur in areas where people have poor personal hygiene habits and poor hygiene conditions. Improving self-protection awareness, unhealthy living habits, and personal health conditions and strengthening regular disinfection of public health facilities and public places can reduce the occurrence of contact-transmitted parasitic diseases. Commonly used monitoring indicators are contact history, health habits, and disinfection of health facilities.

5.2.1.7 Vertical Transmission via the Placenta

A fetus can be infected by parasites, such as *Toxoplasma gondii*, through the placenta, causing vertical transmission, which is seriously harmful to the health of the fetus.

5.2.2 Transmission Factors and Their Characteristics

The transmission and prevalence of parasitic diseases depend on the connection and continuity of three chains: the source of infection, the route of transmission, and the susceptible population. Changes in any chains may affect the prevalence and spread of parasitic diseases. However, the connection of these three links is often influenced and restricted by natural, biological, and social factors. Transmission factors can promote or inhibit the epidemic process of parasitic diseases by influencing the three chains.

5.2.2.1 Natural Factors

Natural factors are the natural environmental factors that affect the spread and incidence rates of parasitic diseases, including geographical, meteorological, ecological, and other factors such as elevation, latitude, temperature, humidity, rainfall, light, and soil type. Natural factors have an important influence on the occurrence and development of parasitic disease epidemics. The geographical environment affects the breeding and distribution of intermediate hosts. For example, the intermediate hosts of *Paragonimus*, crabs and slugs, are found only in mountain streams, so *Paragonimus* is endemic mostly in hilly and mountainous areas. Climatic conditions affect the growth and development of parasites in the outside world and the breeding of intermediate hosts or vector insects, as in the hatching of schistosome miracidium and the escape of cercariae. In addition to the need for water, such conditions are related to temperature, light, and other environmental conditions. In addition, the appropriate temperature increases the chance for the population to contact infested water, thus helping the prevalence of schistosomiasis.

The regional and seasonal characteristics of parasitic diseases are closely related to natural factors. For example, there are endemic areas of kala-azar in northern China and endemic areas of schistosomiasis in southern China. The incidence of some parasitic diseases increases seasonally every year to a certain extent. The reason for the seasonal increase is related mainly to temperature and the density of vector insects. Natural factors can directly affect the ability of pathogens to survive in the external environment; for example, ancylostomiasis is rare in arid areas. Some natural ecological environments have good conditions for the spread of parasitic diseases among wild animals. Human beings can also be infected when they enter these areas; this is called natural epidemic disease or zoonosis.

5.2.2.2 Biological Factors

Some parasites require the existence of intermediate hosts or arthropods in their life cycle. The existence of these intermediate hosts or arthropods determines the prevalence of the parasitic disease. Common parasitic vectors are mosquitoes, flies, lice, mites, sandflies, cockroaches, beetles, and fleas. Different biological vectors have different living environments and different morphological characteristics, life histories, physiological and ecological habits, and relationships with transmitted diseases. Vector population dynamics, such as density changes, spatial distribution, habits, insect age composition, and oviposition ability, are closely related to the prevalence of vector-borne diseases and the harm they cause.

The distribution of *Oncomelania* snails, the intermediate hosts of *Schistosoma japonica*, in China does not exceed 33.15 N northward, so there is no endemic areas of schistosomiasis in northern China. The vector of malaria is anopheles, and in areas without *Anopheles* distribution. There is no prevalence of northern and northwestern China due to mosquito cannot survive in cold weather. Changes in natural factors also affect the migration patterns of vector insects and animal hosts. *Aedes* can live in areas only up to 1000 m above sea level. However, due to the warming climate, *Aedes* can be found in some countries in South America at altitudes of 1350–2000 m.

5.2.2.3 Social Factors

Social factors include the social system, economic situation, religious beliefs, cultural and educational levels, medical treatment and health care, modes of production, and living habits. Social, natural, and biological factors often interact to affect the prevalence of parasitic diseases. Since natural and biological factors are generally stable and social factors are often variable, social stability, economic development, improvement of medical and health conditions, and improvement of people's educational level often play an active role in controlling the prevalence of parasitic diseases.

The influence of social factors on the prevalence of parasitic diseases works through all three chains of transmission. With the rapid development of the economy and society, population mobility is increasing, and the risks of the transmission and outbreak of parasitic diseases from imported infection sources are also increasing. At the same time, improvements of housing and environmental health conditions, the popularization of health knowledge, living standards and disease prevention and control measures, and improved medical services all have restrictive effects on the prevalence of parasitic diseases. The influence of social factors on routes of transmission is most obvious. For example, the elimination of snails, the provision of safe drinking water, and the treatment of feces to render them harmless have a great influence on the control or elimination of schistosomiasis and hookworm.

5.3 Surveillance Components and Methods

Cutting off the transmission route is one of the basic measures for preventing and controlling parasitic disease epidemics. Strengthening the monitoring of routes of transmission of parasitic diseases is of great significance for correctly identifying different routes of transmission and taking reasonable measures to cut them off. Monitoring techniques of parasitic transmission routes can be divided into the following categories.

5.3.1 Epidemic Surveillance Network

Various transmission methods, transmission carriers (water, soil, food, etc.), and high-risk behaviors (eating *Pomacea canaliculata* snails, working barefoot, etc.) that cause the transmission of parasitic diseases are included in the traditional epidemic situation monitoring network, and the dynamic monitoring of various transmission factors is the key to improve the surveillance network. The monitoring of transmission routes of parasitic diseases mainly involves monitoring the population that may be threatened by a pathogen or the environment, transmission carriers, and animal hosts that may carry the pathogen before the occurrence of a parasitic disease epidemic.

Surveillance of transmission routes is divided into routine surveillance and sentinel surveillance. Routine surveillance sites are used mainly to carry out epidemic surveillance and emergency surveillance of parasitic diseases caused by various transmission routes. Sentinel surveillance is carried out in monitoring sentinel points that located in hospital and community to actively monitor outpatients and schoolchildren, parasitic diseases in community residents and related symptoms, and related transmission carriers and high-risk behaviors.

5.3.2 Serological Surveillance

The immunofluorescence test (IFA), ELISA, and other immunological techniques are used to detect specific antibodies during the transmission of parasitic diseases. Immunological technology is used to detect specific antibodies that may exist in humans, animals, insect vectors, and other organisms and to collect relevant data. For example, in areas where trichinosis is likely to break out, regular serological screening is conducted among local residents to determine the level of trichinosis-specific IgM and IgG antibodies in the population in order to dynamically grasp and determine the prevalence of trichinosis in a certain period and specific population.

5.3.3 Laboratory Surveillance Network

Laboratory detection techniques for parasitic diseases can provide etiological information and more direct evidence that can play an unexpected role in the surveillance of parasitic diseases when combined with epidemiological information. At present, with the acceleration of economic globalization and the rapid increase in the flow of people brought about by the great development of transportation, the parasitic disease epidemics have begun to spread across regions and borders more rapidly. Moreover, network analysis can be performed by combining the test results of laboratories with epidemiological monitoring information so that the current epidemic situation and spreading trend of parasitic diseases can be correctly judged on a large scale and across regions. Therefore, the combination of laboratory network detection results and field epidemiological monitoring information is the priority direction of parasitic disease surveillance.

In developing laboratory network surveillance, it is necessary for all kinds of test methods to have stable results, strong repeatability, standardized processes, and simple operations in order to standardize their operation and promotion in different laboratories and make the results and information obtained by different laboratories comparable. Therefore, to participate in parasitic disease network monitoring, laboratory units must use unified technologies, methods, equipment, reagents, procedures, and analytical methods and standards in order to ensure the quality of laboratory tests.

5.3.4 Vector Surveillance

Vector biological surveillance can be used to understand the species, density, geographical distribution, and seasonal growth and decline of vector insects and is the focus of epidemic prevention and prediction as well as risk assessment of biologically transmitted parasitic diseases. Through continuous and systematic ecological monitoring of vectors, potential outbreaks and sudden outbreaks of arthropod-transmitted parasitic diseases can be detected in a timely manner.

For example, in mosquito vector monitoring by the ultraviolet lamp trapping method, the light source of the mosquito trap is an ultraviolet lamp, and an electric fan forms a certain negative pressure below the light area to trap the mosquitoes by using alternating current, thereby overcoming the defects of insufficient suction and instability of the mosquito "manual hour" monitoring method. During monitoring, the lamp is hung 1.5–2.0 m above the ground, and adult mosquitoes can be caught if they are within 10 cm of the mosquito trap lamp. Other light sources should not be arranged around the hanging mosquito trap lamp to avoid the tendency of mosquito dispersion and interference of the monitoring results. The ultraviolet lamp trapping method has a long and continuous monitoring time and can trap mosquito species in different time periods. Therefore, it accurately reflects the situation of the monitored mosquito species. This method is suitable for outdoor monitoring and long-term system monitoring.

5.3.5 Employment of 3S Technology in Parasitic Disease Surveillance

With the improvement of technology and the development of software, 3S technology, which combines the geographic information systems (GIS), remote sensing (RS), and global positioning system (GPS), has been widely applied and developed in the background investigation, monitoring and control, prediction and forecasting of various diseases, such as schistosomiasis, malaria, leishmaniasis, etc. as well as research on transmission pattern of parasitic diseases. Taking schistosomiasis as an example, *Oncomelania hupensis* has a wide distribution area and a complex breeding environment. With continuous changes in the environment, the range of snail habitats may experience great changes. RS and GIS can be conveniently applied for timely and effective updating of snail distribution information by identifying and monitoring snail breeding sites. High-resolution satellite images can be used to clearly distinguish the grass shoals suitable for the growth of snails and reveal the growth and decline of these breeding sites in a timely manner. This provides a basis for timely and effective countermeasures, thus greatly improving the efficiency of schistosomiasis control. With the support of GIS and RS analysis software, the unsupervised classification method is used to classify satellite images by collecting images of water levels, precipitation, and dry seasons in the same period over different years in the study area to calculate the increase or decrease in the grassy beach areas and thus to determine the increase or decrease in snail habitats. Unsupervised classification of remote sensing images is used to distinguish different features of the site environment, and the main areas where snails are distributed can be determined in combination with a site snail survey data. Image synthesis and reevaluation of remote sensing images in flood and dry seasons in schistosomiasis-endemic areas of the lake region can be used to distinguish water bodies and major landforms and identify snail distribution sites.

5.4 Management of the Epidemic Surveillance System

Surveillance is one of the important measures for preventing and controlling parasitic diseases. Timely and scientific information is of great importance for prevention and control decision making, to effectively control and eliminate the hazards of sudden parasitic disease epidemic patterns and to ensure public health security. Parasitic disease epidemic management involves dynamically managing the epidemic data of parasitic diseases by establishment of databases according to certain standard requirements to correctly grasp the dynamic distribution and epidemic trends of parasitic diseases, so that identify specific epidemic patterns as early as possible. The management of parasitic disease epidemic surveillance is important to guarantee the quality of epidemic monitoring, the analysis of recent epidemic information, and the prediction and early warning of epidemic trends. Aforementioned advantages of surveillance provide a scientific basis for formulating corresponding countermeasures for eliminating transmission of parasitic diseases.

5.4.1 Systematic Management of an Outbreak

5.4.1.1 Establishment of a Surveillance Network

To improve the surveillance network on parasitic disease epidemics in all monitoring points, entering the causes of parasitic disease transmission, including transmission routes, transmission carriers (water, soil, food, etc.), and high-risk behaviors (eating snails, barefoot work, etc.), into the monitoring network is key to parasitic disease epidemic surveillance in the traditional epidemic monitoring network. Parasitic disease epidemic surveillance is conducted mainly before the occurrence of parasitic disease epidemics for people who may be threatened by pathogens or to identify the possible presence of pathogens in the environment, transmission carriers, or animal hosts.

5.4.1.2 Surveillance Data Collection and Database Establishment

Relevant epidemic indicators, influencing factor indicators, and other indicators, such as reporting units, regional names and codes, population/society/economic data, physical and geographical data, infectious disease case data, and vector monitoring data, which will be included in the monitoring work, will be used to establish a data table with a unified format or a number of subtables related by monitoring point codes according to different monitoring components and will be reported in a unified way. After the data are reviewed and summarized by CDCs at all levels, a surveillance database will be created.

5.4.1.3 Analysis and Utilization of Surveillance Data

The analysis and utilization of surveillance data include access, inquiries, statistics, reports, etc. CDCs at all levels involved in the surveillance can carry out comprehensive analyses of the monitoring data at different levels, judge epidemic trends, regularly produce monitoring reports, evaluate the effects of intervention measures in local areas, and propose countermeasures and suggestions for prevention and control. The relevant monitoring results can be reported to the health administrative department at various levels and fed back to lower-level CDCs.

5.4.1.4 Surveillance Data Management

The archiving and electronic management of parasitic disease surveillance data involve the classification, cataloging, and centralized storage of parasitic disease surveillance reports, compilations, analysis charts, written materials and electronic documents, and other information sources according to the requirements of relevant technical archive management regulations. Electronic files of monthly and annual backups of epidemic monitoring data should be copied to CD-ROMs for archiving and preservation, and epidemic monitoring data should be compiled regularly (annually if possible); original materials such as paper epidemic reports and epidemic registration books should be properly stored.

5.4.1.5 Forecasting and Early Warning

On the basis of standardization, the database of epidemic information can be completed quickly by using complicated statistical analysis methods or mathematical models to predict epidemic development trends in real time, identify the characteristics of special emergencies, and release the appropriate epidemic warning information in real time, thus providing a scientific basis for rapid emergency treatment of the epidemic. At present, rapid prediction and early warning can be achieved for the epidemic management of schistosomiasis and malaria. For example, with the support of the geographic information system and a distribution model, early-warning information was provided in a timely manner for the malaria outbreak in the Huanghuai Plain of China in 2006, which enabled timely tackling the epidemic and laid a good foundation for the region to successfully enter the malaria elimination stage.

5.4.2 Management of the Laboratory Network

Laboratory detection of parasitic pathogens can provide direct pathogenic information and play an important role in monitoring the epidemic situation of parasitic diseases. After the network management of laboratory testing is realized, scattered laboratory testing technologies can be unified methodologically and standardized in terms of equipment and reagents, operational steps, and scientific production of data and statistics, thus ensuring the surveillance quality. The overall process of quality control and management should be well managed before, during, and after the establishment of the laboratory network.

5.4.2.1 Pre-establishment Management

According to the epidemic characteristics of parasitic diseases in a certain area, top-level design should be performed with clear objectives. It is necessary to provide key technical index requirements for laboratory network construction, such as formulating laboratory construction procedures, clarifying access standards and evaluation methods, and establishment of laboratory evaluation expert groups to achieve unified planning, operational schemes, standards, evaluation measures, and management procedures.

5.4.2.2 Management During Establishment

According to the technical index requirements of laboratory evaluation, infrastructure construction must be well managed, necessary instruments and equipment allocated, professional and technical personnel appointed, and internal laboratory management and quality control systems established. The equipment and configuration of laboratories at all levels should adhere to the principles of science, rationality, practicality, and economy. While meeting the basic functions, the network should be standardized, intelligent, and humanized.

5.4.2.3 Management After Establishment

According to the division of responsibilities of laboratories at all levels, subordinate laboratories should be subjected to a quality control comparison. To perform a quality evaluation of on-site monitoring and diagnosis, training in laboratory testing technology should be strengthened, new testing technology introduced in a timely manner, and on-site monitoring ensured and improved (Fig. 5.1).

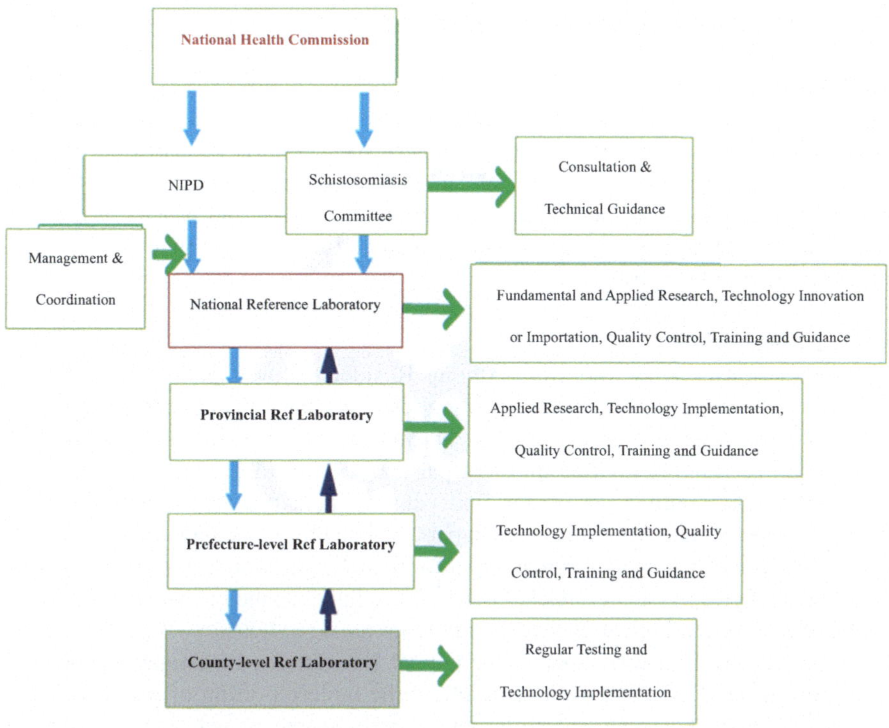

Fig. 5.1 The institutional structure for the management of laboratory networks in schistosomiasis surveillance

5.5 Evaluation of the Effects of Epidemic Surveillance

The surveillance effects of parasitic diseases should be evaluated in a certain area and at a certain stage of prevention and control to revise and improve the surveillance strategies or schemes in this area in a timely manner. As further development or adjustment of new monitoring strategies is crucial to the effective development of parasitic disease prevention and control in a region, it is necessary to continuously evaluate the surveillance effects in the process of prevention and control. The evaluation of epidemic surveillance effects includes investigating missing reports and evaluating the quality of epidemic reports. In terms of methodology, in addition to the commonly used health and economic evaluation methods, more attention has been paid to the application of rapid epidemiological methods in recent years. More new evaluation techniques have been applied in the field surveillance, such as the capture-mark-recapture (CMR) and Delphi evaluation methods.

5.5.1 Investigation of Underreporting

5.5.1.1 Purpose

Investigating the underreporting of parasitic diseases can help us understand the dynamics of certain parasitic diseases, master the situation of parasitic disease underreporting, evaluate the quality of reports on parasitic diseases in CDCs and medical institutions at all levels, calculate epidemic levels, and provide a reliable basis for controlling parasitic diseases and formulating strategic plans for prevention and control of the diseases.

5.5.1.2 Methods

1. Investigation of Underreporting Among Residents. Investigating the underreporting of some parasitic diseases among residents can reflect the underreporting of parasitic diseases among patients. During the investigation, the investigated area can be roughly divided into five parts according to geographical location, namely, east, south, west, north, and center. Then, one to two communities (towns) should be selected from each part, and one to four villages selected from each community (town) (if the population is too large, several production teams or resident groups can be selected from each village) as the investigation units so that the investigation can be carried out from house to house.
2. Investigation of Underreporting in Medical Institutions. Investigating underreporting in medical institutions can reflect the underreporting of patients, but it cannot reflect the underreporting of patients who are ill but do not go to the hospital. The outpatient, emergency, and hospitalization registrations of medical institutions can be consulted to check the detected parasitic diseases against the hospital records, and the parasitic disease epidemic reports (including suspected or confirmed cases) of CDCs can be checked at the same time.

5.5.2 Investigation of Report Quality

5.5.2.1 Purpose

Epidemic reporting of parasitic diseases is the premise of controlling epidemics. A quality investigation of epidemic reporting can strengthen the standardized management of epidemic reporting and online direct reporting in various medical and health institutions at all levels; improve the quality of epidemic reporting; facilitate the collection, analysis, feedback, and utilization of epidemic information on parasitic diseases; and guarantee the timely discovery and effective conclusion of parasitic disease epidemics.

5.5.2.2 Methods

1. Complete investigation of paper reports. Fifty paper reports of parasitic diseases from epidemic reporting units (CDCs and medical institutions) for a specified period of time (fewer than 50 paper reports can be investigated based on the actual number) should be spot-checked to evaluate their completeness and accu-

racy. The components of the reports should include at least the name, sex, age, occupation, current address, disease type, onset date, date of treatment, and report date of each patient, and the reports should be considered complete only after all information is added.

2. Investigation of the consistency of paper and electronic reports. The components of paper and electronic reports of epidemic situations are compared for consistency, and the name, sex, age, disease, date of onset, date of diagnosis, and date of report completion for each patient in paper, and electronic reports should be checked to compare their consistency.
3. Investigation of the timeliness of the reports. According to the relevant provisions of the *PRC Law on the Prevention and Treatment of Infectious Diseases*, *Measures for the Administration of Information Reporting on Monitoring Information of Public Health Emergencies and Infectious Diseases*, and *Standards for the Management of Information Reporting of Infectious Diseases*, cases of Class B and Class C parasitic diseases should be reported to the web-based reporting system within the prescribed time. Medical records, case registration information, epidemic report receipt information, and relevant data in the web-based reporting system can be collected to judge whether reports of parasitic disease epidemics are timely.

5.5.3 Evaluation of Surveillance Effects

5.5.3.1 Purpose

Timely evaluation of the surveillance effects of parasitic diseases can help improve the sensitivity of the epidemic surveillance system and revise and improve the monitoring strategies or programs in a timely manner to enhance monitoring, early-warning, and forecasting abilities and promote the prevention and control of parasitic diseases.

5.5.3.2 Methods

1. Capture-mark-recapture. Reporting incidence is one of the most important indicators of the effectiveness of outbreak surveillance and interventions, but it is usually affected by the integrity of outbreak reporting systems. Therefore, when conducting epidemic monitoring or epidemiological evaluation, areas with sound grassroots health organizations and complete and effective epidemic reporting systems must be selected or monitoring points established within a certain range to reflect changes in epidemic situations before and after the implementation of epidemic monitoring and intervention measures. However, in areas where primary health organizations are not sound and not suitable for establishing monitoring sites, the capture-mark-recapture method can be used to grasp the epidemic situation of a certain parasitic disease.

   Two or more surveys should be carried out. Those who are found (caught) to have a parasitic disease or disease characteristics in the first survey should be registered (marked); those who are registered should be found (recaught) in later surveys; and the number of cases and confidence limits should be estimated by statistical processing.

2. Delphi method. The Delphi method provides an anonymous way to express opinions; that is, experts cannot discuss their assessments with each other, as there is no horizontal contact but only contact with investigators. Through several rounds of investigation, experts are repeatedly consulted regarding the questions raised in the survey; their opinions are then summarized and revised until they are finally summarized into a basic consensus that is the result of prediction. As a subjective and qualitative method, the Delphi method can be used not only in forecasting but also in the establishment of various evaluation index systems and the determination of specific indexes.

Using the Delphi method for team communication can avoid some possible shortcomings of group decision-making. People with the loudest voices or the highest status have no chance of controlling the group because everyone's views are collected. In addition, managers can ensure that important points are not ignored when soliciting opinions to make decisions. This method is widely representative and reliable. The implementation procedure of the Delphi method generally has four aspects: sending the initial questionnaire to team members to collect participants' views on a certain topic; sending a second questionnaire (with other people's opinions) to team members and asking the participants to evaluate other people's opinions according to several specific criteria; sending a third questionnaire to team members (including the evaluation results, average evaluation, and consensus provided by the second questionnaire) and asking participants to revise their original views or evaluations; summarizing the fourth questionnaire (including all evaluations, consensus, and remaining issues) so that the organizer can comprehensively deal with the results.

5.6 Case Study

5.6.1 Cause Analysis of a Kala-azar Outbreak in Jiashi County

5.6.1.1 Proposing the Issue

After September 2008, the national infectious disease network reporting system showed that the number of reported cases of Kala-azar in the Xinjiang Uygur Autonomous Region had increased sharply. As of November 26, 2008, 215 cases of Kala-azar had been reported by CDCs and medical institutions at all levels in the Xinjiang Uygur Autonomous Region through the web-based reporting system of infectious diseases. From January 1 to November 26, 2008, 127 cases were reported in Jiashi County, Kashgar, mostly in infants under 2 years old.

5.6.1.2 Analysis of the Causes of the Outbreak

1. Meteorological factors. The changes in monthly average relative humidity, monthly average precipitation, and monthly average wind speed in Jiashi County from 2005 to 2008 were analyzed. In 2007–2008, the average relative humidity,

average precipitation, and average wind speed in Jiashi County were low. The decrease in humidity, precipitation, and wind speed all affected the density and distribution of sandflies and the distribution and activities of suspected animal infection sources. The comprehensive effect of these factors may have promoted the Kala-azar epidemic in the local area.

2. Vector biological factors. In 2008, the climate in Jiashi was drier than usual, and the wind speed was lower, which increased the density of sandflies. At the same time, the monitoring data of sandfly density in this area showed that in July 2008, the sandfly density was significantly higher in Jiashi County than that in other areas. In addition, the area of newly cultivated or semidesert agricultural areas in this area had increased in recent years, resulting in increased contact opportunities with sandflies and exposure of people, which in turn led to an increase in the number of patients.
3. Socioeconomic factors. The economic backwardness of the region and the poverty of its inhabitants, as well as the poor nutritional status of children, especially newborns, and their poor immune status and reduced immunity to the disease, increased the risk of morbidity.
4. Personal protective factors. In view of the living customs and hot weather in the region, the vast majority of residents did not have the habit of using mosquito nets, and infants tended to sleep in unprotected cradles, which led to an increased chance of exposure to sandfly bites.
5. Production and lifestyle. Local residents had no habit of sending children to nurseries. As most local residents were engaged in farming, parents usually took their children with them to the farm and did not take them home until evening. Near evening, the density of sandflies on farms was quite high, which increased the bite rate among infants and young children.
6. Animal protection factors. Changes in insect-retaining hosts and the number of various wild animals contributing to the epidemic played an important role in the development of this Kala-azar epidemic. In the vast desert and semidesert areas of Jiashi County, there were a large number of Tarim rabbits and other rodents. Because the Tarim rabbit had been designated as a national secondary protected animal in the List of National Key Protected Wild Animals, the number of local Tarim rabbits had increased rapidly, which had a great impact on the transmission potential of natural Kala-azar foci.
7. Host animal factors. Changes in the activity range of hosts and various wild animals linked to the epidemic also played an important role in natural Kala-azar foci. In recent years, the precipitation in Jiashi had decreased, and the climate was relatively dry. As a result, many wild animals that had originally lived far from residential areas had moved closer to residential areas for food and water. The expansion of wildlife activities and their gradual aggregation in the vicinity of residential areas increased the probability of wildlife bringing pathogens to residential areas and increased the natural infection probability of vector sandflies and the infection probability of the population.
8. Prevention and control factors. The investment in local Kala-azar prevention efforts had long been unable to meet prevention and control needs. Due to a shortage of funds, a shortage of prevention and control personnel, backward

prevention and control technologies, sparsely populated areas and inconvenient transportation, as well as the onset and distribution characteristics of desert-type Kala-azar and difficulty in controlling vector sandflies, prevention and control measures, such as patient check-ups, health education, vector monitoring and sandfly eradication spraying, had basically stagnated for many years.

5.6.1.3 Recommendations

The epidemic severity of parasitic diseases is closely related to biological and social factors. These factors should be fully taken into account when analyzing and preventing epidemics in order to comprehensively control the epidemic efficiently and prevent the occurrence of diseases in real time.

5.6.2 Establishment of the National Geographic Information System for Surveillance of Natural Factors of Schistosomiasis Transmission

5.6.2.1 Proposing the Issue

The national geographic information system (GIS) for schistosomiasis was established to explore the possibility of predicting endemic areas of schistosomiasis by establishing models using meteorological parameters.

5.6.2.2 Surveillance Components and Methods

1. Establishment of the GIS Database
   (a) Basic geographic information. Using the "Digital Map of the World" (1:1 million) produced by ESRI, the digital map was processed on ArcView3.0a software and overlapped with AVHRR satellite remote sensing images to form an intuitive map. Both layers of maps were marked with an accurate coordinate system. The former is a vector file map with counties as the unit, and the latter is a image file with the corrected normalized difference vegetation index (NDVI) as the background. In this study, the part of the Yangtze River in southern China was selected for analysis.
   (b) Meteorological data. Based on data from the FAOCLIM database published by the World Food and Agriculture Organization, comprehensive data from meteorological stations in southern China over 30 years (1940–1970) were extracted, including the lowest temperature in January, monthly average temperature, amount of rainfall, number of rainy days, and potential evapotranspiration (PET) index.
   (c) AVHRR remote sensing satellite data. There are five data channels in the images of the advanced very-high-resolution radiometer (AVHRR) satellite, and the resolution of each grid is 1.1 km. The first and second channels were selected to obtain the corrected vegetation index according to the formula NDVI = (CH1 + CH2)/(CH1 − CH2). At the same time, Channel 4 was selected for the ground temperature index. All image processing was performed using ERDAS8.03 software.

   (d) Elevation data. Using the elevation distribution from "Digital Map of the World," the elevation map of southern China was created with the ArcView 3.0a software package for further overlap analysis.
2. Prediction Model. Because it takes an average of 60 days for *Schistosoma japonicum* to develop from larvae miracidia to larvae cercaria which is infected stage inside snails, the appropriate temperature of *S. japonicum* in snails is 18–28 °C, and the diffusion of *S. japonicum* is closely related to water. The modified Malone formula based on temperature and potential evaporation index (surface water balance system) was used to calculate the transmission index of *S. japonicum*.
3. GIS Spatial Data Analysis
   (a) Spatial analysis. With ArcView3.0a as the supporting software, the spatial analysis method was used to divide the distribution of the propagation index at each observation station. The propagation index was divided into different geographical locations by different colors, and areas with strong or weak propagation were indicated by lines and colors.
   (b) Overlap analysis. The NDVI synthesis map of the epidemic season in 1995 (March–October) was used as the main background data together with the distribution map of the schistosomiasis transmission index, snail distribution index, lowest temperature in January distribution map, temperature distribution map of AVHRR Channel 4, and elevation distribution map, supported by ArcView 3.0a software, and the geographical distribution map of the epidemic area was displayed with certain category values.

5.6.2.3 Surveillance Results

1. Distribution of schistosomiasis transmission index. The transmission index of *S. japonicum* was higher in southeast China than that in northwest China, and the index value >900 basically coincided with the endemic areas of schistosomiasis in southern China (except that the area in the eastern region extended slightly to the north). The spatial distribution results showed that the northernmost area was not beyond the northern boundary of Jiangsu Province, and it was in the area south of 34 °C north latitude.
2. Multilayer overlapping distribution. The overlapping map showed that the high-risk areas in the red area were basically consistent with areas with a high incidence of schistosomiasis in the Yangtze River basin. Western Sichuan and western Guizhou were both nonendemic areas, which indicated that the epidemic range of schistosomiasis was closely related to temperature, elevation, rainfall, and other factors.

5.6.2.4 Recommendations

The epidemic range of schistosomiasis is closely related to temperature, elevation, rainfall, and other factors. The meteorological data model and satellite remote sensing data can be used to predict potential endemic areas of schistosomiasis. AVHRR remote sensing data can be used to accurately and quickly predict the epidemic range and intensity of schistosomiasis.

Syndromic Surveillance

6

Jun-Fang Xu and Zhuo Chen

In recent years, with the emergence of new infectious diseases and the reemergence of existing infectious diseases, the traditional diagnosis-based disease surveillance system no longer meets the requirements of emergency responses to public health events. Syndromic surveillance can fill this gap to a certain extent.

Symptom surveillance refers to monitoring the frequency of specific clinical syndromes (such as fever, respiratory symptoms, and diarrhea) in specific populations without any assistance from a specific diagnosis; that is, monitoring based on non-specific symptoms is a detection, evaluation, and reporting mechanism for possible emergencies based on early warning and investigation of disease occurrence. Symptom surveillance aims to establish a threshold for the number of cases with early symptoms, which will occur earlier than the threshold established by traditional surveillance methods, allowing the scale and speed of disease outbreaks to be tracked.

The traditional disease surveillance system is based on hospital diagnosis and laboratory tests. The symptom surveillance system automatically, continuously, and systematically collects and analyzes data on the occurrence frequency of specific clinical syndromes, such as various data on preclinical symptoms, doctor diagnosis results, drug sales records, and even employee attendance. The system performs the analysis using different statistical methods, identifies abnormal signals and generates early warnings in a timely manner, and feeds the abnormal signals back to public health personnel using charts and other information to master the epidemic situation represented by the syndromes and enhance the

J.-F. Xu (✉)
Hubei Institute for Nationalities, Enshi, Hubei, China

Z. Chen
National Institute of Parasitic Diseases, Chinese Center for Disease Control and Prevention (Chinese Center for Tropical Diseases Research), Shanghai, China

X.-N. Zhou (ed.), *Surveillance and Management of Parasitic Diseases in China*,
https://doi.org/10.1007/978-981-97-1241-0_6

early-warning capability of public health surveillance. At present, there are relatively mature symptom surveillance systems with working objectives that include public crisis response, early warning of emerging infectious diseases, outbreak and epidemic detection of key infectious diseases, and safety protection for large-scale activities.

6.1 Sentinel Hospital Configuration and Management

Symptom surveillance is a method of monitoring the symptoms of a specific population by selecting some primary hospitals as sentinel hospitals. Sentinel hospital surveillance is a monitoring system for specific clinical syndromes that uses random or nonrandom methods to select a primary hospital and a certain number of reporting units or reporters covered by it as monitoring sentinels.

Sentinel surveillance can compensate for the defects of a high underreporting rate and low monitoring quality in conventional surveillance. It has the advantages of low cost, high efficiency, and strong purpose. It can select representative surveillance sentinels to serve several specific purposes according to the epidemic characteristics of specific parasitic diseases, special prevention and treatment requirements, and existing weak links. For example, it can be used to obtain case samples from medical institutions serving a particular population or area. In malaria surveillance, malaria epidemic information is collected through the targeted establishment of monitoring sentinel sites to correct nationwide malaria epidemic data. It is also possible to establish surveillance sites for drug resistance of *P. falciparum* in endemic areas and mosquito surveillance sentinels in typical areas.

6.1.1 Configuration

For the sentinel surveillance system based on hospitals, representative hospitals should be selected as surveillance points to observe the occurrence, increasing and decreasing trends, and distribution of diseases in these hospitals over a long period. The cost of sentinel hospital monitoring is relatively low, and the reports are relatively high quality. If the configuration of sentinel sites is reasonable and representative problems (such as population, region, health status, and medical resources) can be solved, the data obtained through such monitoring can not only describe change trends of diseases and detect outbreaks and epidemics but also calculate the overall incidence level and use it to assess the flow of medical resources, thus helping public health emergency agencies allocate limited resources. Therefore, to maximize surveillance effectiveness, symptom surveillance systems should cover most health facilities in designated areas, including representatives of ethnic and socioeconomic zones; otherwise, large outbreaks may not be detectable.

For any type of surveillance, the most important aspect is monitoring population representation. For example, in the sentinel hospital surveillance of *Angiostrongylus cantonensis* infections, three representative hospitals from ten hospitals in Dali City, Yunnan Province, where the incidence of *A. cantonensis* infections is high, are selected as surveillance sentinels. Only cases from these hospitals can be counted as cases for the monitoring project, and they are not subjected to restrictions on the locations of patients' residences. That is, in addition to local patients, patients from other counties and cities or even other provinces and cities can be considered as angiostrongyliasis cases. Similarly, if residents of the area are treated in hospitals outside the scope of the local monitoring network, cases occurring among those residents are not recorded by the surveillance system.

If surveillance sites cover the population in most affected areas, the corresponding information that needs most attention (e.g., geographical location, age, and sex) can be identified. Many examples from retrospective studies have shown that symptom surveillance can provide better early detection, early warning, and rapid response to epidemic outbreaks (emerging infectious diseases, unexplained diseases, and other clustered adverse public health events) than that from traditional disease reporting systems.

6.1.2 Management

The responsibilities of departments and personnel at all levels in sentinel hospitals should be clarified, and a clear and detailed surveillance plan, working system, and specifications should be formulated. In the process of monitoring and implementation, training should be carried out continuously to cover changes in personnel. Public health departments, such as CDCs, should carry out technical guidance, inspection, and supervision of the sentinel hospitals.

Management of all aspects of information reporting, analysis, and feedback also needs to be strengthened. Examples include raising clinicians' awareness of the importance of surveillance and understanding and accurately judging relevant components of surveillance, especially the definition of cases; registering relevant information according to the surveillance requirements; timely reporting according to the requirements of case reports and data transmission; and improving the quality of data reporting and the hospital's HIS system with the establishment of reporting quality control measures.

6.2 Surveillance Indicators and Classification

Symptom surveillance focuses on the occurrence of specific clinical symptoms (such as fever, respiratory symptoms, diarrhea, etc.), rather than disease-specific diagnostic information. Syndrome refers to certain clinical symptoms caused by interrelated organ diseases or dysfunction. The collection of such clinical data is the basis of the symptom surveillance system.

There are many sources of syndrome data that are related to the monitored diseases and syndrome types. To analyze abnormal statistical data, it is necessary to classify symptom data representing health event indicators. The definition and classification of syndrome draw on lessons from many mature experiences; then, according to China's national conditions and new situation requirements, epidemiologists, senior clinical doctors, and health and quarantine administrators can be organized to formulate them. For example, when monitoring the symptoms of heating-related diseases, the International Classification of Diseases (ICD-9) can be used to classify the symptoms. There are several defined symptoms, including, hemorrhagic symptoms, gastrointestinal symptoms, nervous system symptoms, cardiovascular system symptoms, skin rash, local skin lesions, fever with unknown cause, respiratory syndrome, special infection, and serious diseases or death possibly caused by infection. In the global surveillance project of emerging infectious diseases, four types of symptoms that should be monitored first are defined: (1) cases in patients 5–65 years old, severe acute respiratory distress syndrome (ARDS) with fever of unknown cause ≥38.3 °C; (2) cases in patients over 5 years old with encephalitis or meningitis and body temperature ≥ 38.3 °C; (3) clinical symptoms similar to those of cholera, plague, yellow fever, malaria, tuberculosis, and other diseases as well as any serious diseases caused by antibiotic-resistant bacteria or protozoa; and (4) aggregated deaths of animals.

For example, in recent years, the number of reported cases of *Angiostrongylus cantonensis* infections in China has increased. To effectively reduce the incidence and control outbreaks, the China CDC has carried out a pilot program of symptom monitoring and early warning of *Angiostrongylus cantonensis* infections in some areas of China since 2009 and monitored the clinical symptoms of meningitis, such as headache, hyperalgesia, and neck stiffness, which has provided much experience and a scientific basis for a timely understanding of the epidemic dynamics of the disease and formulating prevention and control strategies and measures.

There are clinical and nonclinical monitoring data in the symptom monitoring system. To use clinical data, first, the relationship between public health personnel and doctors is established by implementing symptom monitoring, and second, when an abnormal trend is found, the investigators trace it to individual patients through clinical data. Nonclinical data can supplement clinical data and describe the situation of people outside the hospital to confirm the trend indicated by clinical data. Another task is to normalize and standardize the classification of clinical and management data by readjusting data from different sources and processing information related to symptom monitoring (see Table 6.1 for symptoms of common parasitic diseases).

Table 6.1 Syndromes of common parasitic diseases

Syndromes	Symptoms	Suspected parasitic pathogens
Influenza-like syndrome	Fever, aversion to cold, night sweat, muscle pain	*Plasmodium*, *Babesia*
Febrile respiratory syndrome	Cough, dyspnea, chest distress, chest pain	*Schistosoma*, *Leishmania*, *Filaria*, *Toxoplasma*, *Amoeba*, *Paragonimus*, *Echinococcus*, *Strongyloides stercoralis*, *Ascaris lumbricoides*
Febrile meningitis syndrome	Dizziness or headache, hyperalgesia, stiff neck	*Angiostrongylus cantonensis*, *Filaria*, *Amoeba*, *Taenia solium*, and *Cysticercus cellulosae*
Digestive tract syndrome	Poor appetite, nausea, vomiting, abdominal distension, abdominal pain, diarrhea, pus-blood mature, chyluria	*Cryptosporidium*, *Fasciolopsis*, *Paragonimus*, *Schistosoma*, *Plasmodium*, *Filaria*, *Toxoplasma*, *Amoeba*, *Clonorchis sinensis*, *Trichinella spiralis*, *Taenia solium* and *Cysticercus cellulosae*, *Echinococcus*, *Ascaris lumbricoides*, *Trichuris*, *Enterobius vermicularis*, flagellate protozoan, *Trichinella*
Skin rash or skin lesion syndromes	Dermatitis or pruritus, skin nodules or ulcers	*Schistosoma*, *Leishmania*, *Filaria*, *Amoeba*, *Trichinella* spiralis, *Paragonimus*, *Echinococcus*, *Strongyloides stercoralis*, *Enterobius vermicularis*, *Trichinella*
Lymphadenitis syndrome	Lymphadenectasis, lymph varix	*Leishmania*, *Filaria*

Source: Human Parasitology, Human Parasitic Diseases Primary Prevention and Control Series—parasitic zoonoses

6.3 Data Collection and Analysis

6.3.1 Collection

At present, the application of parasitic disease symptom surveillance data in China is still in the initial stage. The data collected by the symptom surveillance system are mainly the symptom information reported to the public health institutions of a group (class) or multiple groups (classes) before the diagnosis of a specific parasitic disease rather than the diagnostic information. Therefore, syndrome data are the basis of symptom monitoring system data; they predate clinical diagnosis and are mainly related to the types of syndromes to be monitored and target diseases.

6.3.1.1 Data Sources of Parasitic Disease Syndromic Surveillance

At the individual level, data sources are mainly parasitic characteristics, patient (or infected) behavior, and related service supply behavior. From the population perspective, they are mainly population susceptibility, prevalence of parasitic diseases (incidence, prevalence, infection rate, etc.), and population migration status. From

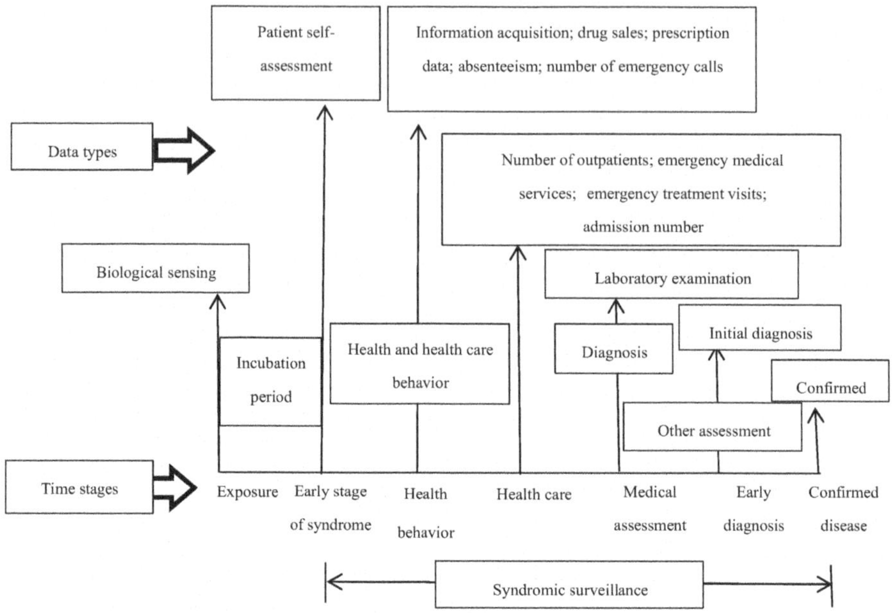

Fig. 6.1 The process of collecting syndromic surveillance data

the vector perspective, they are mainly the breeding environment and vector infections or illnesses. Three types of data sources are frequently used in practice for symptom surveillance: first, clinical data, such as outpatient visits, diagnostic logs, hospital admissions, clinical laboratory tests, and radiological or imaging examinations; second, other data, such as healthcare hotline information, drug use records, absenteeism records, animal diseases and deaths, and vector information; and third, basic information about individuals, including the sex, age, address, zip code, and contact number of the patient.

See Fig. 6.1 for the types of surveillance data in different time stages. Based on various data sources, the effectiveness of symptom monitoring data can be evaluated in terms of the timeliness and characteristics of the information they provide. Table 6.2 provides the classification of different data.

6.3.1.2 Methods of Collecting Syndromic Surveillance Data

The data collection method depends mainly on the monitoring purpose, disease symptoms, relevant original data records, information management methods, and levels of medical institutions. For the elimination programmes of malaria and schistosomiasis, China mainly adopts active monitoring and reactive monitoring to collect data and implements early-warning case monitoring; for case monitoring and early warning of foodborne parasitic diseases, such as *Angiostrongylus cantonensis*, it adopts reactive monitoring. Therefore, the collection methods of parasitic disease symptom monitoring data differ due to disease prevalence, data supply and transmission requirements, special computer network, etc.

Table 6.2 Data sources based on data timeliness and characterization

	Characterization	
Timeliness	Strong	Weak
Quick	Chief complaints from emergency and outpatient services Admission records Laboratory examinations Triage nurse calling calling for ambulance Prescription drug data	Drug sales Absences from work or school
Slow	ICD-9 Laboratory test results Clinical reports	General information (local or regional events)

As symptom surveillance data are mainly nonspecific, it is difficult to ensure their accuracy. When collecting data, attention should be paid to the following aspects: first, the representativeness of data, i.e., whether the monitoring data represent the target population, especially in terms of regional distribution, is very important for the outbreak monitoring of target parasitic diseases; second, the integrity of data, i.e., collecting the vast majority of monitored target information and inputting it into the information management system; and third, the stability of data, i.e., ensuring data quality during the monitoring process.

6.3.2 Analysis

The analysis of symptom data is based mainly on the frequency of various symptoms. The statistical analysis of monitoring data is carried out automatically, including determining whether there is aggregation in time and space, assessing demographic characteristics, and calculating the odds ratio (OR) value. The main purpose is to find or detect the "signal" of disease outbreak or abnormal increase in the incidence of a syndrome and to provide a timely early warning. It is necessary to establish baseline levels or thresholds for identifying and judging abnormal symptom group values in advance. Statistical analysis methods of different times, spaces, or time-spaces can be used to generate early warnings of abnormal numbers and present them to public health personnel in the form of charts so that they can grasp an epidemic situation and provide a timely early warning.

6.3.2.1 Common Statistical Charts and Graphs

1. Bar Chart. A bar chart uses the bars of different lengths and the same width to indicate values in a statistical index that are independent of each other. The bar scale of a bar chart must start at 0, and the width of each bar must be equal. The interval is generally half the width of or equal to the width of the bar. The order of the bars can be arranged according to the index values or the natural grouping order. Figure 6.2 shows the occurrence of nonspecific symptoms in a certain region for a certain period.

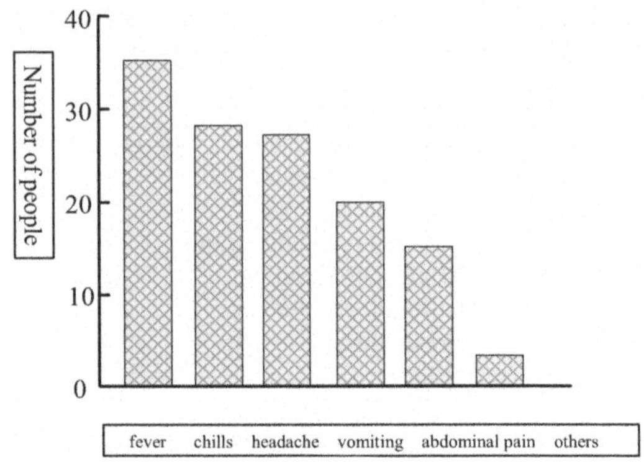

Fig. 6.2 Occurrence of main symptoms in a certain period

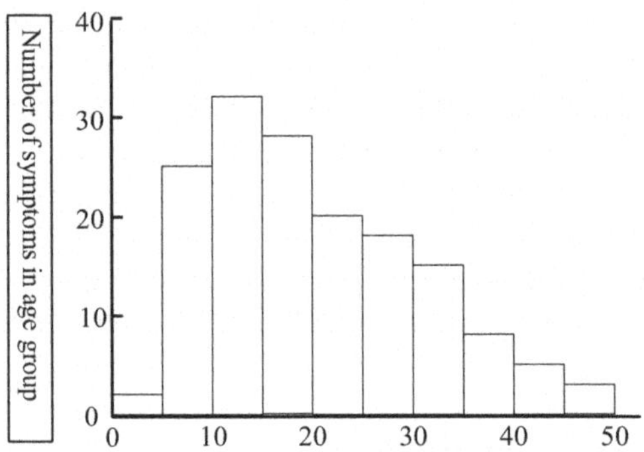

Fig. 6.3 Age distribution of a symptom in a certain period

2. Histogram. A histogram describes the frequencies of each group with the area of each straight square. The sum of the areas is equivalent to the sum of the frequencies of each group; thus, this type of chart is suitable for representing the frequency distribution of numerical variable data. The vertical axis of the histogram is the frequency, and the horizontal axis is the value of the numerical variable. If the distances between groups are not equal, the frequency is converted into equidistant data and plotted. The frequency is divided by the group distance to obtain the frequency of the unit group distance, which is taken as the height of the straight square, and the group distance is the width of the straight square. Another way to represent the frequency distribution of numerical variable data is to divide the observation frequency of each group by the total observation frequency to obtain the frequency of each group segment, and the frequency density obtained by dividing the frequency of each group segment by the group distance is taken as the height of the histogram. The plotted histogram is called the frequency histogram, which represents the frequencies of each level with the area of each straight square, with the sum of the areas being 1. Figure 6.3 shows the age distribution of febrile symptoms at a given time.
3. Line Graph. The line graph is used to express numerical changes by the rise and fall of line segments and is suitable for describing the trend of a statistic that changes with another continuous numerical variable. It is most commonly used

to describe the trend of statistics changing with time. Usually, the vertical axis is the statistical index, and the horizontal axis is time or another continuity variable. If the horizontal axis and the vertical axis are arithmetic scales, the graph is called an ordinary line graph. The vertical axis generally starts at 0; otherwise, special marks or explanations are needed. When the vertical axis is a logarithmic scale, the chart is called a semilogarithmic line graph, which is especially suitable for comparing the changing speeds of different indexes. Different indicators or levels can be represented by different line segment styles (dotted lines or solid lines) or colors, and the marked points of each measured value are connected by straight lines, which cannot be smoothed into curves. Figure 6.4 shows the change trend of fever cases in a certain place for a certain period.

6.3.2.2 Statistical Analysis Methods

Usually, a symptom surveillance system involves a variety of statistical analysis methods, of which there are three types: (1) Time series analysis: for example, for a static cyclic regression model optimized by training data and predetermined parameters, the parameters are based on linear functions learned from historical data, and seasonal influences can be adapted to time series analysis. (2) Spatial distribution analysis: for example, simple geometric shapes can be used to scan the whole area based on likelihood ratio measurement, Bayesian modeling, and a generalized linear mixed model, taking into account various parameters such as region, week, day, holidays, and seasonal changes. (3) Spatiotemporal distribution analysis: an extended statistical model of spatial scanning can be used to search all subregions for possible multiple likelihood ratio tests in space and time to identify specific characteristic groups (age, diagnosis, related diseases, etc.). This section introduces the following three statistical analysis methods. For the analysis methods of the model, please refer to Chap. 9:

1. Exponentially weighted moving average (EWMA). EWMA is an important statistical quality control method that was first developed by Robert in 1959 and has been applied to syndromic surveillance systems in recent years. Its main aims

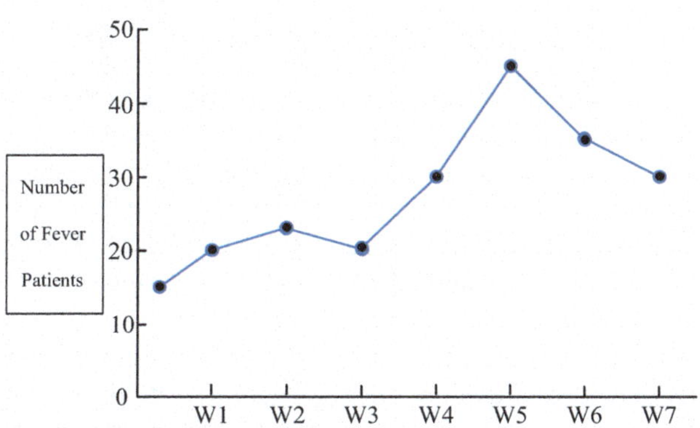

Fig. 6.4 Trends of fever cases in a certain period

are to set up an early-warning database using report data of historical cases, establish an early-warning model using the control chart method, determine the epidemic reference standard using the expert consultation method, and select the appropriate early-warning threshold by calculating and comparing the sensitivity, specificity, and positive predictive value and plotting the receiver operator characteristic curve. In China, some scholars have applied the control chart method to provide early warnings of seven infectious diseases, and the results have shown that this early-warning method has high sensitivity, specificity, and efficacy. The basic operational formula is as follows:

$$S_t = \alpha y_{t-1} + (1-\alpha) S_{t-1}$$

where α is the weight ($0 < \alpha \leq 1$), y_t is the observed value at time point t ($t \geq 3$), and S_t is the average observed value after exponential weighting.

2. Cumulative sum (CUSUM) control chart. The CUSUM control chart is modeled on the long-developed industrial process surveillance statistical methods for monitoring abnormalities in the number of syndromes or medical records. In 1999, it was used to monitor abnormalities in the incidence of rare diseases. The method is currently applied by many foreign surveillance systems, such as the US Early Reporting System for abnormal events, which applies this method to set three different cumulative sum control chart statistics divided into three grades, light, medium, and heavy, as a reference basis for daily monitoring of data abnormality. Early warning of abnormality is carried out through different formulas and set thresholds (double or triple standard deviation) followed by the accumulation of the number of variations in past sequence data, which is used to monitor outbreaks with small changes in the number of cases. The CUSUM control chart can select different cumulative sum control chart statistics according to different data and can modify statistical equations according to different needs or stratifications by different risk factors; however, the problem of epidemiological grouping must be carefully considered when using the CUSUM method. The fundamental equation is as follows:

$$S_t = \max\left(0, S_{t-1} + \left(\left(X_t - \left(\mu_0 + k\sigma_{xt}\right)\right) / \sigma_{xt}\right)\right)$$

where St is the cumulative variation at time point t, X_t is the number of observations at time point t, μ_0 is the mean of the observation interval, σ_{xt} is the standard deviation of the observation interval, and k is the abnormal reference value. This method creates three different statistics (CUSUM1, CUSUM2, and CUSUM3) according to different sensitivities. CUSUM3 (C3) is the highest, followed by CUSUM2 (C2), and CUSUM1 (C1) is the lowest.

3. Scan statistic. The scan statistic was proposed by Naus in 1965 and was initially applied to identify the aggregation of one-dimensional point processes; later, it was gradually extended to explore the aggregation of things in two-dimensional space and three-dimensional space-time. In public health, this method is used for early detection of disease outbreaks and early detection and monitoring of poten-

tial public health risks; in parasitology, the scan statistic is often used to analyze whether there is a tendency or trend of aggregation of parasitic diseases and related vectors in temporal, spatial, or spatiotemporal distributions. According to the different temporal and spatial dimensions of the data, they can be divided into temporal scan statistics, spatial scan statistics and spatiotemporal scan statistics, which are widely used in syndromic surveillance.

The temporal scan statistic can be used to identify one or several clusters on the time axis, and this method has been applied to the detection of schistosomiasis aggregated areas, early warning of the distribution of infected *Oncomelania*, and the characteristic analysis of the spatiotemporal aggregation and distribution of malaria in high-incidence areas. The scan statistic sets the number of symptoms occurring within observation time t as N, determines the occurrence time of each symptom number (point process), defines the scanning window length as w, scans the whole observation period t beginning from time point t_i ($0 \leq t_i \leq t$-w), and records the maximum value of the syndromic number of each window S_{max}. If N is a known constant, the null hypothesis is that the occurrence of symptoms has a uniform distribution. If the scan statistic is greater than the number of cases within the scan window time length, it indicates that the occurrence of symptoms has the possibility of temporal aggregation. The spatial scan statistic is an extension of the temporal scan statistic from one-dimensional time to two-dimensional space; thus, it is used to identify whether the occurrence of symptoms has spatial aggregation and the approximate location of the aggregation. The data can take the form of point data (syndromic location) or regional data (the number of syndromes within each region). Compared with the one-dimensional temporal scan statistic, the calculation of the two-dimensional spatial scan statistic is more complicated, and the hypothesis test process mainly involves the establishment of a likelihood function, the solution of the maximum likelihood ratio, and the calculation of the p-value using the Monte Carlo method. The spatiotemporal scan statistic is an extension of the spatial scan statistic from two-dimensional space to three-dimensional space-time and is used to identify spatiotemporal aggregation and the approximate aggregated location and time. The spatiotemporal scan statistic also uses the Monte Carlo method to calculate the p-value and determine whether there is spatiotemporal aggregation. It allows for retrospective study of past data and prospectively and regularly monitors the occurrence and development trends of diseases.

6.4 Validation and Utilization of Results

6.4.1 Signal Probing of Early Warning

The syndromic surveillance system is a new type of public health surveillance system developed according to the concept and theory of syndromic surveillance. The working principle is the detection of the characteristic signals of the occurrence of public health events against background noise. Signal detection is the process of inputting syndromic surveillance data and transforming them into signals, analyzing and processing them, and finally outputting a result judging whether an event is occurring. The simplest way to determine this result is to compare the signal intensity with a specific threshold value (Fig. 6.5).

In parasitology, syndromic surveillance of parasitic diseases is used to collect data related to early symptoms of parasitic diseases over a given period, converts them into signals by computer, and successively calculates the expected incidence of symptoms (or events) of parasitic diseases in any period of time. If the incidence of one or more symptoms (or events) deviates from the baseline and exceeds a certain threshold, the system automatically raises the alarm, judges whether an outbreak or epidemic of a parasitic disease is occurring, and identifies high-risk populations, regions, and periods. Syndromic surveillance is used to collect relevant data from the stage when patients start to show mild or nonspecific symptoms. Although these symptoms or phenomena are nonspecific, their frequency may increase rapidly over a certain period or be concentrated in specific geographical areas, showing temporal or spatial aggregation. For example, early in the cryptosporidiosis epidemic, a large number of patients went to pharmacies to buy antidiarrheal drugs on their own, leading to an increase in the sales of these drugs. To avoid too many false-positive early-warning signals, syndromic surveillance must be used to conduct in-depth research on signal detection sensitivity and specificity to optimize the surveillance techniques. In response to numerous small outbreaks or minor symptom clusters indicated by early-warning signals, emergency response priorities can be determined in advance to prioritize the investigation of suspected larger outbreaks or serious diseases based on limited emergency response capacity.

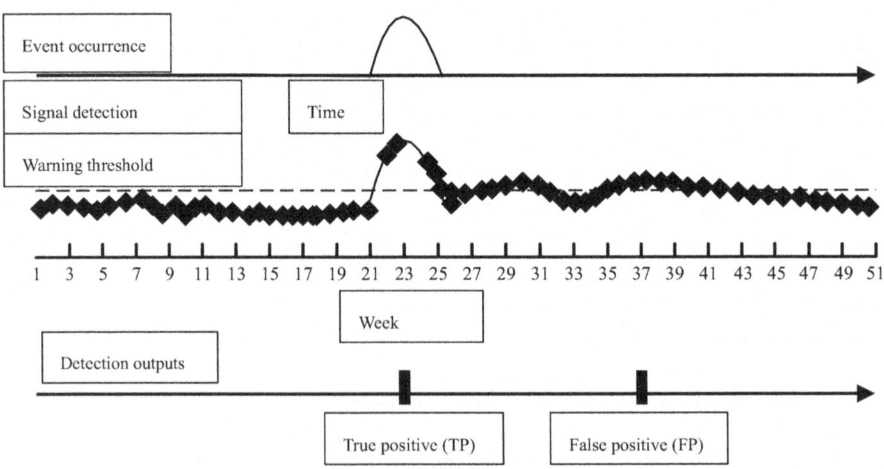

Fig. 6.5 Principles of signal detection

6.4.2 Verification of Early-Warning Signals

Once an early warning of an abnormal signal is issued, field investigation and further laboratory diagnosis are required. During field investigations, epidemiological methods are used to identify pathogenic factors, modes of transmission, susceptible populations, exposure factors, and environmental determinants. Epidemiologists must scientifically propose a rational survey design based on the abnormal signals detected by the syndromic surveillance system and then provide rigorous scientific data for later analysis. Field surveys include both population-level and individual-level surveys. The population-level survey covers mainly medical and health institutions, including individual clinics, hospitals, and schools, whereas the individual-level survey focuses mainly on individuals. A good field survey should (1) be conducted in a timely manner, (2) identify sources of infection or other high-risk factors as soon as possible and evaluate the situation, (3) use reasonable descriptive or analytical epidemiological methods, (4) determine the source of infection or cause of disease with sufficient evidence of causal association, and (5) propose real-time control and intervention measures.

The epidemiological case survey questionnaire is one of the tools used to collect data for field research studies. A well-developed case survey questionnaire can fully reflect the essential components of the survey and the quality of the research plan. Therefore, epidemiological case survey questionnaires are the basis of various epidemiological studies.

6.4.2.1 Principles of Designing Questionnaires

The following basic principles must be adhered to in designing the case survey questionnaire: (1) highlight the characteristics of the symptoms and corresponding diseases under investigation, (2) ensure that the scope of the survey items is reasonable and appropriate, (3) use standard questions as much as possible and try to use objective and quantitative indicators and avoid repeated questions, (4) ensure that the order of the survey items is logical, and (5) conduct a presurvey.

6.4.2.2 Questionnaire Structure

The structure of the case questionnaire generally includes the cover page, questionnaire, and linkage section. The components of each section are as follows: (1) The cover page contains an informed consent form and explains the purpose of the survey; (2) the questionnaire requests the respondents' basic personal information and behavioral habits and presents the items or indicators that need to be investigated; and (3) the linkage section mainly includes interlanguage or tips language.

6.4.2.3 Examples of Questionnaires

1. Case Questionnaire for Suspected Acute Schistosomiasis

Province: _____Prefecture (city): _____County: _______Township: ________ Village: ________

The national standard codes of Province, Prefecture, and County

I. Case investigation:

1. Report date: ______year_____ month _____day
2. Survey date: ______year _____month _____day
3. Case name: ____________ Contact name: ___________
4. Sex: (1) Male (2) Female (9) Unknown
5. Residence address: ____.
6. Floating population or not: (1) Yes (2) No (9) Unknown
   If yes, then from:
7. Date of birth: ______ year ____ month ____ day (or age: ___ years or ___ months old)
8. Fever: (1) Yes (2) No (9) Unknown
   If yes, (1) Fever date: _____ year _____ month _____ day
   (2) Temperature: ______°C
9. Papules: (1) Yes (2) No (9) Unknown
   If yes, (1) Date of papules: _____ year _____ month _____ day
   (2) Character of papules: _______________
10. Cutaneous pruritus: (1) Yes (2) No (9) Unknown
11. Blisters: (1) Yes (2) No (9) Unknown
12. Whether exposed to infested water: (1) Yes (2) No (9) Unknown
   If yes, (1) Way of contact with infested water:
   (2) Time of contact with infested water: __ year __ month __ day __ o'clock
13. Have you ever been infected with schistosomiasis: (1) Yes (2) No (9) Unknown
   If yes, (1) Number of infections: _____times
   (2) Date of last infection: ____ year ___ month ___ day
   (3) Whether treated: (1) Yes (2) No (9) Unknown

II. Laboratory examination:

14. Collection of serum specimens: (1) Yes (2) No
   (1) Collection data: ____ year ___ month ___ day
   (2) Laboratory results: (1) Positive (2) Negative (9) To be determined
15. Collection of fecal specimens: (1) Yes (2) No
   (1) Collection data: ____ year ___ month ___ day
   (2) Laboratory results: (1) Positive (2) Negative (9) To be determined

III. Classification:

16. Results of case classification:
   (1) Confirmed (2) Excluded (3) To be determined (4) Unknown

Evidence of confirmed disease:
(1) Laboratory confirmation (2) Epidemiological linkage (3) Clinical confirmation

Exclusion basis:

Case investigators:

Specimen collectors:

Reporters of laboratory results:

2. Epidemic Focus Questionnaire of Acute Schistosomiasis

Province: ______Prefecture (city): ______County: _______Township: ______ Village: ________

The national standard codes of Province, Prefecture, and County

I. Basic information

1. Report date: ____ year ___ month ___ day

2. Survey date: ____ year ___ month ___ day Investigators:

3. The total population of the village of epidemic focus: ________________

Among them, permanent population: _____floating population:

4. Major crops:

5. Whether there are cattle in the village: (1) Yes (2) No

If yes, how many cattle are there:

6. Is the village an epidemic village: (1) Yes (2) No

If yes, epidemic type: (1) Type I village (2) Type II village
(3) Type III village (4) Type IV village

7. The cumulative number of schistosomiasis cases in the village:

Among them, acute schistosomiasis patients: _____:
advanced schistosomiasis patients: _____;
chronic schistosomiasis patients: _____

II. The distribution survey

8. Number of positive snail points in the village:

9. Situation of snail search: historical snail areas: _____m^2; existing snail areas: _____m^2

Snail areas checked: _____m^2; areas with snails: _____m^2;
positive snail density: _____pcs/m^2

No. of searched snail frames:_____; no. of live snails caught: _____; no. of positive snails: _____pcs

Maximum *Oncomelania* density: _____/m^2; mean of live snail density: _____/m^2

10. Snail control situation: snail control area: _____m^2

6.4.3 Decision-Making Applications

The early-warning role of syndromic surveillance is based on signal detection theory and decision-making theory. Decision-making theory can provide decision-makers with a systematic method for analysis and problem-solving, encourage them to discover and explore various potential countermeasures and possible natural states, and form a comprehensive and systematic management approach by making full use of available scientific knowledge and technical means. In public health, indicators such as sensitivity, specificity, positive predictive value, negative predictive value, and timeliness of signals are adjusted by adapting detection methods to different purposes and requirements. For example, the alarm threshold for symptoms or events can be lowered to improve timeliness and sensitivity, but at the same time, this change can reduce specificity and positive predictive value. Decision-making theory and related analytical methods are useful for determining the optimal value of each indicator (sensitivity, specificity, positive predictive value, etc.) in practical applications and for assessing the benefit of correct alarms and the damage of false alarms. In the practical application of syndromic surveillance results, interpretation of the results is influenced by the following aspects.

6.4.3.1 Determination of the Baseline

It is very important to analyze, judge, and grasp the normal indicator levels to exclude the possibility of the interference of spontaneous changes in data and scientifically determine the baseline of early warning to give a full play to the role of early warning in surveillance. For example, before the arrival of the flu season, people often stock up on certain drugs, leading to increased drug sales. If the consumption psychology and behavioral patterns of these people are not understood and the relevant early-warning baseline is not corrected in time, misjudgment will occur and false alarms will be issued. Therefore, long-term surveillance is necessary to grasp the regularity of natural variations in data and exclude relevant confounding factors to the greatest extent possible in order to improve the scientificity and accuracy of the baseline.

6.4.3.2 Determination of the Alarm Threshold

The scientific determination of alarm threshold indicators is an important factor in the success of early-warning surveillance systems. If the alarm value is low, the surveillance system will become overly sensitive, increasing the rate of false alarms for symptoms (or events). Therefore, determining the appropriate sensitivity and specificity is the key to ensuring the validity of the data analysis results.

6.5 Case Studies

6.5.1 Syndromic Surveillance of a Group Infection Case of Acute Schistosomiasis in Xiantao City in 2005

6.5.1.1 Proposing the Issue

Zijing Village of Maozui Town in Xiantao City is a mildly epidemic schistosomiasis area. The village has 600 households and a total population of 2600, of whom 1560 are permanent residents. It contains 78 cattle, and rice is grown as the main crop. Most of the residents live along a machine-plowed ditch that is overgrown with weeds. Cow dung and scattered cattle can be seen along the ditch, and the water source of the ditch is the Tongshun River where snails can be found.

In mid-June 2005, villagers of Groups 10 and 11 in Zijing Village who were planting rice seedlings in the rice fields beside the machine-plowed ditch frequently came into contact with infested water. Many people reported skin pruritus and rashes after contact with the infested water, which did not attract attention. On July 20, 28 workers and some elderly people in Zijing Village had persistent high fevers of unknown causes. The village doctor thought they had caught colds, but repeated treatments were ineffective. The Schistosomiasis Control Station of Maozui Town learned of this situation and reported it to the Schistosomiasis Control Office and the Schistosomiasis Control Institute in Xiantao City. Serum immunological examination of schistosomiasis indicated that 26 of 28 people were positive for IHA, and their leukocytes and eosinophils were elevated.

The incidence situation of schistosomiasis was as follows: among the 26 febrile patients, there were 10 males and 16 females, of whom the oldest was 80 years old and the youngest was 28 years old. The age groups were 9 patients aged 28 to 39, 12 patients aged 40 to 59, and 5 patients aged 60 or older; all were farmers by occupation. The serum immunological examinations of 26 patients were positive, and fecal examination showed that 5 were carrying schistosome eggs. According to the national diagnostic standard of schistosomiasis GB15977-1995, these five patients were confirmed as cases of acute schistosomiasis, and the other 21 were diagnosed as clinical cases of schistosomiasis.

6.5.1.2 Application of Syndromic Surveillance

Schistosomiasis japonica is a parasitic disease with a feature of endemic and seasonable transmission. The retrospective investigation of this epidemic showed that the epidemic could have been predicted if syndromic surveillance had been carried out in June when skin papules and pruritus appeared among local residents. Therefore, in epidemic areas and during the season of high incidence of schistosomiasis, syndromic surveillance should be applied in advance to prevent and control acute schistosomiasis outbreaks and to ensure that early warning can occur before routine surveillance time Δt (for details, see Fig. 6.6).

6.5.1.3 Recommendations

Surveillance of certain clinical syndromes in specific populations can be used to identify and detect possible epidemics at an early stage, improve the timeliness

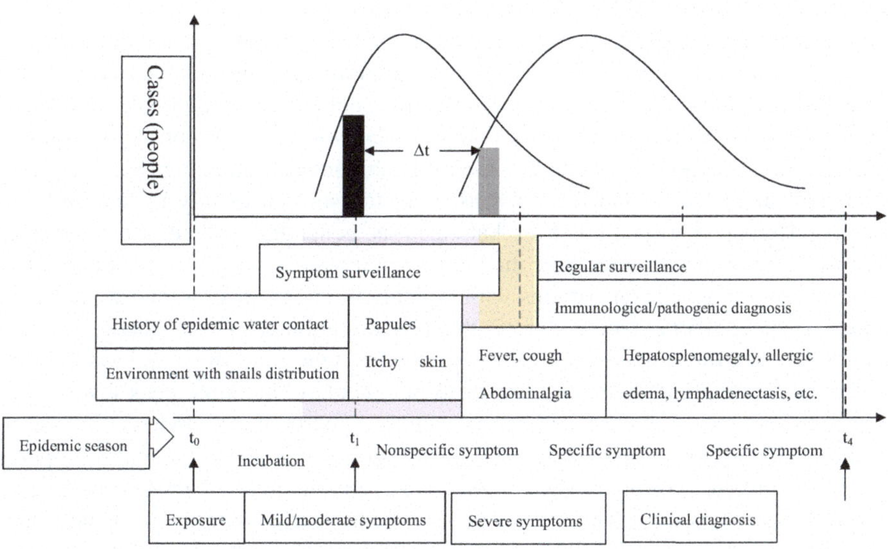

Fig. 6.6 Application of symptom surveillance for acute schistosomiasis outbreak

of response to public health emergencies, and minimize the hazards of such events. Syndromic surveillance, as a new method of public health surveillance, should be applied scientifically and effectively. It should be integrated with epidemiological surveillance, laboratory diagnosis results, disease clinical data, and human behavioral information, focusing on the systematic integration of traditional and nontraditional surveillance data sources to meet the needs of public health early warning.

6.5.2 Syndromic Surveillance of a Kala-azar Outbreak in Jiashi County in 2008

6.5.2.1 Proposing the Issue

Based on the preliminary investigation of a kala-azar outbreak in Jiashi County, Kashgar Prefecture, Xinjiang Uygur Autonomous Region in November 2008, an expanded investigation of the search for cases in key areas of the kala-azar outbreak was carried out from December 16 to 22, 2008. This investigation of the key areas revealed that the epidemic situation of kala-azar in Jiashi County was severe, and the probability of a kala-azar outbreak was again high. In this area, there was still a lack of active case surveillance and follow-up mechanisms, which made it difficult to detect cases in time to avoid deaths of infants and toddlers. Additionally, the proportion of cases with nonstandard treatment was high, which led to the recurrence of kala-azar and then posed a threat to the health and lives of infants and toddlers. The phenomena of underreporting and rereporting continued to occur, affecting the ability to grasp the real outbreak situation and manage cases.

6.5.2.2 Application of Syndromic Surveillance

Kala-azar is characterized by prolonged irregular fever; hepatosplenomegaly, especially splenomegaly, emaciation, anemia, leukopenia, and thrombocytopenia; and increased gamma globulin. Sources of infection include infected people and dogs. Canine visceral leishmaniasis is caused by *Leishmania canis*. According to studies in China, the distributions of canine visceral leishmaniasis and human kala-azar are parallel in many areas, and these diseases are transmitted mainly by sandflies as vectors. The incubation period of kala-azar is generally difficult to determine because it is difficult to determine the date when a patient was bitten by a naturally infected sandfly. Kala-azar patients with typical clinical manifestations are not difficult to diagnose; for example, fever is the most common symptom, usually accompanied by sweating. Infants may have convulsions. One-third of cases have a fever type similar to typhoid fever, and one-third of cases have a double-peak fever type. Hepatosplenomegaly occurs, and splenomegaly is particularly obvious. The high burden of kala-azar disease hampers local socioeconomic development and openness to the outside world. Therefore, syndromic surveillance of kala-azar is beneficial for disease control and prevention, and the early warning of syndromic surveillance is issued earlier than the routine surveillance time Δt (for details, see Fig. 6.7).

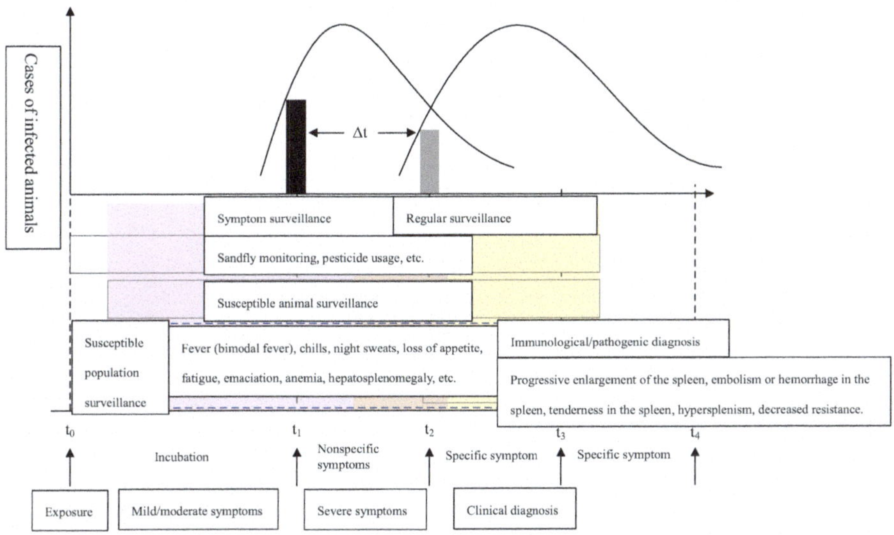

Fig. 6.7 Application of symptom surveillance for Kala-azar outbreak

6.5.2.3 Recommendations

Early detection of cases and scientific judgment of epidemic diseases play important roles in improving emergency response and treatment efforts in epidemics. Traditional disease surveillance usually relies on clinical diagnosis and laboratory diagnosis, which are not only time-consuming and incomplete but also cannot meet the requirements of emergency response to public health emergencies. Syndromic surveillance aims to identify early abnormal "symptoms" in public health surveillance. It is based on information technology and integrates a set of continuous and systematic collection, analysis, and early-warning methods that can provide timely and valuable information to the state, local authorities, health departments, and hospitals and can carry out real-time surveillance of parasitic diseases, and thus gain time for public health response. In addition, as wild animals can be hosts of various parasitic and infectious diseases, disease surveillance of these animals is very important in cooperation with veterinary sector. Syndromic surveillance can be used to monitor the infections and deaths of wild or domestic animals, and deal with epidemics among them in a timely manner, thus reducing the hazard to humans.

Ecological and Environmental Surveillance

7

Kun Yang

Ecological and environmental surveillance is a scientific process of using various technical means and methods to systematically measure and observe the type, structure, function, and components of an ecosystem within a specific temporal or spatial area in time or space. The specific aim is to monitor the ecological environmental situation and identify changes in it by using instruments, field surveys, and other means to evaluate and predict the impacts of human activities on the ecosystem, to provide a scientific basis for the assessment and regulation of eco-environmental quality and eco-environmental management, and to provide a decision-making basis for the rational use of resources and the prevention and control of disease.

7.1 Overview

7.1.1 Basic Tasks

The basic tasks of ecological and environmental surveillance are (1) monitoring changes in the composition, structure, and function of the ecological environmental system caused by human activities and scientifically evaluating the impacts of human activities; (2) monitoring the impacts of changes in ecosystem composition, structure, and function on disease vectors and disease epidemics; (3) monitoring the impacts of various control measures on the composition, structure, and function of the ecosystem; and (4) studying the change rules and development trends of various ecological problems through the accumulation and analysis of surveillance data by establishing mathematical models that lay the foundation for prediction and impact evaluation.

K. Yang (✉)
Jiangsu Institute of Parasitic Diseases, Wuxi, Jiangsu, China
e-mail: yangkun@jipd.com

X.-N. Zhou (ed.), *Surveillance and Management of Parasitic Diseases in China*,
https://doi.org/10.1007/978-981-97-1241-0_7

7.1.2 Surveillance Components

The main components of ecological and environmental surveillance are as follows:

7.1.2.1 Surveillance of Meteorological Factors

This type of surveillance mainly includes air temperature, accumulated temperature, extreme temperature, humidity, wind direction, wind speed, dust storms, amount and distribution of rainfall, evaporation, soil temperature gradient, sunshine, and radiation budget, etc.

7.1.2.2 Surveillance of Hydrological Factors

This type of surveillance mainly includes rainfall intensity, surface runoff, peak discharge and flood volume, runoff coefficient, groundwater reserves, groundwater runoff, chemical composition of dissolved matter in runoff, and groundwater level, etc.

7.1.2.3 Surveillance of Soil Factors

This type of surveillance mainly includes soil nutrients and availability components (N, P, K, S), pH value, and exchangeable acid and its composition, etc.

7.1.2.4 Surveillance of Biological Factors

This type of surveillance mainly includes plants, animals, microbes, biodiversity, and vectors and focuses on animal species, population density, habitat, and other indicators.

7.1.2.5 Surveillance of Ecological Factors

This type of surveillance is mainly the monitoring of macroecosystem succession, such as the land use/land cover changes, large-scale water conservancy projects, and climate change, etc.

7.1.3 Indicator System of Surveillance

The establishment of an indicator system of ecological and environmental surveillance is based mainly on the ecological type and integrity of the system. The establishment principles are as follows: (1) the representativeness, comprehensiveness, and operability of the indicators should be fully considered according to the surveillance components; (2) tests of the same ecological type at different surveillance sites should be carried out according to a unified indicator system to ensure that the surveillance components are comparable; (3) each surveillance site should add specific indicators according to the particular aims of the surveillance project to highlight their own characteristics; and (4) the indicator system should reflect the various levels of the ecosystem and the main ecological environmental problems and should focus on structural and functional indicators.

7.1.4 Techniques and Methods of Surveillance

Ecological surveillance is a complex systemic project with high requirements for environmental surveillance workers. The final result of environmental surveillance is used for the evaluation of environmental quality that enables disease control programs to be proposed. Ecological surveillance serves deeper health management and decision-making departments by supporting ecological environmental planning and ecological design schemes.

Ecological surveillance approaches are as follows: 3S technology and ground monitoring are combined to comprehensively examine ecological quality from macro- and microperspectives, and the network design tends to be integrated, considering changes in global ecological quality and gradually shifting from ecological quality status assessment to ecological risk assessment, which will provide early warning. Additionally, standardization and normalization are emphasized in information management, and geographic information systems are widely used to strengthen cooperation between countries. In summary, with economic growth, increasing populations, resource constraints, and environmental problems are becoming increasingly serious, and it is no longer enough to understand environmental quality simply by monitoring physicochemical and biological indicators. Ecological surveillance is an inevitable trend in the development of environmental surveillance, and the majority of environmental surveillance workers must gradually recognize and master relevant new technologies.

7.2 Surveillance of Pathogens in the Environment

7.2.1 Common Transmission Routes of Parasitic Diseases in the Environment

7.2.1.1 Waterborne Transmission

Many parasites enter the body through water. If a water source is contaminated with the infected eggs or larvae of certain parasites, people may be infected by drinking the water or by contact with it. For example, people may be infected with amebiasis by drinking water contaminated with mature *Entamoeba histolytica* cysts and with schistosomiasis by contacting infested water containing schistosoma cercaria. The characteristics of parasitic diseases transmitted by drinking water are that the case distribution is consistent with the scope of the water supply, and people of different ages, sexes, and occupations can be infected. A characteristic of parasitic diseases transmitted by contact with infested water is that patients have a history of contact with the water, and the incidence of this kind of disease is not only endemic and seasonal but also shows occupational differences.

7.2.1.2 Foodborne Transmission

Human excrement is used as fertilizer in many areas of China, and contamination of vegetables and fruits by eggs in infected human excrement is a common method of transmission. Therefore, eating raw vegetables or unwashed or unpeeled fruits is often an important means of transmission of certain parasitic diseases. Parasites are also contained in fish and meat, so eating these is another important means of spreading certain parasitic diseases. For example, people eating raw or undercooked pork containing infected larvae may be infected with *Taenia solium* or *Trichinella spiralis*, and people eating raw or undercooked fish containing encysted metacercariae can be infected with *Clonorchis sinensis*. The main characteristic of food-borne parasitic diseases is that the patients eat the same food, and people who do not eat this food do not get sick.

7.2.1.3 Soil Transmission

Some directly developing nematodes, such as roundworms, whipworms, and hookworms, need to develop eggs or larvae in soil. Human infection is related to contact with soil. Some parasite eggs are highly resistant to the external environment; for example, roundworm eggs can survive in shallow soil for several years and accumulate in surprising numbers over time.

7.2.1.4 Airborne or Droplet Transmission

Some parasites can be transmitted by air or by droplets in the infection period; for example, pinworm eggs and *Acanthamoeba* parasite can float in the air and enter the human body with breathing to cause infection.

7.2.2 Stool Surveillance Techniques

Stool can be used to detect roundworms, whipworms, pinworms, hookworms, *Clonorchis sinensis*, *Paragonimus*, *Schistosoma*, *Fasciolopsis buski*, *Taenia solium*, *Taenia saginata*, *Hymenolepis nana*, *Entamoeba histolytica*, *Giardia*, etc. Below, schistosomiasis is used as an example to introduce stool surveillance.

7.2.2.1 Surveillance Purposes

The purpose of surveillance is to investigate the situation of grazing prohibitions and pollution by livestock feces in susceptible areas (or environments with snails) in highly epidemic areas of schistosomiasis and to evaluate the risk of schistosomiasis transmission.

7.2.2.2 Determination of Surveillance Scope and Quantity

Each county should select two townships with severe epidemic situations, and each township should randomly select two villages for monitoring. The selected sites should be (1) high-risk environments with frequent human and animal activities or areas where infectious vectors or fecal distribution have been detected within the previous 3 years and (2) within 1000 m of the epidemic village. For example, monitoring sites can be selected in susceptible areas within and outside dykes and dams for epidemic villages with the limnetic type of schistosomiasis, and monitoring sites can be selected on ditches, farmland, or barren slopes near hill-type epidemic villages.

7.2.2.3 Surveillance Time and Components

In June and July of each year, researchers should investigate grazing prohibitions and the type, quantity, density, and infection situation of livestock feces as well as the density and infection situation of *Oncomelania* snails.

7.2.2.4 Determination of the Snail Sampling Frame

Within each snail surveillance site, three sampling frames should be selected, and each sampling frame should have an area of 100 m^2 (10 m × 10 m). The snail sampling frames should in principle be selected within areas of livestock activity; when there is no herd in the monitoring site, sampling frames should be selected according to cow dung distribution, *Oncomelania* snail distribution, and the frequency of human and animal activities.

7.2.2.5 Livestock Feces Pollution Situation

First, the quantity and type of livestock within each monitoring site should be determined. Then, the quantity of feces in each sampling frame should be counted, the nature of the feces (fresh or old/dry) should be recorded, and all fresh feces should be sampled (200 g per pile) and tested for hatched snail eggs.

7.2.2.6 Surveillance Indicators

Surveillance indicators are the species and quantity of livestock in each monitoring environment, the number of piles of livestock feces in each sampling frame (100 m^2), and the positive rate of fresh cattle feces in each sampling frame (100 m^2).

7.2.3 Surveillance Techniques for Pathogens in Water

Amoebiasis and cryptosporidiosis can be detected in water bodies. Cryptosporidiosis is used as the main example to introduce monitoring techniques for pathogens in water.

7.2.3.1 Surveillance Purposes

The purpose of surveillance is to investigate the situation of water pollution in zones susceptible to severe cryptosporidiosis epidemics and to evaluate the transmission risk of cryptosporidiosis.

7.2.3.2 Surveillance Scope

All monitoring sites should be selected based on the principle of long-term water quality trends. They should be divided into sites with a centralized water supply and those with a surface water supply:

1. Monitoring sites with a centralized water supply
   (a) Urban municipal water supply: All municipal water supplies should be included in the monitoring network. For example, each municipal water plant should set up one monitoring site with finished water. Additionally,

urban areas of provincial capitals should set up 40 monitoring sites with terminal tapwater, urban areas of prefecture-level cities should set up 20 monitoring sites for terminal tapwater, and counties should set up 10 monitoring sites for terminal tapwater. The proportion of terminal tapwater monitoring sites in schools should not be less than 10%.

(b) Urban self-built water supply: At least three self-built water plants should be included in the national monitoring network. Self-built water plants should be selected as much as possible with different water sources and water purification processes. Each self-built water plant sets up one monitoring site for finished water and at least one monitoring site for terminal tapwater.

(c) Urban secondary water supply: At least ten secondary water supply facilities should be included in the national monitoring network. At least one school secondary water supply facility should be included.

(d) Rural centralized water supply: Twenty rural centralized water supply plants should be selected for inclusion in the national monitoring network in each county containing agriculture. Centralized water supply plants with different water sources should be selected as much as possible. Each rural centralized water supply plant should set up one monitoring site for finished water and one monitoring site for terminal tapwater.

(e) Water supply of self-built facilities in rural schools: At least five rural schools that use self-built facilities should be included in the national monitoring network. School water supplies should be selected as far as possible with different water sources. A monitoring site should be set up at each school's self-built water supply facility.

(f) Dispersed water supply in rural areas: Four to five monitoring sites for dispersed water supply should be set up in each county containing agriculture and included in the national monitoring network.

2. Surveillance sites for surface water: For surface water sampling, a sampling cross-section is usually set up first, and then sampling sites should then be established on the sampling cross-section. The quantity and location of the sampling sites should be determined according to the width of the water surface (sampling cross-section), the depth of the water, the uniformity level of water quality, and monitoring purposes. The number of sampling sites for rivers is determined by the width of the water surface. For lakes and reservoirs, there is generally one sampling site per 2000 km^2.

3. Collection and preservation of water samples

   (a) Water samples for centralized water supply: Water samples should be collected and preserved in accordance with *Standard Examination Methods for Drinking Water—Collection and Preservation of Water Samples (GB/T5750.2-2006)*. Water monitoring sites should be located as follows: sampling locations for finished water should be located in the water intake after water treatment is completed and before water enters the pipeline. Sampling locations for terminal tapwater should generally be located at users' taps. Sampling locations for secondary water supplies should be located at impounding reservoirs or outlets of water tanks. Sampling locations for school self-built water supply facilities should be located at users' water intakes. Sampling locations of dispersed water supplies should be located in household water tanks.

(b) Collection of surface water: The appropriate sampling time and frequency should be determined according to the sampling purpose. Generally, one to two samples are taken in the low-water season, high-water season, and normal-water season. To grasp seasonal variations in water quality, it is best to sample once a month. Drinking water sources should be sampled no fewer than 12 times a year, and the sampling time should be selected according to specific conditions. Generally, small-sized and medium-sized rivers should be sampled six times a year, with the low-water season, high-water season, and normal-water season each sampled twice. Tidal rivers should be sampled twice in each low-water, high-water, and normal-water season in both the spring-tide period and neap-tide period. Samples should be taken at high tide and low tide on the same day each time and evaluated separately. The water samples from the high-tide period should be collected when the cross-section rises to its highest value at high tide, and the water samples from the low-tide period should be collected when the cross-section retreats to its lowest value at low tide. If there are no such conditions, sampling is not necessary during neap tide. Lakes and reservoirs with special monitoring stations should be sampled no less than once a month and no fewer than 12 times a year. Other lakes and reservoirs should be sampled twice a year and once each in low-water and high-water seasons.

7.2.4 Surveillance Techniques for Soil Pathogens

Parasites such as roundworms, whipworms, and hookworms develop as infected eggs or larvae in the soil, and people can be infected by contact with them. Soil-transmitted helminths such as hookworms, roundworms, whipworms, and pinworms are used as examples to introduce the steps of soil pathogen surveillance.

7.2.4.1 Surveillance Purpose

Understanding the epidemic status and influencing factors of soil-transmitted nematodosis, grasping the epidemic patterns and predicting epidemic trends can provide scientific evidence to support and evaluate prevention and control strategies and establish national and local health cities (towns).

7.2.4.2 Surveillance Scope

The counties (cities, districts) where the monitoring sites are located should have a certain working basis and technical capacity. The monitoring sites should be fixed in one administrative village or several natural villages where the total population is not less than 1500 and the number of people examined each time is not less than 1000.

7.2.4.3 Surveillance Components and Methods

1. Surveillance of population infection: Hookworms, roundworms, whipworms, and pinworms are mainly monitored, although if other parasites are detected, they will be recorded. The people monitored should be above 3 years old. Every year from September to November, the Kato-Katz method (three tests on one fecal sample) should be used to search for parasites. For children aged 3–12 years, scrath cellulose tape method is selected to search for pinworm eggs. According to the monitoring results, the infection rates and intensity of hookworms, roundworms, and whipworms as well as the infection rate of pinworms should be calculated.
2. Surveillance of *Ascaris lumbricoides* egg contamination of soil: At each monitoring site, ten households of residents should be randomly selected. In each household, soil samples should be collected from four types of places: vegetable gardens, toilets, courtyards, and kitchens. One sample should be taken from each type of place in each household, and each sample should be 30 g. Microscopic examination of the samples should then be performed to detect the fertilized or unfertilized *Ascaris lumbricoides* eggs, and the soil culture method should be used to distinguish between dead and live eggs. According to the results, the total number of eggs and numbers of fertilized eggs and after-culture live fertilized eggs of *Ascaris lumbricoides* should be counted.
3. Survey of relevant factors: Natural and social factors such as geographical conditions, temperature, humidity, rainfall, crops, economic level, hygiene conditions, and prevention and control measures of the monitoring sites should be investigated.
4. Monitoring of roundworm infection in children under 14 years of age in cities (towns): For areas declared to be national and local health cities (towns), roundworm infection in children under 14 years of age should be monitored. For cities with a total population of 500,000 or more, the number of respondents should not be less than 1500; for cities with a total population of less than 500,000, the number of respondents should not be less than 1000. Two schools in urban areas and two schools outside urban areas should be randomly selected. Students in all grades under the age of 14 should be monitored for roundworm infection, and at least 100 students in each grade should be checked. The examination components include detecting roundworm eggs in fecal samples by the Kato-Katz method (three tests per fecal sample) and then observing *Ascaris* infection. According to the monitoring results, the *Ascaris* infection rate and infectiousness in children under 14 years of age in the region should be calculated.

7.2.4.4 Quality Control of Microscopic Examination

Professional institutions at high levels should be responsible for rechecking the microscopic examination results from the monitoring sites. At least 30% of the positive samples for each monitoring site should be rechecked (if there are fewer than ten positive samples, all positive samples should be checked), and the false-positive rate should not exceed 10%; at least 5% of negative samples should be rechecked, and the false-negative rate should not exceed 10%.

7.3 Surveillance of Vectors in the Environment

Vector surveillance is important in vector-borne disease control and prevention. Vectors are organisms that can transmit vector-borne disease, generally to humans. Vectors include vertebrates and invertebrates. Vertebrate vectors are mainly rodents, which belong to the mammalian class, and invertebrate vectors are mainly mosquitoes, flies, cockroaches, sandflies fleas in Insecta and ticks and mites in Arachnoidea. Vectors can not only directly affect or endanger human life through bites and contaminated food but also spread important infectious diseases in a variety of ways.

7.3.1 Importance of Surveillance for Vector-Borne Infectious Diseases

Many of the legally notifiable infectious diseases in China are vector-borne diseases, such as plague, hemorrhagic fever, leptospirosis, malaria, dengue fever, endemic typhus, leishmaniasis and filariasis. Some infectious diseases of the digestive tract, such as dysentery and typhoid fever, are spread through mechanical transmission by vectors. The effective control of vectors can reduce harm to the population with the result of economic loss, and the occurrence and transmission of vector-borne infectious diseases can be prevented and controlled. Systematic vector surveillance can provide not only evidence for vector control programs but also prediction and early-warning information for epidemic trends of vector-borne diseases. In the past, vector surveillance was carried out systematically in China, playing a positive role in the prevention and control of vector-borne diseases. However, due to different methods being used in different regions, it was difficult to compare and analyze data. Additionally, some of the original methods were not suitable for using in the current social developments, and the epidemic status of some vector-borne diseases had changed significantly. In recent years, outbreaks of some vector-borne infectious diseases in neighboring countries have become a threat to China. Therefore, strengthening vector surveillance has become an urgent task in disease prevention and control.

7.3.2 Vector Surveillance

The following is an example of mosquito monitoring to introduce vector surveillance in the environment.

7.3.2.1 Surveillance Purpose

The species, quantity, distribution, and seasonal variation of the monitored vectors should be determined to accumulate basic data for predicting and dealing with emergency events and provide evidence for formulating scientific and reasonable vector prevention and control programs. In addition, analyzing correlations between long-term changes in vectors and local infectious diseases can also provide data support for the prevention and control of vector-borne diseases.

7.3.2.2 Mosquito Density Surveillance

1. Adult mosquito monitoring: Adult mosquito density should be monitored at regular and fixed points:
   (a) Mosquito lure lamp
      - Selection of monitoring sites: In each province (municipality, autonomous region), three to five prefecture-level cities should be selected as monitoring sites according to different geographic orientations. In each city, four residential areas, four parks (including street gardens), and four hospitals should be selected. In rural areas, four private houses and four livestock sheds (cowsheds, pigpens, etc.) should be selected. Except for monitoring sites in cowsheds and pigpens, all other monitoring should be performed in the external environment.
      - Monitoring time: Monitoring should be performed two times per month at intervals of 15 days. In the case of wind and rain (wind force above level 5), the monitoring should be postponed. The surveillance should begin 1 month before the activity of overwintering mosquitoes and end after no mosquitoes are caught on two consecutive occasions.
      - Monitoring methods: One mosquito lure lamp should be used in each place, and the monitoring should begin 20 min after sunset and last for 6 h. On the second day, the mosquito collection boxes should be removed to identify the species, sexes, and total number collected.
      - Density index calculation: The formula is:

$$\text{Density}\left(\text{pcs / hour}\right) = \frac{\text{Number of mosquitoes captured}}{\text{Time to catch mosquitoes} \times \text{number of lamps}}$$

   (b) Labor/hour method
      - Selection of monitoring sites: In each province (municipality, autonomous region), three to five cities should be selected as monitoring sites according to geographical orientation. In each city, two residential areas in the city center and three residential areas in the urban-rural fringe area should be selected, and four households should be investigated in each residential area. In each rural area, two natural villages should be selected, and four households should be investigated in each natural village in addition to two cowsheds and two pigpens (or other livestock sheds that are not adjacent). In villages without livestock sheds, the sheds should be replaced by private houses (two households total).
      - Monitoring time and method: Monitoring should be performed two times a month at intervals of 15 days. Electric mosquito traps should be used to capture mosquitoes in the indoor environment of each monitoring site for 15 min 1 h after sunset, and the captured mosquitoes should then be classified and counted. The surveillance should begin 1 month before the

activity of overwintering mosquitoes and end after no mosquitoes are caught on two consecutive occasions.
- Density index calculation: The density index is calculated as number per manual hour.

2. Special surveillance of *Aedes* mosquitoes: The surveillance should target *Aedes albopictus*, and *Aedes aegypti* and should be carried out in various localities according to actual needs. At monitoring sites with a distribution of *Aedes aegypti*, data on the distribution range of this kind of mosquito should be collected during the monitoring process:
   (a) Mosquito ovitrap index method
   - Selection of monitoring sites: A total of 200 mosquito ovitraps should be placed in residential areas, parks, construction sites, hospitals, waste collection stations, and other *Aedes* breeding sites in each city, with one mosquito ovitrap per household in each residential area and one mosquito ovitrap per 10 m in parks and construction sites.
   - Monitoring time: Each monitoring site should be monitored once in the middle of each month. The surveillance should begin 1 month before the activity of overwintering mosquitoes and end after no mosquitoes are caught in two consecutive months.
   - Monitoring method: The mosquito ovitraps should be placed continuously for 4 days, and the trapped adult mosquitoes and mosquito eggs should be checked and collected on the fourth day. The positive rate of mosquito eggs should be calculated, and the average temperature and rainfall during the monitoring period should be recorded.

   (b) Container index method
   - Selection of monitoring sites: Residential areas of four streets should be selected at different geographical directions in each city, and no fewer than 100 indoor and outdoor water accumulation containers should be placed in each street, with two containers placed in each park, construction site, water collection station, and other sites.
   - Monitoring time: Each site should be monitored once in the middle of each month.
   - Monitoring method: The positive rates of *Aedes* larvae and pupae in various water accumulation containers should be calculated. Water accumulation containers should be divided into permanent containers (such as water vats and pools) and temporary containers (such as vases, tires, and discarded bottles and cans) so that the relevant indexes can be counted and calculated accordingly.

7.4 Other Ecological Surveillance Techniques

At present, the epidemic of foodborne parasitic diseases, represented by clonorchiasis sinensis caused by *Clonorchis sinensis* infection, is serious. It is estimated that more than 12 million people are infected with *C. sinensis*. With the improvement of people's living standards and the variety of dietary sources, food safety problems caused by foodborne parasitic diseases have become increasingly noticeable. After infection with *C. sinensis*, mild cases in humans can cause abdominal pain, diarrhea, malnutrition, and hepatomegaly, while severe cases can lead to cirrhosis, ascites, and dwarfism. In addition, after being infected with *Taenia solium* by eating raw or undercooked pork, the infected person may have digestive tract symptoms, which can be complicated by intestinal obstruction and appendicitis. Eating food containing *Taenia solium* eggs may also cause brain cysticercosis, resulting in epilepsy, meningitis, and other serious consequences. There are many influencing factors for the increase in foodborne parasitic diseases. First, it is difficult to change long-formed dietary habits of eating raw or undercooked freshwater fish and meat in some regions, and with the improvement in living standards, people have more opportunities to eat away from home, which leads to an increased chance of infection. Second, although freshwater aquaculture is developing rapidly, quarantine procedures for fish and other food health measures are lagging. Third, the prevention and control of foodborne parasitic diseases have not been carried out systematically. *Paragonimiasis* and *clonorchiasis* are used below as examples to introduce host investigations.

7.4.1 Intermediate Host Survey of Paragonimiasis

7.4.1.1 Selection of Survey Sites

In the endemic areas of paragonimiasis, investigators need catch mallaniid snails (or *Tricula humida*) and freshwater crab (or *Potamon spp.*) or crayfish in selected ditches to examine the infections of those freshwater intermediate hosts.

7.4.1.2 Pathogen Search

1. After the mallaniid snails are broken, the viscera should be put on a slide and then torn open. An appropriate amount of 0.45% saline is then added to the slide. Under the microscope, researchers can look for sporocysts, redia, and cercaria of the parasites.
2. Researchers should dissect crab or crayfish to look for encysted metacercaria.

7.4.1.3 Encysted Metacercaria Separation

A crab is shucked and ground with a mortar; an appropriate amount of 0.45% saline should be added and the coarse residue filtered out with a steel-wire screen. The filtrate should be precipitated naturally and the supernatant poured out (the water should be changed two to three times until the supernatant is clarified) until the sediment is absorbed into the culture dish. Then, encysted metacercaria should be examined under a microscope. Encysted metacercaria are white and spherosome, and the cyst wall has two layers, with a curving larva in the cyst wall. The wavy intestine can be seen on both sides, and the excretory bladder contains black particles.

7.4.2 Survey of the Second Intermediate Host of *Clonorchis sinensis* in Freshwater Fish

7.4.2.1 Microscope with Wet Mount

After species identification of a variety of freshwater fish caught in fish ponds or ditches or purchased at markets, different tissues, such as fish muscle, should be cut into small pieces and then pressed with glass slides to check for metacercariae.

7.4.2.2 Separation Examination of Metacercaria of *C. sinensis*

The freshwater fish, the second intermediate host of *C. sinensis*, should be minced in a meat grinder (or chopped by hand). Every 10 g of fish should be digested with 250 ml of digestive solution (formula: pepsin 9.8 g, 1 N hydrochloric acid 164 ml, sodium chloride 17 g, and water added to 2000 ml) for 4 to 12 h. Then, a copper-wire screen should be used to remove the coarse residue, and the filtrate should be precipitated after repeated changes of normal saline. Finally, when the sediment is absorbed, it should be placed in a petri dish to look for metacercaria under a microscope.

7.5 Case Analysis

7.5.1 Surveillance of Cryptosporidiosis

7.5.1.1 Proposing the Question

Cryptosporidium spp. is a protozoon that mainly parasitic on the digestive epithelium of humans and animals that can cause diarrhea. Immunocompromised hosts, such as infants and immunodeficient patients, especially AIDS patients, may develop chronic, persistent infections that lead to dehydration, malnutrition, and even life-threatening infections. Since the first case of cryptosporidiosis was reported in the United States in 1976, the disease has become a growing concern, and the number of cases has increased dramatically, as the disease has reached more than 90 countries on six continents and been found in an increasing number of AIDS patients. According to epidemiological data, the infection rate of *Cryptosporidium* in China is 0.85%–12.30%. Cryptosporidiosis has become an important public health problem due to the lack of effective prevention and drugs. Frequent outbreaks are caused by water sources and food contamination. Molecular hybridization, polymerase chain reaction (PCR), restriction fragment length polymorphism (RFLP) analysis, random amplified polymorphic DNA (RAPD) analysis, gene cloning, and other molecular biological techniques have been developed and widely used, which has promoted the development of *Cryptosporidium* detection technology, molecular epidemiological investigation, taxonomy, drug design, and other research interventions.

7.5.1.2 Surveillance Results

From January to April 1993, the largest, most severe, and longest-lasting outbreak of waterborne cryptosporidiosis in history occurred in Milwaukee, Wisconsin,

USA. Of the 840,000 people served by the water system, approximately 403,000 developed diarrhea, 4400 were hospitalized, and 69 died. Most of those infected had AIDS. Investigations by a number of relevant departments determined that the outbreak had been caused by *Cryptosporidium* contamination of the municipal water supply system in the southern part of the city. Water samples collected during the event were frozen and preserved, and two ice samples that were tested afterward contained 13.2 and 6.1 *Cryptosporidium* oocysts per 100 L, respectively. The contamination was attributed to *Cryptosporidium* oocysts from cattle feces, slaughtering waste, and domestic sewage entering water sources with stormwater runoff. The supply scale of the water plant was 380,000 m^3/d, and the water source was Lake Michigan. The raw water was treated by coagulation and precipitation with aluminum salt as the coagulating agent, filtration with quartz sand as the filtering material, and chlorination disinfection. The analysis showed that higher turbidity was caused by an inappropriate coagulating agent in the water treatment process and back-flush water recovery.

7.5.1.3 Recommendations

The growth, reproduction, and development of parasites are closely related to the ecological environment and climate warming. Environmental pollution may change the scope of epidemic areas of parasitic diseases. Monitoring the ecological environment to determine transmission routes of cryptosporidiosis is helpful in grasping the transmission dynamics of disease and controlling the development of epidemics.

7.5.2 Surveillance and Evaluation of the Impact of the Three Gorges Project on Schistosomiasis

7.5.2.1 Proposing the Question

There is a possibility that schistosomiasis and *Oncomelania* snails are able to be imported to the Three Gorges Reservoir areas. Surveillance and evaluation of the impact of the project on schistosomiasis can provide a timely understanding of the potential risk factors for schistosomiasis transmission in the Three Gorges Reservoir area.

7.5.2.2 Selection of Surveillance Sites

Because the Three Gorges Reservoir covers a wide area, monitoring sites were selected based on the characteristics of the reservoir area and epidemic features of schistosomiasis. On the basis of the county (district), fixed monitoring sites for schistosomiasis were set up. Considering the geographical distribution of the Three Gorges Reservoir, surveillance sites were set up in different reservoir sections, including areas with frequent human activities and livestock trading, backwater areas that might be suitable for *Oncomelania* breeding after ecological environmental changes, and planning areas for the three pillar industries of animal husbandry, tourism, and aquaculture.

7.5.2.3 Surveillance Components and Methods

1. Monitoring of basic information: Basic information for the monitored counties including population, family per capita annual income, industrial structure, annual (monthly) average rainfall, annual (monthly) average temperature, population flow, and residents' production modes and lifestyle.
2. Surveillance of infection sources: Investigation of the floating population included migrants from schistosomiasis epidemic areas, those returning to their hometowns from epidemic areas, and those migrating to epidemic areas. The general demographic information was investigated by questionnaire, focusing on whether the respondents had come from or been to schistosomiasis epidemic areas (accurate to the county level), whether they had been in contact with infested water in epidemic areas, whether they had a history of schistosomiasis, and whether they had any knowledge of the prevention and control of schistosomiasis. The serological method was used for screening, and those with positive test results were examined for etiology by the Kato-Katz method (three tests per fecal sample). The agricultural and animal husbandry areas around the reservoir were investigated to understand the importation situation of livestock, focusing on the investigation of livestock farms, including the number, variety, breeding methods, and sources of livestock (accurate to the county level). Livestock from schistosomiasis epidemic areas were randomly selected and examined for schistosomiasis infection by the fecal incubation method.
3. Surveillance of *Oncomelania* snail importation and suspicious environments: One to two river beaches were selected from each monitoring site, and the snail situation was investigated by a systematic sampling method every year to determine the sources, quantities, species, and planting sites of plants imported from schistosomiasis epidemic areas. At the same time, snails were sampled in the accumulation sites and planting sites. *Oncomelania* snails input was monitored at wharves and shipyards where ships that had sailed to and from areas with snails docked, and the snail distribution was investigated by environmental sampling every year.
4. Survey of social and economic development: A questionnaire was designed to determine the development and related factors of tourism, animal husbandry, aquaculture, and citriculture in the monitored counties.
5. Surveillance of schistosomiasis infection in the local population: At each monitoring site, a new migrant village was selected to investigate its basic information, including population, income, industrial structure, and farming methods. A questionnaire was used to investigate the situation of drinking water, toilet sanitation, contact with river and canal water, and degree of knowledge of schistosomiasis prevention and control. Serological screening was carried out for permanent residents over 6 years old at the monitoring sites, and those who tested positive were examined by the Kato-Katz method (three tests per fecal sample). More than 300 people were monitored every 5 years to understand the changes in the schistosomal antibody levels in local residents.

6. Monitoring time and cycle: According to the actual situation of each monitoring site, various types of monitoring were implemented. People returning to their hometowns usually do so at the time of the Spring Festival. People from infected areas and migrants entered the Three Gorges Reservoir area at different times. Therefore, people returning to their hometowns were investigated around January every year, and people from infested areas were investigated in the schistosomiasis transmission season. All sites were investigated and monitored once a year.

7.5.2.4 Recommendations

The Three Gorges Reservoir is the largest water-control project in the world to date. After the completion of the reservoir, it is worth paying attention to whether outbreaks and epidemics of waterborne diseases and natural focal diseases, such as schistosomiasis will be introduced by population migration and possible changes in the ecological environment. Carrying out disease surveillance, vector surveillance, and ecological environmental surveillance, such as water quality surveillance, and research on disease prevention and control can provide strong evidence for disease prevention and control in reservoir areas in the future. In addition, training and technology for the prevention and control of relevant diseases and talent recruitment should be performed, and emergency preparedness measures should be implemented.

Health Resource Surveillance and Analysis Techniques

8

Yao Deng

Health resources are a basic factor in the development of health work. They are an important component in human, material, and financial support and guarantee in health activities, such as disease prevention and treatment, public health promotion and education. Health resources have played a key and fundamental role in the reform and development of medical and health care in China. How to fairly and reasonably allocate and effectively use health resources is related to the health of the residents of a country or region and the efficiency of resource utilization, making it a strategic issue of common concern for all countries in the world today.

With the increasing abundance of health resources and the continuous improvement and development of various monitoring technologies, the scope and components of health resource monitoring are also expanding. Based on an overview of the definition of health resources, this chapter introduces the principles and practices of health resource allocation as well as how to evaluate the situation, efficiency, and equity of such allocation. The aim is to guide county health workers to correctly and rationally allocate and utilize health resources in order to maximize their ability to meet the health-care needs of the people.

8.1 Overview

In a broad sense, health resources are the collection of all productivity factors used for health services, including hardware and software health resources. The former include health staffing, health financial resources, and health service equipment, while the latter include information technology and management. In a narrow sense, health resources refers to the human, financial, and material resources needed to provide health services, i.e., hard health resources. Different forms of health

Y. Deng (✉)
Jiangsu Institute of Parasitic Diseases, Wuxi, Jiangsu, China

X.-N. Zhou (ed.), *Surveillance and Management of Parasitic Diseases in China*,
https://doi.org/10.1007/978-981-97-1241-0_8

resources play different roles in the process of health service delivery. Whether a country or region has allocated health resources in a fair, rational, and efficient manner is an objective indicator of its economic strength, cultural status, and health level; this, resource allocation is related to the development of health and is a basic quality of life issue.

8.1.1 Components

8.1.1.1 Human Resources for Health

Human resources for health are the most important component of health resources because of the associated core competencies; additionally, they are a decisive factor in health development because of their function of maintaining and strengthening health systems. This means that people with different educational backgrounds and vocational training can provide health services and contribute their talent and wisdom according to people's health needs. A proper human resources structure plays an important role in the development of local health. It not only reflects the overall level of health services in a country or region but is also the most important factor in maintaining people's health. Whether the quantity of health staffing in a country or region is sufficient—that is, workers are well qualified, the structure is reasonable, the distribution is balanced, and the management is scientific—affects not only the development of health but also the development of the economy and the balance of society. In China's disease surveillance system, the quality of human resources and their work motivation have a greater impact on the quality and efficiency of the health system's operation than resources such as capital, technology, and equipment.

8.1.1.2 Physical Resources for Health

Physical resources for health include health-care institutions and their properties, beds, and equipment, which are the basic conditions for providing health services. The quality and quantity of physical resources for health inevitably affect the quality and quantity of health services as well as their scale and scope. Some researchers believe that physical resources are reflected mainly in the quantity of medical equipment. In recent years, China's stock of large medical equipment has been developing rapidly, and the amount of such equipment has increased significantly. The planning of type B large-scale medical equipment (including CT, MRI, DSA, LA, and SPECT machines) is an important element of regional health planning and symbolizes the comprehensive strength of a hospital.

8.1.1.3 Financial Resources for Health

Financial resources for health care are mainly funds, primarily from financial allocations, the operational income of health units, and operational expenditures, which are often expressed as total health costs. Compared with human and physical resources, financial resources are more representative because they have greater mobility and substitutability in the market economy.

According to channels of health investment, total health costs include government health inputs, such as government budget allocations (recurrent funding, special funding) and health inputs from state-owned, collective, and joint venture enterprises; they also include health inputs from individual residents as well as donations and investments from overseas compatriots and aid funds from the World Bank.

According to categories of health investment, total health costs include investment in health infrastructure, health-care expenses, medical security expenses, medical education and scientific research expenses, and health administrative expenses. Adequate financial resources can provide a reliable guarantee for human and physical resources. Gradually increasing health investment and increasing the proportion of total health costs in the gross national product are of great significance to the development of medical and health care.

8.1.2 Characteristics

8.1.2.1 Finiteness

The health resources of a region are always limited. There is always a gap between the health resources available in society and people's actual needs for health care. The shortage of health resources in China is very prominent, and there is a significant gap between the current health service conditions in China and those in developed and some developing countries.

8.1.2.2 Selectivity

Health resources can be used for various purposes; for example, medical personnel can carry out various health-care service activities, and health funds can be invested in different ways. Therefore, any use of health resources should take into account the issue of opportunity cost. The formulation of strategic goals and plans for health development and the allocation of health resources should be based on the comparison of the effects, benefits, utility, and costs of health resource use. Limited health resources should be invested in projects with the greatest benefits or effects so that the available health resources can be used to obtain the greatest economic benefits.

8.1.2.3 Diversity

Improving the health of the population is the overall goal of health work. However, many specific objectives, such as medical treatment, prevention, health care, rehabilitation, family planning, environmental protection, medical education, and scientific research, are involved in achieving this general goal. Since people's health-care needs are diverse, random, and variable, the rational allocation and effective use of health resources are conducive to promoting the coordinated development of various systems within the health sector and meeting diverse health-care needs at different levels.

8.2 Health Resource Allocation

Health resource allocation is the core component of health resource research and planning. Health resource allocation refers to the distribution and transfer of health resources within the health industry (or sector). There are five main aspects of health resource allocation: the establishment of health institutions, hospital beds, health staffing, health equipment, and health funding allocation. The classification of health resources can be practically expressed as human, physical, and financial resources for health. The monitoring and evaluation indicators of health resource allocation are also divided into three aspects: human, physical, and financial health resources. Whether the allocation of health resources is reasonable is often evaluated in terms of efficiency and fairness.

8.2.1 Allocation Basis

8.2.1.1 Economic Evaluation

1. Cost-benefit analysis (CBA). Cost refers to the consumption of all resources, such as human, physical, and financial resources, invested in the whole process of implementing a health project planning, diagnosis, or treatment program and is measured in monetary terms. Benefit refers to the total benefits achieved after the implementation of a health project plan, diagnosis, or treatment program and is also measured in monetary terms. Cost-benefit analysis means selecting the best option by calculating the expected benefits and costs of a program and serves as a reference for decision-makers when making choices and decisions. Cost-benefit analysis requires that both the costs and benefits of an option be measured in monetary terms and evaluates whether the benefits outweigh the costs so that the option with the lowest possible costs and highest possible benefits will be selected.

   Cost-benefit analysis can provide a basis for evaluation and decision-making from the perspective of health economics when planning, evaluating, and making decisions in the medical and health sectors. Therefore, it helps reduce or even prevent mistakes and possible waste or loss in decision-making, enables available health resources to be better allocated and utilized, and plays a positive role in effective disease prevention and control and maintenance of population health. There is still some controversy about the application of cost-benefit analysis in the health field. It is limited for various reasons, for example, the fact that many distributional problems based on allocation decisions have still not been satisfactorily solved. In contrast, cost-effectiveness analysis and cost-utility analysis can be used more directly in policy analysis.
2. Cost-effectiveness analysis (CEA). CEA, using a specific clinical treatment objective (physiological parameters, functional status, life extension years, etc.) as a measure, calculates the cost per unit of treatment used for different protocols or therapies. The results of CEA are expressed not in monetary units but rather in health outcomes or clinical treatment outcomes, such as changes in the num-

ber of patients saved, cure rates, years of life prolonged, blood pressure reduction values, and other indicators.

Cost-effectiveness analysis does not indicate the outputs of prevention programs in monetary terms. It tends to measure outputs in physical units, such as gains in ultimate health, lives saved, disability averted, and reductions in morbidity or in intermediate products that do not directly measure health gains but are believed to be associated with health improvement, such as the number of cases detected, immunization levels gained, and reductions in risk factors (e.g., reductions in blood pressure or cholesterol levels).

Comparative criteria are needed when comparing the results of cost-effectiveness analyses. Comparisons are more credible when the output metrics are similar. In some extreme programs, cost minimization can be directly compared for the same health risk or disease condition; i.e., the lowest cost with the same effects is preferred.

In addition, CEA is relatively flexible. It allows comparison of projects with different forms of costs and benefits but the same objectives, such as years of life saved. CEA results are often combined with other methods, such as disease burden and benefit volume.

3. Cost-utility analysis (CUA). CUA is an extension of cost-effectiveness analysis. It is applied to the measurement of the value of different health outputs. Quality-adjusted life years (QALYs) and disability-adjusted life years (DALYs) are commonly used as units to be measured. QALYs reflect both the number of years of life and the quality of life, so the cost per QALY can be used to measure the effectiveness of the program:
   (a) Disease burden: This is very different from cost-effectiveness analysis. Disease burden studies must identify which conditions or populations are most affected by disease, and prevalence is not the only criterion. The total benefit is maximized if treatment and prevention services focus on the population groups that lose the most QALYs or DALYs.
   (b) Ability to benefit: The main concern of this approach is distribution. It measures the impacts on individuals in terms of the composition of disease burden (disease prevalence) on the demand side and in terms of QALYs or DALYs lost, including the efficiency of treatment and accessibility, on the supply side.

8.2.1.2 Management Measures

1. Purchaser-provider contracting. Contracting and the purchaser-provider split are different organizational structures with the purpose of separating allocative efficiency from technical efficiency. The development of purchaser-provider contracting has been notable in public finance and management over the past 20 years.

   There is currently a large body of literature on the purchaser-provider split based on the experiences of the UK and New Zealand. For example, in the UK National Health Service (NHS) operating system, the most common management system uses a contractual form to organically link purchasers and providers. Purchasers (district health boards or individuals) purchase various types of health services from hospital consortia or GPs (fund holders) on a contractual basis, depending on the actual situation in their district or for a specific individual. There are three types of contracts: comprehensive contracts, cost contracts, and disease-specific cost contracts. The components of the contract include the components of the health service, the volume of the service, the way the service is provided, and the quality of the service.

2. Health benefit groups and diseases based on economic models. Health benefit groups and diseases based on economic models are two similar planning tools proposed by the Executive Directorate of the NHS in the UK and the Melbourne Health Program Evaluation Centre. For example, the population is divided into five groups: no risk, at risk, requiring primary health care (outpatient), acute illness (requiring outpatient or inpatient care), and requiring long-term care. The type of service available is defined separately as the amount of resources (cost) required at each level of service. For the no-risk group, no resources are consumed; for those at risk but asymptomatic, only preventive services or screening are needed; for patients at the next level, primary health care is needed, etc.

8.2.2 Evaluation Indicators

8.2.2.1 Total Allocation of Health Resources

This refers to the quantity and quality of personnel, beds, institutions, equipment, etc.

8.2.2.2 Structure of Health Resource Allocation

1. Vertical Structure of Health Resource Allocation. This is, for example, the allocation of health resources among medical institutions at all levels, such as primary, secondary, tertiary, and community medical institutions.
2. Horizontal Structure of Health Resource Allocation. This is mainly the allocation of health resources of different categories (such as the ratio of personnel and beds), the regional structure of health resources (such as the allocation of resources between urban and rural areas), and the professional structure of health resources (such as the allocation of resources between medical treatment and prevention and general practice and specialties, such as internal medicine, surgery, gynecology, and pediatrics).

8.2.2.3 Allocation of Human Resources for Health

This includes staff occupations (such as the proportions of medical personnel and medical technicians), education (such as the proportions of postgraduate, graduate, college, and junior college), title structure (such as proportions of chief physician, deputy chief physician, and attending physician), and other structures (such as age and sex ratio).

8.2.2.4 Health Cost Allocation

1. Percentage of health costs to GNP or GDP. This indicator usually reflects the total amount of health costs in a country or region and the extent to which the necessary health resources are provided to ensure the coordinated development of health and the social economy. On the whole, since the reform and opening up, the total health costs in China have shown a trend of annual increase, but there is still a gap compared with developed countries.
2. Health cost per capita. This is an indicator of the level of health resources in a country or region and reflects whether the allocation of health resources among regions and populations is reasonable. Factors such as sex, age, income, education level, and medical system have a great influence on the per capita health cost.
3. Cost allocation ratio among medical institutions. This indicator reflects the reasonableness of the allocation of health resources among medical institutions at different levels. Different levels of medical institutions consume different amounts of health resources due to the different numbers and levels of patients treated and the scope of services provided. Generally, the higher the level of a medical institution, the higher the average cost per consultation. Therefore, health resources should be allocated among medical institutions at all levels based on a specified ratio.
4. Ratio of resource allocation between outpatients and inpatients. This indicator reflects whether the allocation of health resources within medical institutions is reasonable. Generally, the utilization rate of outpatient services is high, but the average cost per visit is low, while the utilization rate of inpatient services is low, but the average cost per visit is high. Therefore, the total outpatient actual expenditure is lower than the total inpatient actual expenditure.
5. Ratio of resource allocation among medical treatment, prevention, maternal and child health care, and family planning. The vast majority of Chinese health resources are allocated to medical services; however, prevention and maternal and child health-care services are more efficient than medical services. Therefore, the ratio of resource allocation among medical care, prevention, maternal and child health care, and family planning should be reasonably determined according to the health of the population, the demand for health services, and the efficiency of health services while taking into account socioeconomic and cultural levels.
6. Ratio of the cost of rural health services to the cost of urban health services. This reflects the fairness and rationality of the allocation of health resources between urban and rural areas.

8.2.2.5 Evaluation of the Efficiency of Health Resource Allocation

Efficiency refers to using limited health resources to obtain the maximum outputs. The ratio of inputs to outputs is commonly used as an indicator of efficiency. The larger the ratio, the greater the output obtained per unit of input and the more efficient the use of resources. Commonly used indicators are outpatient visits per doctor day, bed days per doctor day, average inpatient days, bed utilization rate, average costs per outpatient visit, average costs per inpatient visit, and average prescription costs.

The judgment criteria of these indicators depend on specific situations, and they reflect only the efficiency of specific health service units in the use of funds, staffing, equipment, and technology with the goal of maximizing, minimizing, optimizing, and appropriately measuring and evaluating the degree of utilization of health resources.

8.2.2.6 Evaluation of the Fairness of Health Resource Allocation

The fairness of health resource distribution is one of the important bases for measuring whether the allocation of these resources is reasonable. At present, the phenomenon of expensive and difficult-to-obtain medical treatment in the medical and health service market is rooted in the poor social equity of medical services and low efficiency of medical resource allocation in China. Currently, the Lorentz curve and Gini coefficient are often used to evaluate the equity of the distribution of population and geographical health resource allocation. For example, first, the number of health human resources per 10,000 people in a certain place should be ranked from lowest to highest. Then, using the cumulative population percentage of the place as the horizontal axis and the cumulative percentage of health human resources as the vertical axis, the Lorentz curve of the distribution of health human resources can be plotted by population, and the Gini coefficient of the distribution of health human resources can be calculated by the population of a certain area. Another example is that the Gini coefficient can be calculated by ranking per capita health-care costs in each region from lowest to highest and using the cumulative population percentage as the horizontal axis and the cumulative percentage of health-care costs as the vertical axis to draw a Lorentz curve indicating the fairness of the distribution of health-care costs in each region.

1. Lorentz Curve. The Lorentz curve was proposed by statistician Lorentz in 1905. He divided the total population into ten ranked groups according to income, from low to high, with each ranked group accounting for 10% of the population, and then calculated the proportion of each group's income to the total income. Finally, using the cumulative percentage of population as the horizontal axis and the cumulative percentage of income as the vertical axis, he plotted a curve reflecting the disparity of income distribution among residents. This is called the Lorentz curve.
2. Gini Coefficient. The Gini coefficient is an indicator of the equality of income distribution calculated from the Lorentz curve and is used to measure the inequality of income distribution among countries, regions, ethnic groups, and industries. It is also used in many countries to measure how and to what extent fiscal policy affects the distribution of income among the population. The Gini coefficient takes values between 0 and 1, with larger values indicating greater inequality, 0 indicating absolute equality in income distribution and 1 indicating absolute inequality in income distribution. The calculation formula is as follows:

$$G = \sum_{i=1}^{n} W_i Y_i + 2\sum_{i=1}^{n} W_1 (1 - V_i) - 1$$

where *W*i is the proportion of population by region (or area) to total population (or area); Y_i is the proportion of health resources owned by each region to total health resources; Vi is the cumulative number of Yi from i = 1 to i, $V_i = Y_1 + Y_2 + Y_3 + \ldots\ldots + Y_i$; and i is the ranked number of health resources owned.

8.3 Surveillance of Burden of Disease

The rational allocation of health resources is an important element of health resource monitoring, as it not only has a great impact on the supply of health resources but also has a bearing on the development of health. As the discrepancy between people's expectations of the accessibility of medical services and the actual availability of health resources continues to widen, allocating limited health resources faces rational and political challenges. The disease burden, as a relatively reasonable and fair, comprehensive, and objective evaluation indicator, provides a more scientific and objective basis for health policy-makers to determine how to prioritize the allocation of limited resources. Monitoring the dynamics of the burden of disease in a country or region can provide a preliminary evaluation of established resource allocation plans and measure the effectiveness of health resource allocation.

8.3.1 Definition and Components

Disease burden, also known as "cost of disease," is not yet a unified concept. Some studies have expressed it as "the consequences and effects of the loss and harm of disease." "Consequences" refers to the outcome of the disease (death, disability, recovery) and the previous losses caused by the disease, such as personal, family, and social losses; "effects" refers to the biological, psychological, and social harm caused by the disease.

The study of burden of disease should include the patient group burden, the family burden, and the social burden (Fig. 8.1). DALY, a widely used indicator for

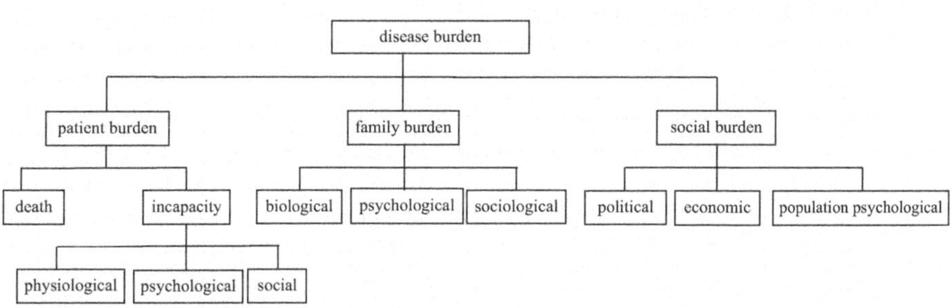

Fig. 8.1 Components and framework of burden of disease

disease burden assessment, is a good indicator of the burden on the patient population. The family burden of disease refers to the adverse effects of disease on the family, including the effects on family economics, daily life, recreation, relationships among family members, and the health of family members; the social burden of disease refers to the psychological, physical, and social effects of disease on people other than the patient and family members, including the effects on social and economic stress, public psychology, business trade, government image, and social stability. The methods used to comprehensively measure the family and social burden of disease are still in the exploratory stage.

8.3.2 Evaluation Methodology

8.3.2.1 Evaluation of Patient Burden of Disease

1. Concepts. As a negative composite measure of health status, quality of life adjustment and life expectancy loss, DALY is often used to evaluate disease burden at the patient level, which refers to the total number of healthy life years lost from onset to death, including years of life lost (YLL) due to premature death and years lived with disability (YLD) due to the disease. DALY is a composite of nonlethal health outcomes of disease and premature death, with one DALY representing 1 year of healthy life lost. It is one of the summary measures of population health (SMPH) that determines the difference between the actual health status of a population and an ideal situation (where everyone is in perfect health and lives to their standard life expectancy).
2. Calculation Method
   (a) Basic formula: DALY includes YLL due to premature death and YLD due to disease-induced disability, and the formula is written as follows: DALY = YLL + YLD. Premature death owing to YLL is relative to the life expectancy of a certain age group and can be calculated by referring to the standard life expectancy table when the death rate of each age group from a disease is known. The basic formula for YLL is YLL = N x L, where N is the number of deaths and L is the standard life expectancy at a given age. The disability part of YLD is relative to the ideal quality of life in a healthy state, and the severity of the impact of a disease on health is expressed by the disability weight (the value range is 0~1, with 0 representing the best health condition, i.e., complete health, and 1 representing the worst health condition, i.e., death). The basic formula is YLD = $I \times \text{DW} \times L$, where I is the number of incident cases during the study period, DW is the disability weight, and L is the average duration of disability (usually measured in years).

      These two basic formulas allow a preliminary calculation of years of healthy life lost by etiology, age, and sex groups. However, from a health economics perspective, the time lost must be further discounted by its social value. It is generally believed that present time is more valuable than future time, just as the value of money now and in the future differs, so future time needs to be discounted to the present value (usually with reference to the discount rate of banks). The social value of different age groups is also different, and it is generally believed that the social value of young adults is higher than that of elderly people or children. Therefore, the complete calculation includes two social value parameters, i.e., age weights to assign productivity to age groups and discount rates to assign present value to future time. The scientific and rational nature of the DALY formula has been thoroughly discussed, and there is basically no objection. At present, the controversy focuses on the assignment of health status, age, and time, especially the subjective arbitrariness of the health status assignment process, which is considered to be too great:

(b) Determination of disability weights: To establish a temporal link between death and nonfatal health outcomes so that the two can be combined in a comprehensive measurement method, it is necessary to assign values to time spent on disability loss (determined by the degree and duration of disability) for different health states and to use empirical weights to express the relationship between disability and full health. Depending on the source of the weights (different composite measures), they can be classified as disability weights, QALY weights, health state valuations, health state preferences, or health state utilities. The values of these weights generally range from 0 to 1, with 0 corresponding to death and 1 corresponding to perfect health. DALY (with 0 representing perfect health and 1 death) is the opposite of QALY, as DALY measures health loss (negative) and QALY measures healthy life years (positive). Measurement of health status differs from health state valuations. The former refers to a multidimensional description of an individual's quality of life, while the latter refers to an assignment of a value to evaluate the time lost to disability for a given health state. In conducting health state evaluations, it is important to consider how health states are described, the type of subject, and the evaluation method.

Parasitic diseases are one of the major categories of neglected tropical diseases. The disease burden evaluation of these diseases is essential for a proper understanding of their public health significance. Currently, disease burden has been evaluated for a few parasitic diseases, such as schistosomiasis and clonorchiasis, and more definite disability weights have been derived. In the global burden of disease evaluation, the disability weight of schistosomiasis was previously underestimated and was considered to range from 0.005 to 0.006 for all age groups, which is comparable to the health impact of facial leukoplakia. A subsequent meta-analysis of schistosomiasis disability weights by some researchers concluded that those weights were 4–30 times those of the estimated global disease burden. Chinese scholars studied the disease burden of chronic schistosomiasis in the middle and lower reaches of the Yangtze River and concluded that the disability weights were 0.095–0.246. Schistosomiasis, an important foodborne parasitic disease in Asia, has not been included in the WHO global burden of disease assessment. The weight of disability for this disease was estimated to be 0.022~0.118 by the Monte Carlo simulation method.

8.3.2.2 Evaluation of Family Burden of Disease

The family burden can be divided into subjective burden, which refers to the psychological feelings of family members caused by the disease, and objective burden, which refers to the patient's disease becoming a disruptive factor in family life and the resulting adverse events. Family burden is universal, complex, and multidimensional, including impacts on family economics, daily life, recreation, relationships among family members, and the physical and mental health of family members, and is an important component of the burden of disease. Studies have shown that caring for patients places a significant burden on families, exposing them to many detrimental health effects, including increased stress, reduced subjective well-being, poorer health, and reduced self-worth. Objective and subjective burdens are prevalent in patients' families, and these burdens can have an impact on all dimensions of the family and result in family dysfunction. Currently, studies measuring family burden in a multidimensional way are still limited to a few conditions, such as psychiatric disorders, and the methods used are mostly scale measures, the most common of which is the Family Burden of Disease Scale (FBS).

Currently, in the field of parasitic diseases, some researchers have evaluated the reliability and validity of applying the FBS to the evaluation of the family burden of advanced schistosomiasis. The results showed that the FBS has excellent reliability, validity, and responsiveness in the evaluation of the family burden of advanced schistosomiasis patients in China and is suitable for such evaluation. Families of advanced schistosomiasis patients have different degrees of burden in various dimensions, such as family economics, daily life, recreational activities, relationships among family members, and the physical and mental health of family members, among which the family economic burden is most significant. The severity of the family burden is influenced by the prevalence of the epidemic area and family income and expenditure.

8.3.2.3 Evaluation of Social Burden of Disease

The social burden of disease refers to the negative impact of disease on various aspects of society, including not only the decline of the social function of the patient group caused by disease and the economic pressure of disease on society but also the impact of disease on the psychology of the society, social, and economic trade and commerce, the image of the government, and social stability, such as the sociopolitical and economic impact caused by an infectious disease epidemic. The measurement of social burden has been more widely explored; for example, some researchers have studied the social burden of stroke patients during hospitalization by converting the time lost by accompanying caregivers into an estimation in terms of person-years of lost work. However, the Delphi method has mainly been used in studies of the social burden of disease. For example, in a study of the burden of Alzheimer's disease, Parkinson's disease, and cerebrovascular disease in Beijing, the social burden was divided into six dimensions, the impact on health resources, social labor productivity, community psychology, business trade, government image, and social stability, and each dimension was assigned a weight between 0 and 10. The sum of the scores for each dimension was the social burden. In addition, other researchers have used brainstorming and the Delphi method to construct an evaluation index system for the social burden of advanced schistosomiasis and applied it to field surveys. The social burden of disease can also be measured by the social burden index (SBI).

8.4 Case Studies

8.4.1 Evaluation of the Disease Burden of Chronic Schistosomiasis in Dangtu and Hanshou Counties

8.4.1.1 Proposing the Issue

Schistosomiasis is endemic in 74 countries in the world and still seriously endangers the health of people in some third-world countries; more than 600 million people worldwide are threatened by schistosomiasis. However, in the World Bank's Global Burden of Disease (GBD) study in 1990 and 2000, the DW of schistosomiasis in Japan was seriously underestimated. Therefore, it is necessary to correctly evaluate the disability weight (DW) and YLDs of schistosomiasis to provide a scientific basis for the formulation of national schistosomiasis control program and the rational allocation of health resources.

8.4.1.2 Research Methods and Steps

1. Site selection and data sources. Two schistosomiasis-endemic counties, Dangtu and Hanshou, in the middle and lower reaches of the Yangtze River were selected as survey sites, and data on chronic schistosomiasis were collected from 91 administrative villages in the 2 counties.
2. Establishment of chronic schistosomiasis cases was confirmed and clinical cases were determined according to the diagnostic criteria for schistosomiasis in Japan (GB15977-1995).
3. Calculation of DW. The European Quality of Life Assessment Scale (EQ-5D+C) was used to measure the patient's health status, the VAS of the scale was used for an overall assessment of the patient's quality of life, and the individual disability weight was deduced from the patient's VAS score. The formula was as follows:

$$DW = 1 - VAS/100$$

4. Calculation of YLD

   (a) Prevalence and Incidence Rate: The prevalence rates of 91 administrative villages in the two counties were calculated by age group and sex based on the positive blood test rate and the expected proportion of cases among those with positive blood tests in each group of the population. The study adopted the same schistosomiasis prevalence projection strategy as that of the GBD and WHO; i.e., assuming a 1-year disease duration, the prevalence rate was approximately equal to the incidence rate based on prevalence = incidence rate × disease duration.

The specific calculation formula was as follows:

$$\text{Population prevalence} = \frac{\text{Number of positive blood tests}}{\text{Number of people in this population receiving blood tests}} \times \frac{\text{Number of people identified as chronic cases}}{\text{number of blood test positives who underwent further testing}} \times 100\%$$

(b) YLD: The calculation formula was as follows:

$$YLD = IDW \left\{ \begin{array}{l} KCe^{(ra)} / (\beta + r)^2 \left[\begin{array}{l} e^{-(\beta+r)(L+a)} \left[-(\beta + r)(L + a) - 1 \right] - \\ e^{-(\beta+r)a} \left[-(\beta + r)a - 1 \right] \end{array} \right] + \\ (1 - k)(L / r)(1 - e^{-rL}) \end{array} \right\}$$

where I is the number of morbidities, DW is the disability weight, L is the mean duration of disability, a is the age of onset, r is the time discount rate (the GBD standard is 0.03), the parameter K determines whether to choose ($K = 1$) or not choose ($K = 0$) the age weight, C is the correction constant for the age weight (the GBD standard is 0.1658), and β is the parameter of the age weight equation (the GBD standard is 0.04). The study set the duration of chronic schistosomiasis to 1 year so that there was no time discounting problem and YLD was calculated without the discount rate or age weight ($r = 0$, $K = 0$).

(c) YLD rate and per capita YLD loss value

- The YLD rate was expressed as YLD per 1000 people and was calculated as follows:

$$\text{YLD per 1000 population} = \frac{\text{sum of YLD for this population}}{\text{total population of the target population}} \times 1000,$$

  with the unit being "person-years/1000 people."
- The per capita YLD loss value of cases reflected the per capita burden of disease for different patient groups and was calculated as follows:

$$\text{per capita YLD loss value} = \frac{\text{sum of patient YLD values}}{\text{total number of patients}}$$

- Rate Ratio and Rate Difference. The rate ratio and rate difference were used to compare the magnitude of the difference in YLD between different populations. The calculation formula was as follows:

$$\text{rate ratio} = \frac{\text{YLD rate in the observation group}}{\text{YLD rate in the control group}};$$

$$\text{rate difference} = \text{YLD rate in the observation group} - \text{YLD rate in the control group}$$

5. Projection of all county population YLD. The population of each county was divided into 60 subgroups, i.e., age (5 groups) × sex (2 groups) × prevalent village type (6 groups). Of these, age was divided into 5 groups: 5 to 14, 15 to 29, 30 to 44, 45 to 59, and over 60 years. The populations of endemic and nonendemic villages were standardized according to the age and sex composition ratios of the whole population of the two counties, and people younger than 5 years old were excluded. The YLD for a population was calculated as follows:

$$\text{YLD values for a population} = \frac{\text{YLD rate for this population} \times \text{total population}}{1000},$$

with the unit being a person-year.

8.4.1.3 Study Results

1. Information on disease detection. Approximately 40 and 51 administrative villages were selected as sample villages in Dangtu and Hanshou counties, respectively. A total of 59,765 people in the 2 counties were examined serologically, and 3405 were positive, with a total positive blood test rate of 5.70% in the 2 counties. In Dangtu County, a total of 33,339 people were examined, and the number of positive blood tests was 2025, for a positive blood test rate of 6.07%; in Hanshou County, a total of 26,426 people were examined, and the number of positive blood tests was 1380, for a positive blood test rate of 5.22%. Of the 3405 positive serological screening cases in the 2 counties, 2843 individuals underwent fecal tests, with 290 positive results. The total fecal positive rate of the positive blood tests in the 2 counties was 10.20%: 3.99% in Dangtu and 17.96% in Hanshou. The infection rate was 0.24% in Dangtu and 0.94% in Hanshou, with Hanshou being higher than Dangtu ($P < 0.0001$).
2. Case confirmation. The number of chronic schistosomiasis cases was determined to be 1466, of which 1225 were clinical cases and 241 were confirmed cases.
3. Disability weights. The mean disability weight for the total patient group was 0.192, and the age-specific disability weights were 0.095 (5–14 years), 0.160 (15–44 years), 0.210 (45–59 years), and 0.246 (60 years and older). One-way ANOVA results showed significant differences between age-specific disability weights ($F = 61.07$, $P < 0.0001$). Additionally, the Bonferroni t test showed significant differences between comparisons of pairs of age-specific disability weights (all $P < 0.05$), and the differences increased with age.

4. Sample YLD
   (a) Different Populations: The YLD rate was 3.70 person-years/1000 people in Dangtu County and 7.28 person-years/1000 people in Hanshou County. The YLD rate was higher for males than females in all age groups. YLD rates increased with age, with the lowest rate in the 5–14 group (0.30 person-years per 1000 people) and the highest rate in the 60-plus group (11.96 person-years per 1000 people), with the latter being 39.87 times higher than the former (11.96/0.30). The YLD rate was 1.93 times higher for males than for females (1.75 to 3.07). The difference between the male and female rates ranged from 0.29 to 7.32 person-years/1000 persons and increased with age.
   (b) Different Epidemic Areas: YLD rates in both counties generally increased with the increase in epidemic intensity. Among them, the YLD rate was 0.25 person-years/1000 people in nonendemic villages in Dangtu County and 13.14 person-years/1000 people in Class I villages, but the YLD rate was higher in Class III villages (only one) (7.79 person-years/1000 people) than in Class II villages (4.23 person-years/1000 people). Hanshou County had the smallest YLD rate, 0.89 person-years/1000 people, in Category V villages, followed by nonendemic villages with 0.90 person-years/1000 people and Category IV villages with 1.68 person-years/1000 people. There was a gradual increase to 18.60 person-years/1000 people in Category I villages.
5. Projection of YLD for the Whole County Population
   (a) Total YLD: The total expected number of cases for the population aged ≥5 years in Dangtu County was 5350, with a total YLD of 1056.26 person-years, including 323.29 person-years for females and 732.97 person-years for males; 40.1% of YLD was concentrated in Class I villages (423.35 person-years). The total expected number of cases for the population aged ≥5 years in Hanshou County was 20,925, and the total YLD was 3967.43 person-years, with 1302.84 person-years for females and 2664.59 person-years for males; 37.39% of the county's YLD was concentrated in Category I villages (1483.35 person-years). The total YLD in Hanshou County was 2911.17 person-years higher, or 3.76 times higher, than that in Dangtu County. YLD was higher in both counties for males than for females by 1771.43 person-years, and YLD in Class I villages accounted for 37.9% of the disease burden of chronic schistosomiasis.
   (b) Per capita YLD: The per capita YLD of cases in Dangtu County was 0.197 person-years, with 0.177 person-years in nonendemic villages and 0.195, 0.203, 0.189, 0.206, and 0.177 person-years in endemic villages from Class I to Class V, respectively. In Hanshou County, the per capital YLD was 0.190 person-years, with 0.186 person-years for nonendemic villages and 0.188, 0.186, 0.195, 0.215, and 0.181 for endemic villages from Category I to Category V, respectively.

8.4.1.4 Recommendations

This study for the first time confirmed by field survey that the DW of schistosomiasis was severely underestimated in the GBD. The total DW of schistosomiasis should be 0.192, and the age-specific DW should be between 0.095 and 0.246. The results of population and regional evaluations of chronic schistosomiasis YLD will help to identify priority intervention areas and targets and support scientific and fair health decisions.

8.4.2 Laboratory Capacity Survey of County-Level Schistosomiasis Control Institutions at National Surveillance Sites

8.4.2.1 Proposing the Issue

Improving laboratory testing capacity at the grassroots level is one of the major measures for the construction of disease prevention and control systems. Since 2004, China has focused on strengthening the capacity of provincial, local, and county disease prevention and control center laboratories; however, many county and municipal schistosomiasis control institutions in schistosomiasis-endemic areas have not been included in the plan. A survey of the current situation of laboratory resource allocation and schistosomiasis diagnosis level in county-level schistosomiasis control institutions can provide an understanding of the current situation of diagnostic laboratories in grassroots schistosomiasis control institutions in China and provide a reference for future schistosomiasis control health resource allocation and laboratory capacity building in grassroots schistosomiasis control institutions.

8.4.2.2 Research Methods and Steps

1. Survey site selection. Among the 80 national schistosomiasis surveillance sites nationwide, 40 surveillance sites with the top 50% random number rankings were selected using the Excel random number table method, and the laboratory of the county-level schistosomiasis control institution where each surveillance site located was used as the survey site.
2. Questionnaire survey. By means of a questionnaire survey, basic information was collected from the schistosomiasis diagnostic laboratories of county-level schistosomiasis control institutions, including information on laboratory personnel (age, sex, education, title, etc.), laboratory conditions (housing area, instrument, and equipment configuration), and common laboratory diagnostic methods for schistosomiasis (types of diagnostic methods, annual use, etc.).
3. Detection capacity survey. In conjunction with the national survey of residents of schistosomiasis surveillance sites, serum samples and modified Kato tablets from residents of national surveillance sites were randomly selected. Then, the samples were reviewed by provincial schistosomiasis diagnostic reference laboratories or provincial (municipality directly under the central government) schistosomiasis control (disease prevention and control) institutions and diagnosed by testing experts to assess the detection capacity of the investigated laboratories' serum immunology (IHA) and fecal pathogenesis (modified Kato method).

8.4.2.3 Study Results

1. Laboratory distribution. The laboratories performing schistosomiasis diagnosis belonged to three types of institutions: 9 county (city)-level disease prevention and control centers, 27 county (city)-level schistosomiasis control clinics/stations, and 4 county (city) schistosomiasis control hospital laboratory departments.
2. Laboratory personnel. The 40 laboratories had a total of 196 personnel, and the average number of staff in each laboratory was 4.90 ± 2.73. The average age was 40.93 ± 9.56 years, and 69.39% (136/196) were >35 years old. Male and female personnel accounted for 57.14% (112/196) and 42.86% (84/196), respectively, and the mean age of men (43.18 ± 9.16 years) was significantly greater than that of women (37.94 ± 9.30 years) ($t = 3.94$, $P < 0.01$).

   The percentages of laboratory personnel with less than a bachelor's degree, a bachelor's degree and a higher degree were 86.22% (169/196), 13.27% (26/196), and 0.51% (1/196), respectively, and the differences in the educational composition of the three types of laboratory personnel were not statistically significant ($\chi^2 = 7.29$, $P > 0.05$). Junior, intermediate, and senior titles accounted for 56.63% (111/196), 39.29% (77/196), and 2.04% (4/196) of the total number of survey respondents, respectively, except for four people whose titles were unknown; the proportion of junior titles was significantly higher in the schistosomiasis control stations/institutes than in disease prevention and control centers and schistosomiasis control hospitals ($\chi^2 = 28.04$, $P < 0.01$).
3. Laboratory room area and equipment configuration. The areas and equipment configurations of the laboratories varied widely, with the average areas of laboratories for schistosomiasis diagnosis at CDCs, schistosomiasis control stations/institutes, and schistosomiasis control hospitals being (91.00 ± 57.49) m^2, (52.81 ± 40.08) m^2, and (90.00 ± 54.77) m^2, respectively. The analysis showed that the average area of laboratories in schistosomiasis control stations/institutes was significantly lower than that of laboratories in CDCs undertaking schistosomiasis diagnosis ($t = 2.14$, $P < 0.05$), while the difference from the area of laboratories in schistosomiasis control hospitals was not statistically significant ($t = -1.66$, $P > 0.05$).

   There was a large difference in instrumentation among laboratories, and the main items of equipment with >50% laboratory configuration were microscopes (100%, 40/40), centrifuges (92.50%, 37/40), refrigerators (70.00%, 28/40), water baths (65.00%, 26/40), dissecting microscopes (57.50%, 23/40), thermostats (57.50%, 23/40), oscillators (57.50%, 23/40), and biochemical analyzers (50.00%, 20/40). Other auxiliary testing and analysis equipment, such as enzyme markers, hematology analyzers, electrolyte analyzers, urine analyzers, spectrophotometers, and pH meters, represented <50%.
4. Detection capability survey. The compliance rate between the IHA test results and the retest results of the basic laboratories was 95.89% (95.24% to 99.41%), and the kappa value was 0.91 (0.89 to 0.99), indicating that the results of the two tests were highly consistent and that the difference was not statistically significant ($P > 0.05$). When the results of retesting by the provincial reference laboratories/schistosomiasis control institutions were used as the reference standard,

the leakage rate of positive *Schistosoma* antibody samples tested by the various types of laboratories ranged from 0% to 6.72%, with higher rates in the schistosomiasis control stations/institutes and CDC laboratories, but the differences between the various types of laboratories were not statistically significant ($P > 0.05$).

A total of 1695 modified Kato tablets were sampled from the 40 surveillance sites, and the compliance rate of the 2 tests was 99.53% (99.39% to 100%) with a kappa value of 0.97 (0.97 to 1.00), indicating that the results of the two tests were highly consistent. When the test results of the provincial reference laboratory/schistosomiasis control institution laboratory were used as the reference standard, the leakage rate of positive schistosomal pathogenic specimens by all types of laboratories ranged from 0% to 2.31%, with leakage occurring mainly in the schistosomiasis control station/institute laboratories, but the differences between the various types of laboratories were not statistically significant ($P > 0.05$).

8.4.2.4 Recommendations

The average age of laboratory personnel in the primary schistosomiasis control institutions is increasing, with lower education and a low proportion of personnel with senior titles. This indicates that the laboratory testing personnel of schistosomiasis control institutions have an unreasonable structure and low overall quality; thus, these laboratories cannot yet meet the needs of schistosomiasis prevention and control. Laboratory areas vary greatly among counties, and the configuration of instruments varies widely among laboratories, but the overall equipment level is low. All county laboratories have good detection capability for IHA and the modified Kato method, but there is a certain false-negative rate for positive serum and modified Kato tablet specimens.

Establishment of the Early-Warning Mechanism

9

Guo-Jing Yang

Early warning can be defined as follows: before the occurrence of disasters or catastrophes and other dangers that must be borne in mind, an emergency signal is sent to the relevant departments to report a dangerous situation according to laws summarized in the past or the observance of probable precursors to avoid the occurrence of hazards without knowledge or preparation and thus to minimize the losses caused by those hazards. Prevention beforehand is better than remediation afterward, as it can reduce the loss of life and property and improve people's ability to survive. Since early-warning technology is first applied to events such as disasters, the concept of early warning is first applied to disaster warning, which includes natural disaster and human disaster safety warnings.

The early-warning technology discussed in this chapter is intended to protect people's health from epidemics by analyzing trends of disease outbreaks and providing early warning before an outbreak or epidemic in order to maximize various effective preventive measures, thus reducing the loss of life and property and improving people's quality of life.

9.1 Overview

9.1.1 Definition and Characteristics

The early-warning mechanism refers to an early-warning system composed of institutions, systems, networks, and measures that can sensitively and accurately predict an epidemic or outbreak of parasitic disease and provide a timely warning. Its role is to provide advance feedback, enable a timely response, prevent risks before they

G.-J. Yang (✉)
Jiangsu Institute of Parasitic Diseases, Wuxi, Jiangsu, China

X.-N. Zhou (ed.), *Surveillance and Management of Parasitic Diseases in China*,
https://doi.org/10.1007/978-981-97-1241-0_9

occur, and act as a complete mechanism of information security. The early-warning mechanism has the following six characteristics.

9.1.1.1 Timeliness

The system can detect the occurrence of emergencies as early as possible and gain time for the initiation of response measures. Timeliness should be reflected in the investigation, collection, transmission, analysis, and release of information as well as the adoption of measures.

9.1.1.2 High Efficiency

This refers to the collection of information from as many angles and directions as possible to make predictions as accurate as possible and to avoid unnecessary initiation of countermeasures.

9.1.1.3 Operability

The system should be in line with the current situation of China and the actual situation of different regions, based on the corresponding personnel and material reserves; thus, it should be simple to operate and easy to implement.

9.1.1.4 Expandability

The system should have sufficient space to continuously increase or decrease the events targeted for early warning according to the specific situation and have the ability to continuously adjust and improve them.

9.1.1.5 Social and Corresponding Legal Effects

Since early warning involves many sectors of society, having only legal effects ensures that it can work in a short period of time.

9.1.1.6 Relevance to the Emergency Response System

Early warning and emergency response are two continuous processes. Without accurate early warning, it is difficult to coordinate and smoothly carry out the subsequent emergency response.

Early warning can be divided into natural disaster and human disaster safety warnings. The former refers to the construction of early-warning facilities to reduce human and material losses caused by potential natural disasters based on lessons learned from repeated natural disasters (earthquakes, floods, etc.). The latter refers to preventive methods adopted by different groups of people as a result of various human conflicts in order to minimize the losses caused by these conflicts. It includes early warning of war through defense construction, early warning of economic crises through a degree of economic self-sufficiency, cultural early warning through the creative construction of culture, and internal early warning through the establishment of internal security mechanisms.

The internal early warning of internal security mechanisms is further divided into disease early warning, crime early warning, corruption early warning, insurgency early warning, and population early warning. Thus, disease early warning is a branch of internal early warning in the context of human disaster security. Disease early warning entails predicting the occurrence, development, and epidemic trends of possible diseases based on the laws of occurrence and development of diseases and related factors, using methods such as analytical judgment and mathematical simulation to improve foresight and initiative as well as the effectiveness of disease prevention and control.

9.1.2 Early-Warning Technology and Classifications: The Selection and Use of Early-Warning Indicators

Early-warning indicators are indicators with potential warning value, and the magnitude of fluctuation of the indicators is associated with the prevalence or outbreak of parasitic diseases. In the early-warning mechanism, prediction and forecast are synonymous, as both refer to estimating unknown or nonoccurring events, but there are obvious differences. Prediction is a judgment of the occurrence, development, and trend of unknown events based on known data, occurrences and development patterns of events and related factors, and the use of analytical judgment or mathematical simulation and other methods. Therefore, the elements of analysis can be in the spatial or temporal domain, and the events can have not occurred or already have occurred. Forecasting, on the other hand, is based on the prediction of events that have not yet occurred over a longer period of future time and the use of appropriate forecasting methods to transmit correct prediction information so that appropriate preventive measures can be taken to prevent disease outbreaks or epidemic events. In summary, the ability to correctly apply prediction tools and implement forecasting measures makes up a sound early-warning mechanism.

Based on the combined characteristics of different time phases of parasitic disease outbreaks and development and early-warning theory, the early-warning indicators of parasitic disease outbreaks should include three major categories: indicators of a preoutbreak or pre-epidemic situation, indicators of an atypical symptom period, and indicators of a typical symptom period. Once the fluctuation range of the indicators exceeds the specified alert line, an alert can be issued to initiate the corresponding epidemiological investigation or intervention. Therefore, early-warning indicators must be timely, accurate, and operable.

9.1.2.1 Timeliness

Indicators should detect a parasitic disease outbreak or its precursors as early as possible and reflect the relevant situation before, during, and after the development of the outbreak to gain time for response measures to be initiated.

9.1.2.2 Accuracy

Indicators should be used to predict crises as accurately as possible and prevent unnecessary response initiation. The information reflected in the indicators should be consistent with the actual epidemiological situation of the parasitic disease to the greatest extent as possible.

9.1.2.3 Operability

The data and information needed for the indicators should be easy to collect. The establishment of a sensitive and effective early-warning indicator system is a prerequisite and foundation for the success of the early-warning system.

9.2 Methods of Prediction

Prediction is the science of recognizing and mastering the laws of disease occurrence and development, inferring, and judging future development trends based on the observation and analysis of complex and disorganized disease epidemic or outbreak data or information. Prediction is a practical discipline and is the basis for decision-making. Correct prediction is the premise and basis for correct decision-making.

9.2.1 Classification

There are many different methods for predicting parasitic disease epidemics or outbreaks. In practice, different prediction methods should be used in accordance with different prediction purposes to develop different prediction plans.

Prediction can be divided into short-term (monthly, quarterly, semiannual, and 1 year), medium-term (1–3 years), and long-term (>3 years) predictions. In general, short-term prediction serves to prevent and control parasitic disease epidemics or outbreaks, while long-term prediction serves to develop medium- and long-term prevention and control strategies. The shorter the prediction period is, the greater the accuracy.

Additionally, prediction can be divided into macroscopic and microscopic prediction. Macroscopic prediction is the prediction of the entire epidemiological trend of parasitic diseases on a larger scale, involving the whole situation (such as regional- or national-scale prediction), and microscopic prediction is the prediction of the possibility of an epidemic or outbreak of parasitic diseases in a local area, which is the prediction within the inherited organization (such as the prediction within a county).

Prediction can also be divided into qualitative and quantitative prediction. Quantitative prediction has higher accuracy than qualitative prediction. With the development of prediction methodologies, prediction models are being used more widely. A prediction model describes the prediction objects (such as the prevalence of parasitic diseases) as simply and abstractly as possible and is used to describe the relationship between the prediction objects and related factors of connection, dependence, change, and movement. Prediction models can be qualitative or quantitative. In practice, commonly used prediction methods are qualitative prediction, quantitative prediction, and integrated prediction.

9.2.1.1 Qualitative Prediction

Qualitative prediction is used to determine the trend and intensity of an upcoming disease epidemic through specific analysis of the local parasitic disease occurrence and development pattern and related factors. Qualitative prediction is based mainly on the epidemic control chart method, Delphi method, and brainstorming method.

1. Epidemic Control Chart Method. This statistical tool is applicable to parasitic diseases of various distributions and is used to discriminate between ordinary and special causes of process change. It is most effective for parasitic diseases with seasonal or periodic epidemic patterns (such as malaria and black fever). It has a center line (CL), upper control limit (UCL), and lower control limit (LCL), which represent the median line, the upper alert line, and the lower alert line, respectively. The limits are adjustable, as in the case of malaria projections, and the upper and lower standard deviations are the upper and lower limits. Control charts are also tools for statistical quality control. They are easy to create and are a good early-warning method in disease surveillance.
   (a) Classification of Control Charts. Different types of control charts are used for different data and are based on different probability models. The types of control charts are related to the classification method and can be divided into univariate and multivariate control charts based on the number of variables. According to the type of information, X-single value control charts can be used for data counting. The type of control chart for measurement data is related to the sample size, specifically, the X-single value control chart or single-value polarity control chart (when the sample size is large). Control charts can be divided into variable control charts (such as X-movement control charts, X-polarity control charts, single-value charts, and average movement-polarity movement control charts) and attribute control charts (such as defect rate control charts, defect number control charts, defect number control charts, and unit defect number control charts).

       Process quality control charts used in early warning commonly fall into three categories. First is the Swart quality control chart. This method is most commonly used, but it does not work well when there is a small drift in the process mean. Second is the cumulative and control chart, which is based on the derived likelihood ratio. Third is the exponentially weighted moving average control chart. The latter two have been proven to be effective in detecting small drift.
   (b) Issues of Concern When Control Charts Are Applied to Early Warning
       - Setting control limits. This can affect the determination of abnormal signals. There are three main ways to set control limits: specifying control limits, calculating according to the theoretical value, and calculating according to the formula. There are three curves in the control chart, which represent the upper alert line, lower alert line, and median line, and the alert line can be adjusted. When an apparent abnormality is used as the alert target, the alert line is the control limit. If only an increased level

of occurrence is of interest, the lower control limits can be omitted from the application of the control chart method.

- The calculation of the upper and lower control limits is related to the type of distribution of the information, and the formula used for the calculation is a direct application of conventional probability theory. The setting of control limits using the center line to plus/minus 3 standard deviations is appropriate only when the distribution is normal or when the distribution form is similar to the normal distribution. For other types of distributions that may more frequently respond to these signals, false alerts may still occur even if the process does not change or changes less. The calculation of the control limits can also be based on the approximation of the binomial criterion and the Poisson distribution.
- The calculation of the center line is also related to the type of distribution of the information. When the information does not conform to the normal distribution, the center line is the median line. For control charts, the center line (i.e., the baseline level) is determined based on an original dataset, and data from the prevalence period should be excluded from the calculation. The more stable the reference baseline is, the more accurate the set control limits are. Historical data can be used when baseline measurement methods are not available:
- Setting the temporal window. The temporal window used by the identification system may have an important influence on the identification of outbreaks of different intensities. The setting of the temporal window has not been uniform in previous studies. Some studies have used a specific month and the month before and after the month of observation for the previous 5 years; others have used the average incubation period of the disease and oscillated back and forth; and still others have used 7 days, but a 7-day moving average for the temporal window would see outbreaks at a stable level lasting 2–3 days under certain conditions, which may reduce the signal of an outbreak. Conversely, a shorter moving window may not detect outbreaks that are slowly increasing in number.

2. Delphi Method. Also known as the expert scoring method or expert consultation method, this is one of the most widely used prediction methods worldwide. The Delphi method is a combined quantitative and qualitative prediction and evaluation method that takes an anonymous approach to widely consult experts who evaluate the object of prediction. It was not until the mid-1970s that the Delphi method began to be applied in medicine, and it is now increasingly used in parasitic disease surveillance systems.
   (a) Delphi Method Procedure
   - To determine the prediction problem, the requirements should be determined, the prediction objectives specified, and the selected experts and specialists informed in writing (questionnaire). The selection of experts is key, as experts are generally people who have advanced knowledge and skills in a particular field. The number of experts should not be too large;

approximately 8–20 experts are generally appropriate. Each expert is asked to state what special information is available to analyze the issues and how this information will be used. The experts are also provided with relevant information and are asked to suggest what further information is needed.

- After receiving the notification (questionnaire), the experts, based on their knowledge and experience, make their own predictions about the future development trend of the object of prediction, explain their basis and reasons, and respond in writing to the unit hosting the prediction.
- The presiding prediction unit or the leading group summarizes and organizes the experts' predictions according to their opinions, explains the basis and reasons for the different predicted values (according to the experts' anonymous opinions), and then sends a new questionnaire with more detailed components to the experts, asking them to modify their original predictions in response to the new requests.
- After receiving the second questionnaire, the experts analyze the various predictions and their bases and reasons, make another prediction, and present their own revised predictions and the bases and reasons for their opinions. The questionnaire is repeatedly sent, summarized, and modified until the opinions are basically in agreement. The number of revisions is determined as necessary.

(b) Features of the Delphi Method

- Anonymity: The Delphi method is conducted by means of letter solicitation, and neither party knows who has requested the opinion. This anonymity eliminates the influence of psychological factors, which is conducive to the discussion of problems and the revision of opinions.
- Feedback: The Delphi method of letter solicitation is not a one-time process but a three- or four-round process. This kind of multiple controlled feedback provides a good opportunity for opinion exchange and mutual inspiration and creates favorable conditions for faster agreement.
- Statistical power: The Delphi method can be used to statistically analyze experts' opinions or views during the consultation process and enables the statistical processing of the prediction results.
- Strong representativeness: Since the Delphi method uses the letter solicitation method, many experts can be invited to participate in the prediction process, providing wide representativeness.

(c) Principles to Be Followed in the Delphi Method

- The Delphi method should be fully explained so that the experts understand the importance of the survey.
- When designing the questionnaire, the questions asked should be relevant.
- The questions asked should avoid combined events.
- When designing the questions, the wording must be precise.
- The design of the questionnaire should be as simple as possible, and the views of the leadership team should not be imposed on the questionnaire.

3. Brainstorming. Brainstorming is also known as the talk meeting or B.S. method. It involves organizing experts to hold meetings, gives full play to the creative thinking of the experts through the direct exchange of information and mutual inspiration in a relatively short period, and produces as many programs or ideas as possible. The experience of a successful passenger ship escape during World War II was summarized as the brainstorming method because of the irregularity of the passengers' discussion, which was similar to a storm, before they made a decision. American scholars have summarized the characteristics of this method by defining the acronym BRAIN as follows:

   B—build on others' ideas
   R—reserve judgment
   A—aim for quantity
   I—imagine wildly
   N—no killer phrases

   (a) The principles of brainstorming should be as follows: clear meeting topic, spontaneity with no mutual criticism, and concise speech.
   (b) Brainstorming meeting implementation steps.
      - Premeeting preparation: the participants, moderator, and subject tasks should be clarified, if necessary, and flexible training should be provided.
      - Idea development: The moderator should announce the meeting theme and introduce the reference situation related to the theme; the participants should then break through the inertia of thinking and make bold associations. The moderator should control the time and strive to obtain as many creative ideas as possible within the time limit.
      - Classification and organization of ideas: There are two categories, practical and fantasy ideas. The former refers to visions that can be realized by the current technical process, and the latter refers to visions that the current technical process cannot achieve.
      - Perfection of practical ideas: For practical ideas, brainstorming is used to argue again to further expand the scope of the idea.
      - Redevelopment of fantasy ideas: The brainstorming method is used to further develop fantasy ideas. It is possible to transform creative imagining into a mature practical idea. This is a key step of brainstorming and is also a clear sign of the quality of the method.
   (c) Disadvantages of Brainstorming
      - Since the brainstorming method involves many people, once it becomes impossible to unify their opinions, the phenomenon of the minority giving in to the majority often occurs. Given that most people involved in brainstorming are nonprofessionals, the majority opinion is often wrong.
      - If brainstorming is conducted in a meeting, it is easy for it to disrupt the thinking of designers and planners; decisions made in such a meeting are often not well thought out, and brainstorming tends to lead to blind and disorderly conjectures.
      - Brainstorming is not applicable to some issues that require confidentiality, high-level technology, and professionalism.

9.2.1.2 Quantitative Prediction

Quantitative prediction is the use of statistical methods and mathematical models to perform a statistical analysis of historical data and create a quantitative description of a future development trend, the rate of increase or decrease, and the possible level of development. The mathematical model is used to express the basic laws for predicting the future development of a disease. In quantitative prediction, the idea of a mathematical model is the core of the construction of various prediction methods.

Quantitative prediction methods include regression analysis, time series analysis, gray system prediction, Markov prediction, and artificial neural network prediction. These methods all use the idea of mathematical modeling, and the following basic modeling steps are implied in the process: (1) assumptions of the model: simplify the actual problem by making reasonable assumptions, and then, discard the minor factors and consider the most significant ones. (2) Establishment of the model: translate real-world problems into mathematical problems. (3) Solution of the model: find the answer to the math problem. (4) Testing and application of the model: applying mathematical modeling ideas to disease prediction involves applying solutions of mathematical problems to real-world problems.

Quantitative prediction is more objective, relying mainly on historical data with a focus on the quantitative description of disease development. It is less influenced by subjective factors and can use computer programs to process large amounts of data:

1. Regression Prediction Method. The regression prediction method is used to analyze the association between disease and causative factors, establish a regression equation (regression model), and then predict changes in diseases according to the regression model. According to the number of causative factors considered and the relationship between the disease and causative factors, this method can be divided into the univariate linear regression prediction method, multiple linear regression linear prediction method, and nonlinear regression prediction method. Regression analysis has been widely used in medical statistics to predict diseases, and the technology is mature. It can take into account the joint effects of a variety of factors, but the model error is large, the extrapolation is poor, the amount of data required is large, and the distribution of the sample has high requirements.
2. Time Series Prediction Method. The so-called time series prediction method arranges the historical data at certain time intervals to form a statistical series over time, establishes a corresponding model of data changes over time, and extrapolates the model to the future for prediction. The Box-Jenkins (B-J) method (or ARMA model method) proposed by American statistician George E.P. Box and British statistician Gwilym M. Jenkins is the most important and commonly used prediction tool in time series modeling. The ARMA model requires that the time series used as a predictor be a stable series. However, many medical phenomena exhibit an upward or downward trend over time, constituting a nonstationary time series, so the integrated autoregressive moving average model (ARMIA model) with d-order chi-squared nonstationary time series is

applied: $\Phi(B)\cdot(1 - B)dXt = \Theta(B)\alpha t$. Since most medical health phenomena are not stable, the ARMIA model is commonly used.

3. Markov Prediction Method. Markov prediction divides a time series into several states and predicts the possible state intervals according to the transfer probability between states. It is an interval prediction, and the final prediction is the range of the actual value. A Markov chain is a sequence of random events, a Markov process with discrete states and time, and has the non-aftereffect property (this means that the future state of an event and its probability depend only on the present state of the object and not on the previous state). Its main function is to determine the probability of transfer between states. For example, if a disease is in state s at time t, it will be in state S_k with probability P at time $t + 1$, where p is the probability of the transfer of state S_j to state S_k. After i steps of transfer, the transfer probability is assumed to be P(i)jk, which in turn yields the step state transfer probability matrix $P(i) = P(i)jk\ i, j{-}1,\ldots, n$. Thus, the state transfer probability matrix reflects the probability of disease transfer from a certain state to various states, which can be used to predict future changes in the disease.

9.2.1.3 Gray System Prediction

Gray system theory is an emerging discipline founded by Professor Deng Julong in 1982 to discover and grasp the development pattern of diseases through processing raw data and establishing of gray models and to make scientific quantitative predictions of future development.

Compared with traditional prediction methods, gray prediction requires only a few data points to build a prediction model, and the prediction accuracy is higher for prediction problems over a short time and with little data information and low volatility. The most commonly used prediction model is GM(1, 1), which represents a first-order differential equation of one variable and studies only the effect of time, a comprehensive gray quantity, on a certain disease. Thus, it is often used to infer the morbidity or mortality of a disease at a certain moment or several moments in the future. The GM(1, 1) model is simple, the causative influences are largely disregarded, and the sample size is small. These characteristics also lead to coarse prediction; the epidemiological trend of the disease can be predicted only for the near future, which is suitable for exponential growth prediction, and the fitting prediction result for random and volatile data is poor. While the GM(1, N) model is an analytical model, a factorial model, which represents the first-order differential equation of N variables, does not have full information and is generally not suitable for multifactor prediction.

9.2.1.4 Artificial Neural Network Prediction

An artificial neural network (ANN) is an algorithmic mathematical model that imitates the behavioral characteristics of animal neural networks and performs distributed parallel information processing. It is based on the basic understanding of the neural network of the human brain, abstracts the functions of the human brain from the perspective of information processing by mathematical methods, and establishes a simplified model. The artificial neural network has nonlinear fault tolerance and self-organizing, self-learning, self-reasoning adaptive ability and can continuously change the connection power and topology of the network through learning training on samples so that the output of the network is constantly close to the desired output. It does not need to analyze the specific relationship of related factors but can summarize the complex inner law by itself through training on certain samples to make disease predictions.

The BP neural network is one of the most widely used multilayer feed-forward networks. In epidemiology, it can be used to build disease prediction models. First, the main influencing factors leading to disease occurrence and outcomes are collected; then, the influencing factors and disease outcome are input into the designed neural network model for repeated training until the expected training error is reached; and finally, the established model is used for disease prediction. A BP neural network is a very powerful nontraditional multivariate nonlinear model that can identify complex nonlinear relationships between variables. It can be used to model variables regardless of their type and whether they satisfy the conditions of normality and independence. The design of the BP neural network is directly related to the application effects of the network, so the design of the network is considered in every detail according to the characteristics of the data.

9.2.1.5 Comparative Analysis of Several Quantitative Predictions

Several methods of quantitative prediction are commonly used in disease prediction, and they have their own characteristics. Some methods are more accurate in short- and medium-term prediction, while others are more advantageous in long-term prediction; some are more suitable for stable data, while others are especially suitable for data experiencing volatile changes. Table 9.1 lists the characteristics and shortcomings of several prediction methods, which facilitates further comparison and analysis of the advantages and applicability of each method.

Table 9.1 shows that the various prediction methods differ in terms of the study, the form of data used, the size of the data sample, and the conditions of applicability. In terms of applicability, the regression prediction method is suitable for predicting with large samples and constant development patterns, and the gray model prediction method seeks patterns through the collation of raw data, which is suitable for analysis and prediction under conditions of lack of information or unknown information. In terms of the form of data, gray system theory is modeled using generated data, regression analysis is modeled using raw data, and the Markov method is used to predict future value intervals by calculating transfer probabilities. In terms of data characteristics, the time series method, gray model prediction method, and regression prediction method require stable and less volatile data, and the Markov prediction method is suitable for more volatile data. Based on the applicable time classification, the time series, Markov, and gray model methods are more suitable for near-term prediction, and the regression prediction method is more suitable for long-term prediction. While the neural network method is not limited by these factors, it is a variable structure model that can constantly change its own data structure by learning from the sample and adapting to changes in the sample. However, its network structure is difficult to determine, and the algorithm is complex. It is important for multidisciplinary knowledge of the science of thought, artificial intelligence, and computer science, while for general applications, there are certain difficulties in model building.

Table 9.1 Comparison of quantitative prediction methods

	Characteristics	Shortcomings	Scope of application
Regression prediction method	Able to consider various influencing factors and correlations of disease but requires a large sample size and a good distribution pattern	When affected by random disturbances and other intricate factors, the error is large and extrapolation is poor	Suitable for medium- to long-term prediction
Time series prediction method	Does not need to consider each influencing factor, uses time variables to represent each influence factor in an integrated manner, and is simple and easy to grasp but requires a stable sample series with little volatility	Inability to make full use of correlations among disease influencing factors; poor prediction capability when encountering highly volatile data	High accuracy of short-term predictions
Markov prediction method	This is an interval forecast, and the result obtained is a possible interval; the accuracy of the forecast is high; it is especially suitable for information with high volatility	Since it is an interval forecast, it necessarily leads to low forecast accuracy	High accuracy of short-term predictions
Gray system prediction method	This is a time series prediction method, suitable for predicting with small samples, incomplete information, and exponential growth	Poor prediction results for data with high randomness and volatility	High efficiency of short-term predictions
Artificial neural network prediction method	Nonlinear fault tolerance, self-learning, self-organizing, and self-adaptive; a variable-structure model that can constantly change its own data structure by learning from the sample, thus adapting to changes in the sample	Difficult to determine network structure; complicated algorithm; easy to fall into local minima	High accuracy of both short- and long-term predictions

9.2.1.6 Integrated Prediction Method

Since each individual prediction method has defects, the combined prediction method integrates different prediction methods in a way that gives full play to the advantages of each method and compensates for their shortcomings to achieve better prediction results.

Combined prediction refers to the application of two or more prediction models to predict a certain transmission trends of parasitic diseases using the information provided by various individual prediction models in the form of an appropriate weighted average of the combined prediction model.

9.2.2 Basic Procedures

Regardless of the prediction method used, several steps must be followed when making a prediction. These steps systematically summarize the various aspects of starting, designing, and applying a prediction. If predicting is performed on a regular basis, data should be collected regularly. The actual calculations can be performed by a computer:

1. Determine the purpose of the prediction. This step determines what we want to achieve by making the prediction.
2. Select the prediction objects. This step determines what object or variables we need to predict.
3. Decide the time horizon of the prediction. This step determines whether the prediction is to be made for the short, medium, or long term.
4. Choose the prediction model. This step is to choose a suitable prediction model for the next step according to the characteristics of the object or variables to be predicted and the nature of the prediction.
5. Collect the data needed for the prediction. When collecting the data required for the prediction, ensure that the information is accurate and reliable. All the data in the database need to be cleared before use.
6. Validation of predictive models. This step is to determine whether the model chosen is valid for the prediction we want to make.
7. Make a prediction. In this step, we make a reasonable prediction for the object or variables to be predicted based on the data collected and the prediction model we have selected.
8. Put the prediction results into practical application. In this step, we apply the prediction results in practice to achieve the goal of our prediction.

9.2.3 Points for Attention

9.2.3.1 Accuracy and Cost of Prediction

When choosing a prediction method, there is an obvious trade-off between cost and accuracy. A prediction method with high accuracy is generally more expensive to implement but can achieve a higher accuracy, i.e., a prediction value that deviates less from the actual value, thus ultimately leading to lower production and operational costs. It should be noted that, first, there is no 100% accurate prediction method, so efforts should not be wasted in attempting to achieve absolute prediction accuracy. Second, in any prediction problem, there is a minimum cost interval with reasonable accuracy.

9.2.3.2 Time Frame and Update Frequency of Prediction

Prediction is based on history and the present and oriented toward the future. The time between the present and the future is the time horizon of prediction. Different prediction methods have different time horizons, so special attention should be paid to this point when choosing a prediction method. In addition, the larger the time range is, the less accurate the predicted results. At the same time, no single prediction method can be fully applicable to a specific prediction problem, and the prediction method should be constantly tested according to the actual demand. If the prediction value deviates too much from the actual value, the prediction method should be updated.

9.2.3.3 Stability and Responsiveness

Stability and responsiveness are the two basic requirements for the prediction method. Stability refers to the ability to resist random interference and respond to stable demand. A stable prediction method can eliminate or reduce the influence of random factors and is suitable for predicting problems affected by random factors. Responsiveness refers to the ability to quickly respond to changes in demand. A good responsive prediction method can keep up with changes in actual demand in a timely manner and is suitable for problems that are less affected by random factors. Good stability and responsiveness are the main prediction goals.

9.2.4 Advances in Prediction and Early Warning

The description and prediction of parasitic diseases usually use empirical statistical methods to examine disease incidence trends by collecting previous incidence data and using simple statistical prediction models. In recent years, with the introduction of mathematical and physical results, models exploring the transmission mechanism of parasitic diseases have provided many strong theoretical bases for studies of parasitic disease transmission prediction, which emphasize the explanatory nature of the research and are much more complex than statistical prediction models, such as propagation dynamics models, complex network models, and Bayesian prediction models.

9.2.4.1 Complex Networks

Complex networks are an important tool for the study of complex systems and have recently become a hot research topic in physics and mathematics. A complex network used to study contact behavior between populations is called a social contact network (SCN). An SCN is used to study the transmission process of infectious diseases in a population by describing and analyzing contact behaviors among individuals in the population. SCN analysis has three main parameters. First is the clustering coefficient (C), which is used to describe the aggregation of nodes in the network, i.e., the closeness of the network. Second is the average path length (L), also known as the characteristic path length of the network, which describes the degree of separation of the network. Third is the degree (D), which is an important concept used to describe the properties of individual nodes; the degree distribution plays an important role in the study of parasitic transmission mechanisms.

The software commonly used for social contact network analysis is UCINET, which performs network characterization based on the constructed network topology. Based on the understanding of network topology and transmission rules, the process and characteristics of parasitic disease transmission are analyzed. The application steps are as follows: establishing the corresponding contact network through the population survey, determining the statistical properties of the population contact network structure, predicting transmission behavior in the network on the basis of the known network structural properties and formation rules, and exploring the parasitic disease control strategy on the basis of the predicted network transmission behavior characteristics.

Predicting the behavior of network systems using network models focuses on identifying critical nodes; if these nodes are infected, then the spread of parasitic diseases will be accelerated. Therefore, critical nodes play an important role in controlling the spread of parasitic diseases in a population. Nodes with a larger degree of infection will spread the disease rapidly. As a result, preventing nodes with a larger degree of infection from becoming sources of infection is the first step in preventing the rapid spread of a disease. In addition, local information around the infected nodes is used to control some of the nodes around the infected nodes to control the spread of parasitic diseases in the whole network on a large scale.

9.2.4.2 Bayesian Prediction Model

The Bayesian model is a prediction method that uses Bayesian statistics. Bayesian statistics differ from general statistics in that they not only use model information and data information but also make full use of a priori information.

The statistical prediction method of Thomas Bayes is a time series prediction method with a dynamic model as the object of study. In making statistical inferences, the general model is as follows:

A priori information + overall distribution information + sample information → posterior distribution information

Thus, the Bayesian model not only uses prior data but also adds information such as the experience and judgment of decision-makers and combines objective and subjective factors, making it more flexible for predicting the occurrence of abnormal situations.

Raso et al. (2005) used cross-sectional sampling to collect socioeconomic information on 3818 school-age children aged 6–16 years in Côte d'Ivoire and used pathogenetic methods to diagnose them with *Schistosoma mansoni* infection. He also extracted environmental indicators from remote sensing images, analyzed risk factors for *Schistosoma mansoni* infection in school-age children using Bayesian spatial models, and performed risk prediction and mapping for unsampled areas. The results showed that the prevalence of *Schistosoma mansoni* infection in school-age children was strongly related to age, sex, and household economic status, while the relationship with elevation and rainfall was relatively weak. They also indicated that the risk maps produced could provide a reference for the development and implementation of schistosomiasis control strategies.

Yang et al. (2005) collected cross-sectional survey data from 47 schistosomiasis-endemic counties in Jiangsu Province, China, from 1990 to 1998, extracted the normalized differential vegetation index (NDVI) and land surface temperature (LST) from AVHRR remote sensing data in the same period, and analyzed the data using a Bayesian spatiotemporal model. The results showed that from 1990 to 1992, the spatial autocorrelation of schistosomiasis in Jiangsu Province first decreased significantly and then gradually increased, which may have been related to large-scale praziquantel chemotherapy. The results also showed that the vegetation index was negatively correlated with *Schistosoma japonicum* infection and that surface temperature was positively correlated with it.

Kleinschmidt et al. (2002) explored the temporal and spatial patterns of malaria incidence using Bayesian spatiotemporal models from 1986 to 1999 in two small geographic areas of northern KwaZulu Natal, South Africa. The results showed spatial inconsistency (variability) in the increasing incidence of malaria, suggesting a trend toward geographic expansion of high-risk areas and a period of stability in areas with relatively high malaria incidence before that the incidence increased further. The authors concluded that small geographical incidence maps and trend maps should be smoothed to better explain the spatial distribution of disease incidence and morbidity variability. In addition, Nobre et al. (2005) conducted spatial and temporal modeling and mapping of malaria incidence in Pará State, North Region of Brazil.

9.3 Construction and Evaluation of Early-Warning Systems

9.3.1 Construction Framework

Parasitic disease prediction and early-warning systems consist of three main core modules: parasitic disease surveillance, prediction, and early warning. Parasitic disease surveillance is the basis of prediction and early-warning analysis; parasitic disease prediction can be regarded as a tool to transform data from surveillance data into information for early-warning decision-making through certain methods and models; and parasitic disease early warning is one of the purposes of parasitic disease surveillance and is an important practical application of prediction technology.

With the rapid development of mathematics, computers, and other multidisciplinary systems, the methods of parasitic disease prediction and early warning are becoming increasingly diversified, scientific, and integrated. China's research on parasitic disease prediction and early-warning systems started early, and the system has the following two main aspects.

9.3.1.1 Establishment of a Complete Parasitic Disease Surveillance System

Two main aspects of the parasitic disease surveillance system should be improved and strengthened, including surveillance methods and surveillance techniques. Disease surveillance is an important part of public health and is an important countermeasure for disease prevention and control. Parasitic disease surveillance is the basis for prediction and early warning, and its focus is on solving the problems of underreporting, misdiagnosis, timeliness, and diagnostic standardization. Making full use of surveillance information for in-depth description, analysis, evaluation, and early warning of possible parasitic disease outbreaks can support timely and effective interventions to control the extent of transmission of parasitic diseases in the population and the harm caused by them.

9.3.1.2 Selecting the Optimal Parasitic Disease Prediction Methods and Models

Parasitic disease prediction requires scientific inference and judgment of the future development trends or status of parasitic diseases based on the occurrences, development, and changes of parasitic diseases and related influencing factors. The selection of parasitic disease prediction methods and models should not only be based on the unique characteristics of the disease but also take into account the nature of the relevant data and information. The appropriateness of the prediction method and model selection will directly affect the reasonableness and accuracy of the prediction results, thus impacting the accuracy of early-warning information.

The two best times for early warning of parasitic diseases before the occurrence of an event and at the beginning of an event. The best preventive effect can be achieved by making predictions before an event occurs so that preventive measures can be taken to avoid the event, but this requires many resources (human, financial, and material) for special monitoring. The timing of early warning needs to be carefully considered to determine the threshold value of early warning. If the threshold value is low, the early-warning system can be more sensitive in early identification of abnormal events, but the specificity of the system may be reduced. Therefore, a good grasp of the optimal timing of parasitic disease early warning can improve parasitic disease prediction and early warning.

9.3.2 Indicator System

9.3.2.1 Basic Framework of the Early-Warning Indicator System

Since parasitic diseases often go through several stages in the process of occurrence and development, such as periods of exposure, incubation, atypical symptoms, and typical symptoms, the establishment of an early-warning indicator system for

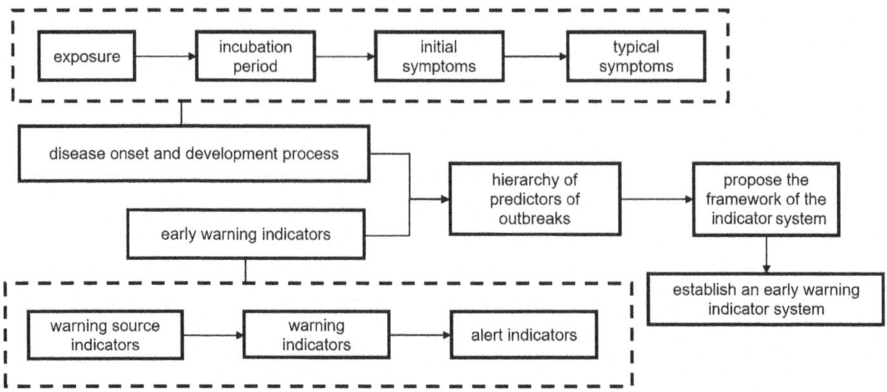

Fig. 9.1 Basic framework of the early-warning indicator system for parasitic disease outbreaks

parasitic diseases should be hierarchical. Through a literature review and expert argumentation, this section proposes a framework for the early-warning indicator system (Fig. 9.1).

9.3.2.2 Early-Warning Indicator Levels of Parasitic Disease Epidemics

Early-warning indicators of parasitic disease epidemics can be divided into three levels: indicators in the early stages of an outbreak or epidemic, indicators of atypical symptom stages, and indicators of typical symptom periods:

1. Indicators in the early stages of an outbreak or epidemic (warning source indicators). These are mainly risk factors, such as sudden changes in climate, a reduction in the immune level of the population, poor hygiene around food and water sources, and outbreaks or epidemics in other areas that are conducive to the outbreak or prevalence of infectious diseases identified by CDC, medical, or other department personnel in the outside environment. In this case, it is necessary to strengthen the monitoring of risk factors and implement targeted interventions, including immunizing susceptible populations, strengthening laboratory monitoring of pathogens, and disinfecting the external environment.
2. Indicators of atypical symptom stages (warning indicators). There are often nonspecific symptoms in the early stages of a parasitic disease outbreak or epidemic, such as an increase in the amount of certain drugs sold in pharmacies, an increase in the number of medical consultation calls, and an increase in the absences due to illness in collective units schools. All of these indicators can be considered precursors of an epidemic. At present, the development of many early-warning systems for infectious diseases in China and abroad involves the construction of symptom monitoring systems in efforts to achieve early detection of epidemics through the monitoring and data mining of certain early, nonspecific symptom indicators.

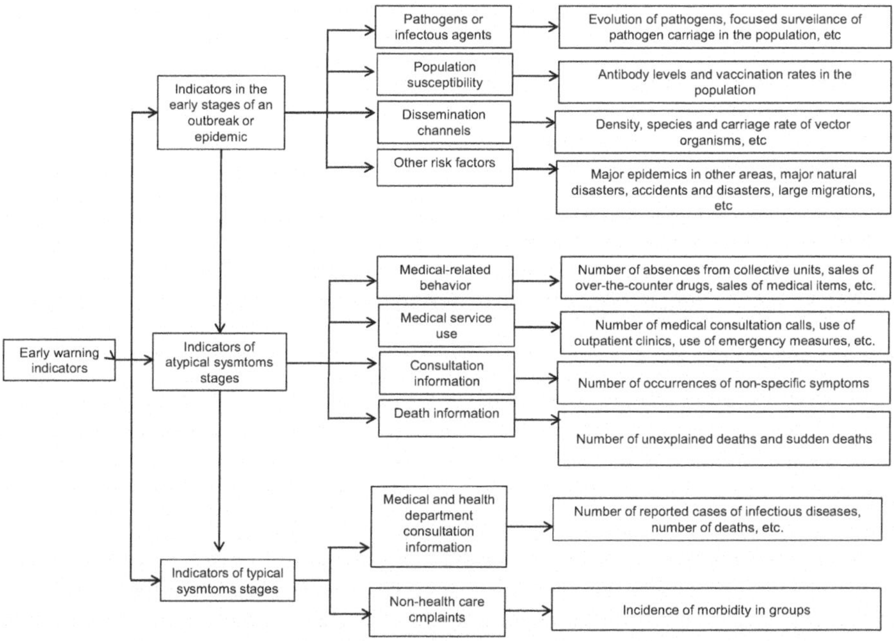

Fig. 9.2 Hierarchical map of early-warning indicators of parasitic disease outbreaks

3. Indicators of typical symptom stages (alert indicators). These refer to the risks of an epidemic for society, such as the number of reported cases of parasitic diseases, the number of deaths, and reports of mass morbidity events.

9.3.2.3 Early-Warning Indicators of an Outbreak

Early-warning indicators for an outbreak of parasitic diseases can be divided into three levels (Fig. 9.2), and specific indicators for each level are proposed:

1. Indicators in the early stages of an outbreak or epidemic. These are pathogen indicators, transmission route indicators, population susceptibility indicators, and other risk factor indicators.
2. Indicators of atypical symptom stages. These are medical-related behavioral indicators, indicators related to the use of medical services, consultation information indicators, and death information indicators.
3. Indicators of typical symptom stages. These are information on medical consultations and treatment in the medical sector and complaints and reports from non-medical sectors.

9.3.3 Systematic Evaluation

An effective evaluation of the early-warning system can improve the ability of relevant decision-making departments to use the system for decision-making. This research suggests that the evaluation of the early-warning system should include the following aspects.

9.3.3.1 Timeliness

The timeliness of the early-warning system is reflected in various aspects between the emergence of outbreak factors and public health interventions, including the timely collection, reporting, and analysis of data, timely determination in the intensity of the outbreak using early-warning methods and the issuance of corresponding warnings, the timely implementation of interventions to control the expansion of the outbreak, and the timely organization of further investigations.

9.3.3.2 Effectiveness

Evaluating the effectiveness of an early-warning system in detecting disease outbreaks and epidemics first requires determining the nature of a disease outbreak. Usually, an outbreak is determined to have occurred when the incidence level is statistically higher than the baseline incidence rate to a certain degree, but in practice, an outbreak is usually determined by the relevant authority only when the prevalence of a disease has reached a level that attracts sufficient attention from the public health community. Theoretically, effectiveness should be assessed according to the characteristics of different diseases after taking into account the statistical aspects of the sensitivity, specificity, and positive predictive value of the system, which should include the following main aspects: the degree of harm generated by an outbreak, the value of early warning and intervention, and the resources needed for intervention and further investigation. For example, when a disease outbreak generates more harm and the value of the timely detection of an outbreak is greater, higher sensitivity is required to achieve effective early warning.

9.3.3.3 Flexibility

Flexibility refers to the ability of the early-warning system to spend the least amount of time, personnel, and resources to adjust the early-warning indicator system to meet the actual needs of the actual situation in different regions and periods. The principle of this adjustment is to achieve a balance between the hazards of the outbreak, the effectiveness of intervention, and limited resources. For example, indicators can be increased or decreased and thresholds adjusted according to local conditions, and at different stages of an epidemic outbreak, data can be added or streamlined or case definitions adjusted as needed.

9.3.3.4 Acceptability

Acceptability refers to the willingness of relevant participants and authorities to contribute to the collection and analysis of data required for early warning, including whether the relevant surveillance data can be authorized for use and the formulation of relevant laws and regulations. For example, if local residents have symptoms indicative of an outbreak but lack evidence for emergency reporting, obtaining and analyzing the data at this time requires the authorization of relevant authorities.

9.4 Case Studies

9.4.1 Establishment of Schistosomiasis Surveillance and Early-Warning System in Jiangsu Province

9.4.1.1 Proposing the Issue

The surveillance of schistosomiasis in China began in the late 1980s, when some scholars began to conduct preliminary studies on the scope of surveillance and surveillance methods in areas that had reached the standard of elimination. By the early 1990s, national schistosomiasis surveillance sites had been established, and schistosomiasis surveillance had gradually become a systematic process of collecting long-term and continuous information on the dynamic distribution of the disease and its influencing factors. Additionally, annual analysis and notification of the disease had been achieved. In the beginning of the twenty-first century, with the rapid development and popularization of computer application technology, remote sensing application technology (RS), network technology, and geographic information systems (GISs), conditions were created for the rapid analysis and processing of surveillance information and timely feedback on high-risk factors to carry out early warning of schistosomiasis. However, how to use these modern high technologies to express schistosomiasis field surveillance information and experimental testing data in three-dimensional form to achieve dynamic real-time updating and off-site sharing at any time is still being explored.

9.4.1.2 Methods and Steps

Based on research on establishing schistosomiasis monitoring and early-warning databases and water infectivity monitoring methods, the real-time operation and expression platform of schistosomiasis monitoring and early-warning systems in Jiangsu Province was constructed using Google Earth map technology and global positioning technology.

The early-warning system uses Google Earth and Picasa3 image management software to establish a monitoring and early-warning site landmark library and a web album management system; import epidemic information, image information, and experimental test results from on-site monitoring into the management system; create information management documents and send them to end users; and realize online real-time querying and expression. The monitoring and early-warning system in Jiangsu Province was successfully constructed through the operation and expression platform and realized real-time querying and sharing in different places with the advantages of intuitive and clear graphic display and easy and fast operation. The total number of views of the site pictures has reached 331 since the initial application, and the information of each monitoring and early-warning site has been viewed an average of 7.36 times. The monitoring and early-warning system has played an important role in the prevention and control of schistosomiasis in Jiangsu Province (Figs. 9.3, 9.4, and 9.5).

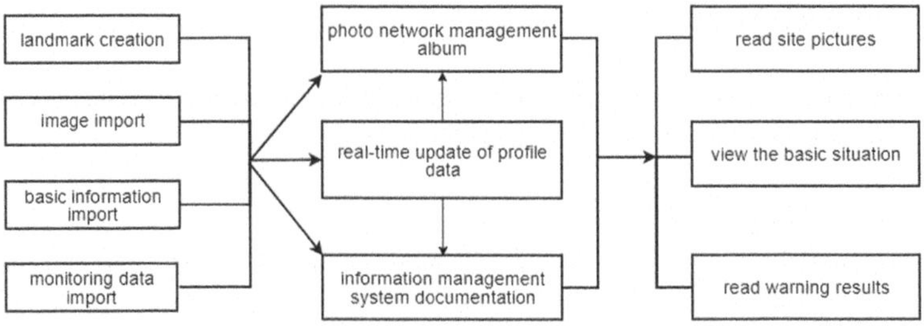

Fig. 9.3 Structure of the operating platform of the schistosomiasis monitoring and early-warning management system

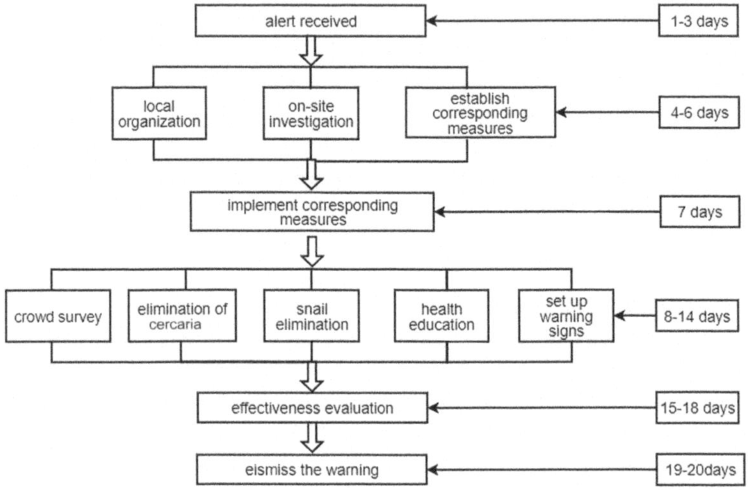

Fig. 9.4 Emergency prevention and control measures for early-warning information of schistosomiasis monitoring and early-warning management system

9.4.1.3 On-site Verification

From 2009 to 2011, Jiangsu Province carried out early-warning sentinel surveillance of schistosomiasis infection in high-risk environments and waters along the Yangtze River for 3 consecutive years. Using the sentinel larvae assay, the researchers set up 45 points along a 200 km section of the river during the May–September flood season and conducted five consecutive sentinel larvae assays each month to establish a dynamic data inventory of river water infectivity. The Google Earth early-warning platform was used to analyze the spatial and temporal distribution and environmental characteristics of the positive sentinel sites during the Yangtze River flood season. Schistosoma infectivity in the Yangtze River flood season was found to be bimodal in distribution. September was the most intensive month for schistosomiasis infectivity, July was the period of alternating generations of

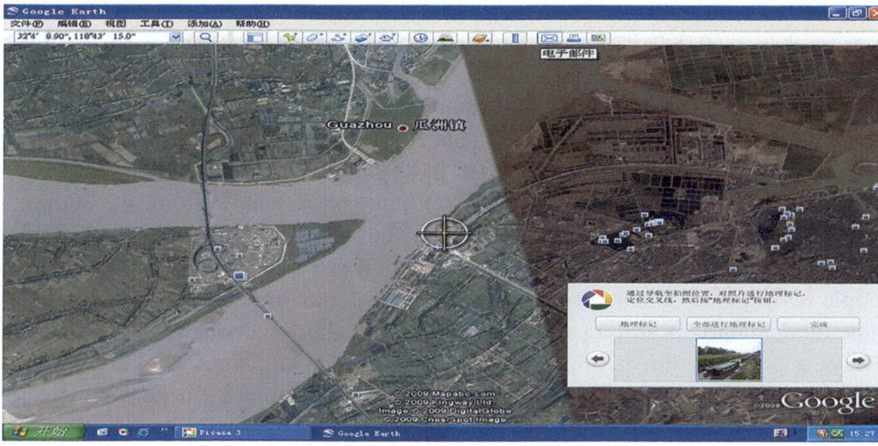

Fig. 9.5 Geographical mark of early-warning point pictures

infective snails in the riverbank area, and August to October was the key period for prevention and control of acute schistosomiasis.

9.4.1.4 Recommendations

At present, China's schistosomiasis surveillance is being fully and actively carried out. Since the purpose of active surveillance is clear, the data authenticity and timeliness are strong, and the match with on-site epidemics is high. The system can provide timely information for providing early warning of schistosomiasis. After schistosomiasis infectivity indicators are obtained, the use of modern and advanced technologies and the establishment of an ideal operational and expression platform for surveillance and early warning is one of the keys to take rapid response and real-time release of early-warning information.

9.4.2 Risk Assessment of the Malaria Elimination Program in China

9.4.2.1 Proposing the Issue

Since the founding of the People's Republic of China, China has achieved remarkable results in malaria control program under the close attention of governments at all levels. The number of malaria cases was reduced from more than 24 million in the early 1970s to tens of thousands of cases in the late 1990s, and the scope of endemic areas has been significantly reduced. With the implementation of *The 2006–2015 National Malaria Control Plan*, the central and local governments at all levels have increased their support of investment in malaria prevention and control, effectively curbing the spread of malaria. As of 2009, the number of reported malaria cases nationwide had dropped to 14,000, and the malaria incidence rate in 95% of the 24 malaria-endemic provinces (districts and cities) had dropped to less than

1/10,000, with only 87 counties (cities and districts) having a malaria incidence rate of more than 1/10,000. This signified that China's malaria prevention and control efforts were well positioned to move from control to elimination. However, there were still many difficulties and challenges in achieving malaria elimination. First, the malaria epidemic is still serious in some of local areas. Second, grassroots prevention and control capacity are weak in most medium- and low-endemic areas of China, and it is still difficult to reach the elimination standard. Third, the elimination of malaria is a brand-new goal in China, and there is no mature experience to learn from. Fourth, international communication is frequent, and the number of people leaving the country for business, tourism, and work is increasing. At the same time, malaria is still a serious problem in neighboring countries, and there is always a potential threat of local transmission from imported infection, which places more pressure on malaria elimination efforts in China. Fifth, some key technical problems around the work of malaria elimination have yet to be solved.

To this end, the project team collected information on the current status of malaria prevalence, potential transmission risk, and the institutional capacity of each endemic county in 24 malaria-endemic provinces (municipalities and districts) in 2010, quantitatively analyzed and assessed the malaria transmission risk and malaria elimination capacity on a county basis, and proposed risk areas in the process of malaria elimination in China, aiming to provide a basis for the comprehensive implementation of malaria elimination strategies.

9.4.2.2 Assessment Methods

1. Collecting information and constructing a database. Data on the malaria transmission risk and malaria elimination capacity of 24 malaria-endemic provinces (cities and districts) in China in 2010 were collected on a county basis, and a database was established. Malaria transmission risk data included malaria incidence rate and incidence of *Anopheles* vector species. The data on malaria elimination capacity included the existence of malaria elimination leading groups, whether malaria elimination plans had been issued, the amount of malaria elimination support funds, the number of malaria control personnel, the number of malaria trainers, the number of work supervisors, the number of mirror inspection stations, the number of active case monitoring sites, the coverage rate of the malaria reporting system, the rate of epidemic site investigation and disposal, and the number of copies of public education materials distributed.
2. Assessing malaria transmission risk and malaria elimination capacity. Using the database of malaria transmission risk and malaria elimination capacity, all quantitative data indicators were first averaged in terms of county population, divided into five categories with equal distance, and then assigned different rank values; all qualitative data were directly assigned different rank values according to their importance. The product of the malaria incidence rate rank value and the vector species rank value in each county was used as the malaria transmission risk index for that county. At the same time, the actual values of each indicator were converted into deviation ratio values to unify the order of magnitude by the deviation standardization process, i.e., deviation standardized value = (variable

value − minimum value)/(maximum value − minimum value). Then, different weights were assigned to each indicator of malaria elimination capacity, and the sum of the weighted rank values of each indicator of malaria elimination capacity in each county was used as the malaria elimination capacity index. Using the malaria transmission risk index and the malaria elimination capacity index as the vertical and horizontal axes, respectively, and the incidence rate in 2010 as the diameter of the circle point, a multidimensional malaria elimination feasibility analysis chart was created to quantitatively assess the malaria transmission risk and malaria elimination capacity of each county.

3. Estimating the malaria transmission risk assessment index. The malaria transmission risk index (MTI) was calculated according to the formula MTI = ATI × RTA, where ATI is the malaria risk index by *Anopheles* transmission and RTA is the regional malaria risk classification. The *Anopheles* mosquito transmission risk index refers to the malaria transmission index of the main local malaria vectors. Based on the results of previous studies, *Anopheles dirus* was assigned a malaria transmission index value of 6, *Anopheles jeyporiensis candidiasis* was assigned a value of 5, *Anopheles* (Cellia) *minimus* was assigned a value of 4, *Anopheles lesteri anthropophagus* was assigned a value of 3, *Anopheles sinensis* was assigned a value of 2, and subgenus *pseudowillmori* was assigned a value of 1. In addition, the regional malaria risk classification was based on the judgment of local malaria transmission conditions, and the country was divided into four categories based on the national malaria classification standards. The risk ratings were assigned as 10 for Class I areas, 7 for Class II areas, 4 for Class III areas, and 1 for Class IV areas.
4. Estimating the malaria elimination capacity assessment index. The malaria elimination capacity assessment index (CI) was calculated based on the formula $CI = R \times 5 + P \times 6 + M \times 4 + ES \times 8 + D \times 6 + A \times 5 + E \times 8 + PS \times 10 + F \times 9$, where R is the epidemic reporting rate, P is the 2010 intervention plan, M is the organizational management, ES is the effective system, D is supervision, A is publicity and mobilization, E is monitoring and evaluation, PS is the prevention input system, and F is financial support.

9.4.2.3 Evaluation Results

1. Information was obtained on malaria transmission risk, disease control capacity, and incidence level in each endemic county. A total of 2147 endemic counties in China were analyzed for malaria elimination risk, and a correlation graph was made to show the relations of the total capacity assessment index and transmission risk index for each county. The results showed that approximately 40% of the counties were below the mean line, which suggested that those counties still needed to strengthen their malaria elimination capacity.
2. Analysis of the correlation between malaria transmission risk, malaria control capacity assessment index, and malaria incidence. A three-dimensional analysis was created based on the correlation between malaria transmission risk, the malaria control capacity assessment index, and malaria incidence in 2147 malaria-endemic counties nationwide (Fig. 9.6). The results showed that the

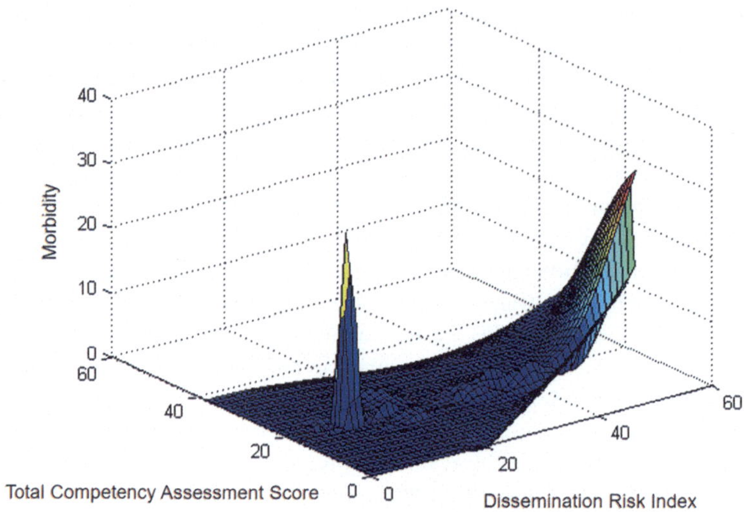

Fig. 9.6 Three-dimensional analysis of malaria transmission risk, disease control capacity, and malaria incidence level in endemic counties

transmission risk index was positively correlated with the incidence rate and the malaria control capacity assessment index was negatively correlated with the incidence rate. However, a higher incidence rate (nearly 10%) was observed in the region with a low transmission risk index (approximately 10) and a medium malaria control capacity assessment index (approximately 20), which may be related to the occurrence of imported malaria cases.

3. Comprehensive classification according to risk level. The endemic counties can be classified in four categories according to their risk level (Table 9.2), and the risk map is drawn with the four categories of endemic counties.

9.4.2.4 Recommendations

China launched a national malaria elimination program in 2010 with the aim of eliminating malaria nationwide by 2020. This goal is very timely and urgently. Since there are some weakness in block the malaria transmission in China, it is urgent to carry out malaria risk assessment on a county basis in implementing the malaria elimination action plan and to analyze and assess the possibility and feasibility of malaria elimination in each place based on gap analysis to scientifically guide the work of malaria elimination. Specifically, the risk assessment of the implementation of China's malaria elimination action plan is intended to achieve the action plan target based on the local malaria transmission risk, the level of control capacity, and related influencing factors.

Over the past 60 years, different methods have been used to classify malaria-endemic areas or risk areas in China. However, these classification methods are based mainly on vector distribution and malaria incidence for a larger qualitative classification of malaria areas in China and have failed to classify transmission risk

Table 9.2 Classification of county-level malaria risk elimination

Categories of endemic counties	Illustrations	Features	Number of counties
Category 1	Marked as red area in Fig. 9.6	These areas have high local vector transmission energy (transmission risk index of 50–60), low malaria prevention capacity (malaria prevention capacity index of approximately 20), and high incidence (>1/10,000)	Yunnan, 9; Guangxi, 5; Henan, Hunan, Hebei, Sichuan, Chongqing, and Tibet, 1 each, for a total of 20 counties
Category 2	Marked as green area in Fig. 9.6	These areas still have a high risk of malaria transmission (transmission risk index of 50–60), moderate to low malaria prevention capacity (malaria prevention capacity index of approximately 10–30), and a lower incidence rate than Category I endemic counties (<1/10,000)	Yunnan, 3; Guangxi, 2; Guizhou, 2; Shaanxi, 2; Guangdong, Jiangxi, Hubei, Sichuan, Gansu, Hebei, Fujian, and Tibet, 1 each, for a total of 16 counties
Category 3	Marked as yellow area in Fig. 9.6	These areas have a malaria transmission risk index of approximately 10 and a malaria prevention capacity index of approximately 20, but the incidence rate is close to that of Category II endemic counties	Yunnan, 15; Guizhou, 14; Hebei, 14; Sichuan, 13; Shanxi, 10; Guangxi, 9; Hunan, 9; Anhui, 9; Jiangsu, 9; Shaanxi, 9; Shandong, 9; Chongqing, 8; Gansu, 8; Jiangxi, 7; Henan, 7; Fujian, 6; Guangdong, 5; Hubei, 5; and Zhejiang, 4; for a total of 170 counties
Category 4	Marked as blue area in Fig. 9.6	Both the transmission risk index and malaria prevention capacity index in these areas are in the range of 0 to 40, but malaria incidence is significantly lower, often below Category I to III endemic counties, and in most counties, the incidence is 0 or close to 0	They are found in all 24 endemic provinces, for a total of 1941 counties

by control agencies, i.e., by administrative jurisdictions (county-based), to guide the implementation of surveillance measures for malaria elimination at the grassroots level. Based on the level of malaria incidence and the level of mosquito vector transmission capacity, this study integrated the transmission index as a basis for assessing the level of malaria transmission risk in a county, which can serve as a guide for local organizations to implement malaria prevention and control measures.

This study attempted to propose a new malaria risk analysis method based on the transmission risk and malaria control capacity of malaria-endemic counties nationwide in 2010 in the hope that it can be used for risk analysis of malaria elimination programs in China and to effectively guide the implementation of malaria elimination action plans in various endemic counties in China. Based on the analysis results, the endemic counties in China that have reached the malaria elimination target can be classified in four categories. This study took into account malaria transmission factors (e.g., infection sources, and vectors), transmission impact factors (e.g., capacity levels of local institutions), and other indicators to more objectively reflect the risks of achieving malaria elimination targets in each endemic county. It can be used to conduct annual malaria risk assessment in each county to identify factors that reduce the risk of malaria so that targeted elimination strategy measures can be taken. It can also be used to specifically guide the implementation of local malaria elimination monitoring and response measures and the evaluation of their effectiveness over a longer period.

References

Kleinschmidt I, Sharp B, Mueller I, Vounatsou P (2002) Rise in malaria incidence rates in South Africa: a small-area spatial analysis of variation in time trends. Am J Epidemiol 155(3):257–264. https://doi.org/10.1093/aje/155.3.257. PMID: 11821251

Nobre AA, Schmidt A, Lopes HF (2005) Spatio-temporal model for mapping the incidence of malaria in para. Environmetrics 16(3):291–304

Raso G, Matthys B, N'Goran EK, Tanner M, Vounatsou P, Utzinger J (2005) Spatial risk prediction and mapping of Schistosoma mansoni infections among schoolchildren living in western Côte d'Ivoire. Parasitology 131(Pt 1):97–108. https://doi.org/10.1017/s0031182005007432. PMID: 16038401

Yang GJ, Vounatsou P, Zhou XN, Tanner M, Utzinger J (2005) A Bayesian-based approach for spatio-temporal modeling of county level prevalence of Schistosoma japonicum infection in Jiangsu province, China. Int J Parasitol 35(2):155–162. https://doi.org/10.1016/j.ijpara.2004.11.002. Epub 2004 Dec 15. PMID: 15710436

Techniques for Risk Assessment 10

Shi-Zhu Li and Ying-Jun Qian

Risk assessment is an important component in public health surveillance and early warning of emergencies as well as an important means of early detection, identification, and assessment of event risks. Timely risk assessment can effectively prevent or reduce the possible impacts and harms caused by outbreak events. At present, the WHO and health agencies in many countries consider risk assessment to be an important measure for dealing with public health emergencies. In 2010, the WHO also included risk assessment as a core competency in its strategy for the prevention and control of emerging and reemerging diseases. During the Beijing Olympic Games, Shanghai World Expo, and major natural disasters, such as the Wenchuan earthquake and Yushu earthquake, China carried out many public health risk assessments and gained some experience. In 2011, in the process of preventing and responding to the outbreak of enterohemorrhagic *Escherichia coli* (EHEC) in Germany, risk assessment, as a red line, ran throughout the emergency response work, playing a very important role in gaining a timely understanding of the epidemic trend and dynamically adjusting the response strategy.

The emergency response law of China, other laws and regulations, and the twelfth 5-year plan for Chinese national economic and social development programs, such as national medium- and long-term development planning, have proposed strengthening source control, especially proposing an emergency response that adheres to the principle of putting prevention first and combining prevention with emergency response. China should establish a risk assessment system for major emergencies, make a comprehensive assessment of possible emergencies, reduce the occurrence of major emergencies, and minimize the impacts of such emergencies.

S.-Z. Li (✉) · Y.-J. Qian
National Institute of Parasitic Diseases, Chinese Center for Disease Control and Prevention (Chinese Center for Tropical Diseases Research), Shanghai, China
e-mail: lisz@nipd.chinacdc.cn; qianyj@nipd.chinacdc.cn

X.-N. Zhou (ed.), *Surveillance and Management of Parasitic Diseases in China*,
https://doi.org/10.1007/978-981-97-1241-0_10

10.1 Overview

Public health emergency risk assessment (hereinafter referred to as risk assessment) refers to the systematic identification, analysis, and evaluation of the related risks that may cause public health emergencies in order to determine the level of public health emergency risk. Risk assessment can provide an objective basis for risk management and decision-making. As the basis of emergency public health risk management, emergency public health risk assessment is an important part of monitoring and early warning. It is an important means of early detection, identification, and assessment of event risk. It is an important way for governments at various levels and administrative agencies to obtain accurate information about public health events, thus playing an important role in incident prevention and decision-making.

10.1.1 Assessment Components

Risk assessment includes the whole process of risk identification, analysis, and evaluation. Risk identification refers to the process of discovering, identifying, and describing risks. Risk analysis refers to the process of recognizing risk attributes and estimating or assigning values to the possibility and severity of consequences. Risk evaluation is the process of comparing the results of risk analysis with risk criteria to determine the risk level. In the public health risk assessment of emergencies, there may be no clear risk criteria. In this case, the risk assessment will be based mainly on the risk analysis results and the possible acceptable risk level to determine the specific risk level.

10.1.2 Risk Management

Broadly speaking, risk assessment is an important part of the risk management of public health emergencies, which includes the following main aspects (Fig. 10.1).

10.1.2.1 Risk Assessment

Conducting hazard, exposure, and environmental assessments and analysis of risk characteristics to determine the risk level of events.

10.1.2.2 Control Measure Identification

Identifying priorities based on the likelihood of control success, the feasibility of implementation, and significant consequences that may occur.

10.1.2.3 Monitoring and Evaluation

Developing and implementing monitoring and evaluation plans, establishing and improving corresponding working systems and mechanisms, and carrying out continuous monitoring and evaluation. This part also includes the information construction of the data platform; the establishment of an information system that monitors

Fig. 10.1 Risk management concepts

data collection, screening, statistics, and preliminary analysis; and the regular summary and analysis of monitoring data.

10.1.2.4 Continuous and Effective Risk Communication

Ensuring communication to gain understanding and support for action from risk managers, risk groups, and other stakeholders.

10.1.2.5 Summarizing Experience and Lessons

Summarizing experience and lessons in risk management and proposing reasonable suggestions and measures.

10.2 Methods and Implementation

10.2.1 Methods

In health emergency management, commonly used risk assessment methods can be divided into unstructured and structured methods. Unstructured methods include mainly the expert consultation method, and structured methods include mainly the Delphi expert consultation method, matrix method, and decision tree method.

10.2.1.1 Expert Consultation Method

The expert consultation method refers to evaluation in the form of expert collective discussion. This method is based on the basic theory and common steps of risk assessment. The problem is discussed by experts participating in the consultation according to the components and relevant information of the assessment combined with their knowledge and experience. Then, relevant risk assessment opinions and suggestions are put forward. The consultation organizer summarizes and organizes the expert opinions in a risk assessment report.

The advantages of this method are that organization and implementation are relatively simple and fast, different experts can exchange opinions fully, and the components considered in the evaluation may be more comprehensive. However, opinions and conclusions can easily be influenced by a small number of "authoritative" experts, and the results may vary depending on the experts involved in the evaluation.

10.2.1.2 Delphi Method

In the Delphi method, according to the determined logic framework of risk assessment, experts independently express their opinions and use a unified questionnaire to carry out multiple rounds of an expert survey. After repeated consultation, induction, and modification, basically consistent opinions are finally summarized and taken as the results of the risk assessment.

The advantages of this method are that the experts' opinions are relatively independent, the experts involved in the evaluation can come from a wide range of professional fields, and the results are less limited by time and space. However, the preparation process is complicated, and there is a long evaluation cycle that requires large personnel and material resources.

10.2.1.3 Risk Matrix Method

The risk matrix method refers to the probability of occurrence of risk factors and the severity of consequences determined by experts. Quantitative grading results on a two-dimensional matrix are calculated in a table by adopting a combination of quantitative and qualitative analysis methods. Then, the possibility of risk occurrence, the seriousness of consequences, and ultimately the risk level are determined.

The advantages of this method are that it quantifies risk, evaluates multiple risks systematically at the same time, compares different risk levels, and is convenient for decision-makers. However, the risk factors to be evaluated must be relatively certain, the experts involved in the evaluation should have a high level of understanding of the risk factors, and the number of personnel involved in the evaluation must reach a certain level.

10.2.1.4 Analytical Flow Chart Method

The analytical flow chart method refers to the establishment of a logical analysis framework of risk assessment, the adoption of a hierarchical logic judgment method, the proper classification of various situations that may be associated with the assessed problem, the categorization of risk factors for each kind of situation, the layer-by-layer measurement and discrimination of risk factors, the analysis of the possibility of occurrence and the severity of consequences of the assessed object or situation, and finally the formation of the risk assessment results.

The advantage of this method is that the related risk factors for different types of events are brought into the analysis and discrimination process in advance, and the analytical process is more logical. Once the logical framework is developed, it is easy to unify the thinking of the participants into an evaluation opinion. However, this method requires strong professionalism and logical thinking abilities in the analysis and discrimination process.

10.2.2 Assessment Components

The WHO recommends the risk matrix method for risk assessment, which includes the assessment of risk factors, exposure, and the environment, but these assessments are not always completed at the same time. Although the three aspects are evaluated separately, some of them overlap (Fig. 10.2).

10.2.2.1 Risk Factor Assessment

Risk factor assessment is the identification of the risks (or potential hazards) and possible health effects of emergencies. Health risks include biological, chemical, physical, and nuclear radiological hazards.

1. Components of risk factor assessment. These include identifying potential hazards, analyzing key risk information (i.e., risk characteristics), and ranking risks when more than one risk factor is present.

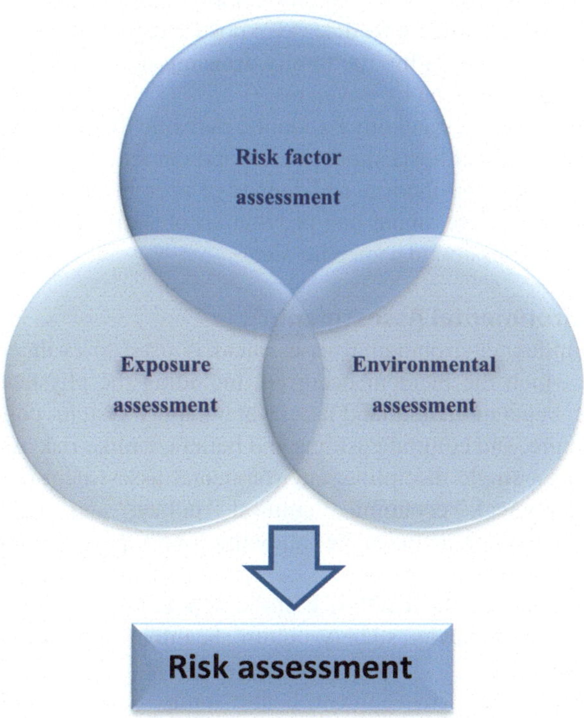

Fig. 10.2 Risk assessment components

When the pathogen is confirmed in the laboratory or the event shows clinical and epidemiological features that are easy to identify, it is easy to identify the risk factors. However, in most other cases, possible risk factors are listed only according to the description of the event, the burden of disease in the affected areas, and the types and distribution of existing risk factors. The more detailed the information in the emergency report is, the more favorable it is for determining the risk factors. With the enrichment of information, more dangerous factors will be eliminated, and they should be sorted according to the possibility that they will occur.

2. Likelihood of risk factors. This can be determined according to the clinical characteristics and natural history of the disease, the rate and trend of development, the affected area and environment, and the affected population.

10.2.2.2 Exposure Assessment

Exposure assessment is the assessment of individual and group exposure to risk factors, including the number of people known or likely to have been exposed and the number of people at risk in the exposed population (e.g., susceptibility to a disease due to lack of immunity). The answers to these questions require information on the mode of transmission, dose response, incubation period, disease lethality, estimates of potential transmission (e.g., basic reproduction number R0), and vaccination status of the exposed population.

For some exposures, such as inhalation of the heavy metal lead, Salmonella infection, or isotope exposure, dose-response relationships are critical. Additionally, exposure time is critical. Infectious disease exposure can occur in a variety of settings, such as transmission within the family, transmission through contact, sexual transmission, and transmission by travelers.

For vector-borne diseases and other zoonotic parasitic diseases, exposure assessment of vectors and animal hosts should be carried out simultaneously, and factors such as the populations, distribution, and density of vectors and animal hosts should be analyzed. Exposure assessment can provide an objective basis for judging the risk of zoonotic transmission in a specific area.

10.2.2.3 Environmental Assessment

As the name implies, environmental assessment is used to evaluate the external environment in which the incident occurred, including the physical environment (such as climate, vegetation, and land use), water supply system, population health status, infrastructure, and cultural customs and beliefs. Unlike risk assessment from the perspective of a single discipline, environmental assessment considers social, scientific, technological, economic, cultural, policy, and political aspects. Environmental assessment includes the following four important components.

1. Which factors increase the health risk of the population in terms of the environment, health status, behavior, society, culture, health infrastructure, and legal and policy framework?
2. Are there factors in the environment, health status, or social and cultural practices that can reduce the risk of population exposure?

3. What is the probability that all suspected cases will be identified?
4. What are effective preventive and therapeutic drugs and supportive therapies?

10.2.3 The Implementation Process

The steps of public health emergency risk assessment are mainly planning and preparation, risk assessment implementation, and risk assessment reporting. Planning and preparation include the determination of topics and components, the selection of assessment methods and personnel, the preparation of basic materials and assessment forms, and the formulation of the implementation plan. The implementation of risk assessment includes risk identification, risk analysis and judgment, grading risk assessment results, and proposing risk control measures and suggestions (Fig. 10.3). Risk assessment reporting includes report writing and utilization.

10.2.3.1 Planning and Preparation

1. Identification of assessment issues. The issues of risk assessment are similar to the research questions. The main considerations in the design of risk assessment

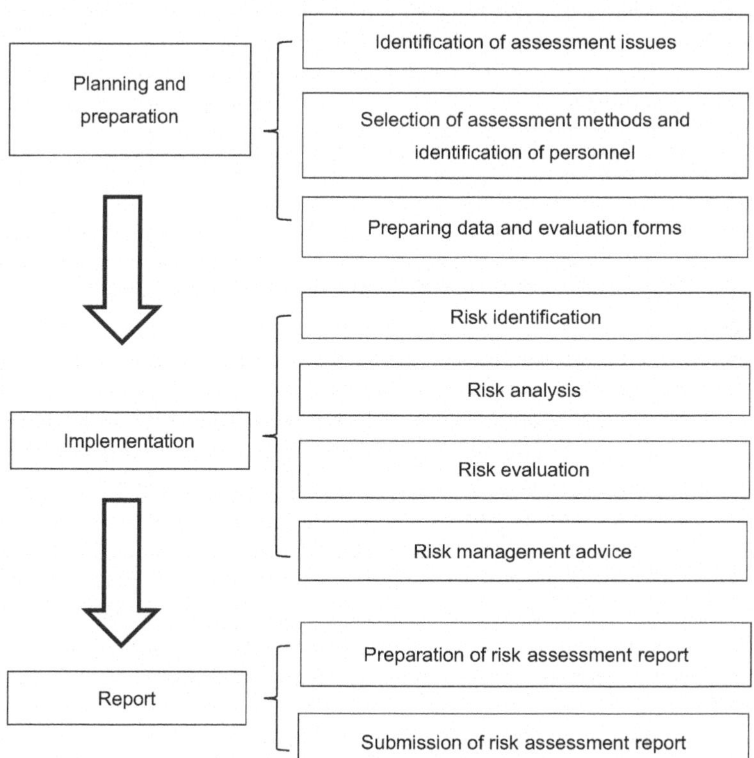

Fig. 10.3 Flow chart of public health risk assessment in emergencies

should be (1) which populations are likely to be affected, (2) possible exposure, and (3) when, why, and how the event will have adverse effects on the population due to exposure to risk factors. In the implementation of risk assessment, these problems may be affected by the following aspects: (1) the risk population; (2) the level of risk assessment, whether provincial, municipal, national, transnational, or global; (3) the inclusion of technical and policy institutions with relevant experience on the risk assessment team (e.g., for a public health emergency of unknown origin); (4) decision-makers, stakeholders, and society accepting the judged risk level (acceptable risk level); (5) the time needed to implement risk assessment; (6) the results of the previous risk assessment and similar historical circumstances; and (7) attention from outside the country (e.g., the international community).

Risk assessment cannot answer all questions at once but should prioritize issues so that critical issues are addressed in less time. The question of concern during a public health emergency is "What public health issues will arise from this event?" That is, what is the risk of exposure to a particular risk factor in a particular location or the risk of exposure to a particular population at a particular time? This problem often raises a number of questions, such as the possible consequences of exposure if no action is taken and the type and severity of public health problems arising from such events. Depending on the risks, a range of scenarios may be considered, for example, what the public health risk of the event is, whether the risk will affect large cities, and how the risk will affect other regions and countries.

The assessment can be divided into daily risk assessment and thematic risk assessment. Daily risk assessment is based on the analysis of surveillance data from different sources, and the assessment issues are determined according to abnormal changes in surveillance data, the characteristics and trends of diseases and public health emergencies, and the degree of concern of the government and the public. The sources of surveillance information usually include public health emergency surveillance systems, various disease surveillance systems, the retrieval of media information related to public health emergencies, and public health service hotlines and information bulletins.

For thematic risk assessment, the assessment themes are information about important diseases and emergencies found in daily risk assessment, information about large-scale activities and various major natural disasters and accidents, and important assessment themes designated by health administrative agencies.

2. Selection of assessment methods and identification of personnel. The risk assessment team decides on the key issues to investigate, which helps define the scope of assessment and ensures that all relevant information is gathered. Clearly defined questions help determine which assessments to prioritize, including literature review, epidemiological investigations, surveillance, counseling, and research.

   Appropriate risk assessment methods should be selected according to the subject and purpose of risk assessment. The expert consultation method is used mostly in daily risk assessment. One or more of the Delphi method, risk matrix method, and analytical flow chart method can be used in thematic risk assessment, as can the expert consultation method or other methods. The number of participants and requirements should be determined according to the purpose, field, and method of assessment.

The participants in daily risk assessment are usually epidemiologic professionals engaged in the surveillance and analysis of public health emergencies, related disease surveillance, and prevention and control. Laboratory testing professionals can also be invited to participate if necessary. In addition, when assessing specific public health emergencies, participants should be selected based on the focus of the assessment agenda. In principle, participants in thematic risk assessment should come from different professional fields related to the theme and have a high level of authority in their professional field. When necessary, relevant experts from outside the health system should be invited to participate, and the number of experts should meet the requirements of the method used.

After confirming the authenticity of an emergency and its possible public health risk, the importance of the event to public health must be identified. The risk assessment team should be established on the basis of the quality and completeness of information. It is essential to establish a risk assessment team of experts with specific professional backgrounds, such as toxicology, animal health, food safety, and radiation protection. They should be included at the outset when the following conditions are met: (1) harm unknown, (2) not caused by an infectious agent, (3) animal disease or death and/or suspected zoonosis, and (4) recall of food or other products due to chemical accidents or radionuclides.

Business and risk communications are integral parts of risk management. A communication mechanism should be established between risk assessment and risk communication. If possible, a risk communication specialist should be a member of the risk assessment team. Ensuring good communication between policy-makers and affected populations can improve the effective implementation of control measures, especially in regard to behavioral changes.

The knowledge and expertise of the team can greatly influence the risk assessment. An understanding of the emergency environment is an important part of risk assessment. The degree of risk of a serious public health event depends on the social, economic, environmental, and political conditions in the affected area and the effectiveness of local health services (such as treatment and public health services). For some hazards, the effectiveness of linkages between health services and other sectoral institutions, such as zoonotic animal health departments, can also influence risk and must be assessed.

3. Preparation of data and evaluation forms. Prior to formal risk assessment, a preliminary analysis of surveillance data should be completed, and relevant studies should be collected. For example, the information that may be involved in the risk assessment of infectious diseases includes pathogenicity, transmission rules, population vulnerability, public concern, emergency response capacity, and available resources. When conducting risk assessments for large-scale events, natural disasters and accidents, data on the natural environment, population

characteristics, health knowledge and behavior, and health-related background information should also be collected according to the characteristics of the theme.

Risk assessment forms, such as the expert questionnaire used in the Delphi method, should be designed and developed according to the risk assessment issues and methods used.

10.2.3.2 Implementation

1. Risk identification. Risk identification refers to the process of identifying and describing risk factors. Only by well-conducted risk identification can risk factors be correctly analyzed, public health risks better assessed, and health emergency countermeasures formulated.
   (a) Risk identification in daily risk assessment: In daily risk assessment, risk identification is often combined with the determination of assessment issues, and the determination process of assessment issues is the preliminary preparation for the implementation of risk assessment. In daily risk assessment, it is very important to determine the key assessment issues. First, the daily risk assessment, especially the comprehensive assessment of the risk of various public health emergencies carried out regularly on a monthly or weekly basis, needs to determine the key issues of the assessment on the basis of comprehensive analysis to ensure the effectiveness of the assessment and ensure its efficiency and pertinence. Second, the assessment components and results of each daily risk assessment may not only have a certain continuity and repeatability but also differ due to seasonal factors, related events, and risk factors. Therefore, risk assessment issues must be redefined before each assessment. In the process of expert consultation and specific assessment, the identified key assessment issues or the comprehensiveness and rationality of the identified risks can be further reviewed, confirmed, and supplemented.

       Daily risk assessment is used to identify the risks of public health emergencies such as infectious diseases, food poisoning, occupational poisoning, and environmental pollution as well as events such as natural disasters, accidents, and large-scale events. The key issues to be included in the assessment should be determined based on the analysis of various types of relevant surveillance information. For example, infectious disease assessment should focus on the following aspects: Class A infectious diseases managed according to requirements of Class A; clustered epidemic or outbreak; the infectious diseases with obvious abnormal distribution or pathogen monitoring; many cases of death or severe infectious diseases with epidemiological connection; rare, new, or imported infectious diseases; infectious diseases that had previously been eliminated; and mass unexplained diseases.
   (b) Risk identification in thematic assessment: The focus is on enumerating and describing the risk elements involved in the assessment theme. The thematic risk assessment of major public health emergencies should focus on categorizing and describing the key information about the event, such as its back-

ground, characteristics, causes, susceptible and high-risk groups, potential consequences, available prevention and control measures, and their effectiveness. For example, when assessing the risk of the EHEC O104:H4 outbreak in Germany, the focus should be on describing the time and place of the event, the infected population, the pathogen, and the characteristics of the disease (the severity of the disease and the mode of transmission), the import and sale of suspected contaminated food in China, and monitoring, treatment, and prevention and control capacity.

For the thematic risk assessment of large-scale activities, the following components should be emphasized: (1) the characteristics of large-scale activities, such as time, location, scale, main activity components and form, number of participants, living environment, and susceptibility; (2) the occurrence of various public health emergencies, such as the types and intensity of infectious diseases, the types and incidence of poisoning, and the occurrence of heatstroke or freezing disasters; (3) imported diseases or other health hazards that may be spread during large-scale events; (4) other public health risks of emergencies that may occur during large-scale events, such as terrorist incidents, natural disasters, accidents, and disasters; and (5) existing health security capacity and measures, such as surveillance capacity, treatment capacity, prevention and control capacity, diet and drinking water security level, and population immunity level. On the basis of these characteristics and related information, various potential public health risks can be listed and described.

2. Risk analysis. Risk analysis is the process of recognizing risk attributes and determining risk level, that is, the risk level of risk elements can be determined by analyzing and comparing relevant data to determine the possibility of risk occurrence, the severity of consequences, and vulnerability.

   For daily risk assessment, the emphasis of the analysis varies with the type of event. For the risk analysis of public health emergencies of infectious diseases, the clinical and epidemiological characteristics (virulence, transmission, seasonality, region, transmission routes, high-risk groups, etc.), demographic characteristics, population susceptibility, impact on the government and the public, population tolerance of risk and government response ability, etc. should be taken into account. In the risk analysis of non-disease-related public health emergencies, such as accidental injury, poisoning, and terrorist events, the nature of the event, the affected area, the severity of the impact on people's health and society, the public's psychological endurance and the government's response ability should be considered comprehensively.

   For thematic risk assessment of large-scale events, natural disasters, and accident disasters, experts can be organized to conduct a qualitative or quantitative analysis of the likelihood of risk occurrence, severity of consequences, and vulnerability.

   (a) Occurrence possibility analysis: The occurrence possibility of infectious diseases, poisoning, accidental injuries, and other secondary and derived public health risks caused by large-scale activities, natural disasters, and accidents and disasters can be analyzed in combination with the event background, monitoring information, historical events, and the resulting harms. The possibility of occurrence can be divided into five grades: very low, low, medium, high, and extremely high, and assigned according to need.

(b) Consequence severity analysis: The severity analysis of the public health consequences of large-scale events, natural disasters, and accidents can be considered on the basis of the geographical scope of risk impact, the number of people affected, the economic losses caused, the severity of the impact on people's health, the degree of damage to important infrastructure or the ecological environmental system, the impact on social stability and government credibility, and the psychological pressure on the public. For large-scale activities, the possible negative impact of risks on the smooth running of the activity should also be considered. The severity of the consequences can be divided into five grades: extremely low, low, medium, high, and extremely high, and assigned according to need (e.g., corresponding to 1–5 points, respectively).

(c) Vulnerability analysis: The analysis of vulnerability to large-scale events, natural disasters, and accidents includes the analysis of risk tolerance and risk control ability. It can be considered in terms of the aspects of population susceptibility, public psychological tolerance, public health awareness, self-rescue and mutual rescue ability, medical rescue ability, technical reserve, health resources and expansion capacity, public health infrastructure, drinking water, food supply, health emergency capacity, etc. Vulnerability can be divided into five grades according to its extent, very low, low, medium, high, and extremely high, and the value can be assigned according to the need (e.g., corresponding to 1–5 points, respectively).

3. Risk assessment. Risk assessment is the process of comparing the results of risk analysis with risk criteria to determine the risk level. In the public health risk assessment of emergencies, there may not be clear risk criteria. In this case, the risk assessment will be based mainly on the risk analysis results and the possible risk level of the comparison to determine the specific risk level. For example, the risk is divided into five levels: very low, low, medium, high, and extremely high. Risks that are rare, have almost no potential impact, and have low vulnerability are classified as extremely low; risks that cannot easily occur, with little potential impact and low vulnerability, are classified as low; risks between high level and low level are set as medium; risks that can easily occur, with great potential impact and high vulnerability, are defined as high; and risks that can extremely easily occur, with great potential impact and great vulnerability, are defined as extremely high. The risk level can also be determined according to the risk assignment result.

   Daily risk assessment mostly adopts the expert meeting method, and the risk level is generally determined by experts after comprehensive analysis and evaluation based on their professional experience, historical monitoring data, and other relevant data.

   If the risk matrix method is adopted, the possibility and severity of the consequences of each risk can be scored, and the risk score of each risk can be calculated. According to the risk score, the risk level is classified and determined. If the analytical flow chart method is adopted, each risk factor can be selected and judged based on the predetermined analytical process and the comprehensive collection, summary, and analysis of relevant information to the greatest extent possible; then, the risk level can be determined intuitively.

4. Risk management advice. According to the risk level and controllability, existing problems and weak links are analyzed, and the risk control strategy is determined. According to the principles of effectiveness, feasibility, and economy, suggestions for early warning, risk communication, and control measures are put forward with the aim of reducing the possibility of risk occurrence and mitigating risk hazards.

10.2.3.3 Reporting

The Centers for Disease Control and Prevention (CDCs) at county levels should be responsible for writing daily risk assessment reports and submitting them to health administrative agencies at the appropriate levels and to the CDC at a higher level. Daily risk assessment focuses on the analysis, assessment, and risk levels of recent events or risks that should be considered within the jurisdiction and includes targeted risk control measures and suggestions. The assessment report should include an introduction, event and risk level, and risk management suggestions. The introduction briefly introduces the components, methods, and main conclusions of the evaluation. In the event and risk level part, the risk level and main evaluation basis of the identified key events or risks are explained, and the occurrence risk and development trend of the events can be described in detail when necessary.

The risk management suggestion part puts forward suggestions for early warning, risk communication, and control measures. According to the needs, it also proposes topics that need special risk assessment.

CDCs at all levels that carry out thematic risk assessment are responsible for organizing the writing of thematic risk assessment reports, submitting them to the appropriate health administrative departments, and informing other relevant CDCs. The main components of the thematic risk assessment report are assessment events and their background, purpose of the report, methods, conclusions and evaluation basis, and risk management suggestions. When necessary, the health administrative agency should submit the relevant evaluation results to the government at the appropriate level, inform the relevant departments and relevant medical and health institutions, and release the results to the public.

CDCs at all levels shall submit the completed risk assessment report to the health administrative department at the appropriate level and the CDC at a higher level in a timely manner and notify relevant medical and health institutions as needed.

10.3 Matters Needing Attention

10.3.1 Risk Assessment Categories and Selection of Methods

Public health emergency risk assessment is divided into routine risk assessment and thematic risk assessment.

10.3.1.1 Daily Risk Assessment

Daily risk assessment is used mainly to analyze all kinds of routinely collected information of public health emergencies, identify potential public health emergencies or public health threats through expert consultation and other methods, carry out preliminary and rapid risk analysis and evaluation, and put forward risk management suggestions. Themes for thematic risk assessment can be identified as needed.

The daily risk assessment mentioned in this program refers to regularly performed risk assessment, which currently is mainly monthly risk assessment. As risk assessment progresses, the frequency should gradually increase. If conditions permit, risk assessment should be carried out daily or at any time for public health emergencies, and related information should be monitored on a daily basis. This kind of risk assessment can be conducted simply by holding a small roundtable meeting or video meeting. The assessment results should be integrated into the analysis of routine outbreak and public health emergency surveillance data. When an assessment identifies an event or relevant information on a potential public health event, a thematic risk assessment should be carried out immediately.

10.3.1.2 Thematic Risk Assessment

Thematic risk assessment is a comprehensive and in-depth special public health risk assessment for major public health emergencies, large-scale events, natural disasters, and accidents and disasters in China and abroad. Thematic risk assessment can be carried out dynamically at different stages before, during, and after an event according to the acquisition of relevant information and changes in the situation, risk duration, etc. A different assessment method is selected for each risk assessment based on factors such as available time, information and resources, and primary assessment objectives. Specific situations include the following:

1. Public health emergencies. Such assessments can be carried out dynamically at different stages of the occurrence and development of the event according to the characteristics of the event and information acquisition.
   (a) Public health emergencies occurring in foreign countries that may pose public health hazards to China
   (b) Public health emergencies occurring in China that may cause public health hazards within their jurisdiction
   (c) Risks found in routine risk assessments that may lead to major public health emergencies
2. Large-scale events. Such assessments can be carried out dynamically at different stages of event preparation and staging.
   (a) Large-scale crowd-gathering activities involving multiple countries, provinces, or cities and lasting for a long time, such as large-scale sports events, trade fairs, and exhibitions
   (b) Large-scale activities for which evaluation is required by the host or local people's government.

3. Natural disasters and accidents. After a forecast of a major natural disaster or after the occurrence of a major natural disaster or accident, timely risk assessment should be conducted in response to the primary, secondary, and other derived public health hazards that may result from the event. Such assessments can be carried out dynamically at different stages before or after a disaster as needed.

10.3.2 Risk Feature Description

Risk levels are determined after risk factor, exposure, and environmental assessments are completed. The process is known as risk characterization. If there is no mathematical model or criteria, they can be graded according to team opinion.

The figure below is the risk matrix. The ordinate represents the possibility of risk occurrence, and the abscissa represents the magnitude of risk. Since most risk assessments are qualitative analyses, risk levels are not determined by quantitative values but are based on the description of risks and possible consequences. Possibility definitions and consequences can be refined according to the scope of assessment. In practice, there are two risk matrices to choose from. In addition to expected morbidity, mortality, and immediate long-term health effects, the panel should consider a variety of other possible risks.

The risk matrix can also reflect changes in risks before and after control measures are implemented. In some cases, this approach is not necessary when there is little information and obvious risk (Fig. 10.4).

10.3.3 Quantitative Risk Assessment

Quantitative risk assessment depends on the level of information and the time and complexity of risk assessment. In some disciplines, such as engineering, risk

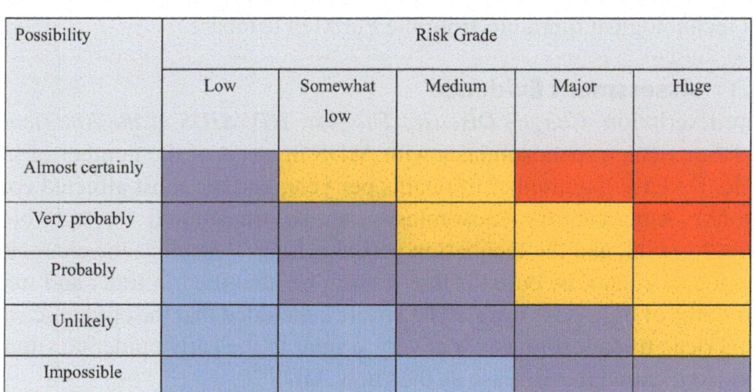

Fig. 10.4 Risk matrix

quantification is feasible. However, qualitative assessment may be the only way to assess the risk of a public health emergency, especially when there is insufficient information. Biological risk assessments take longer, and it is not possible to obtain quantifiable data at every stage. In practice, a comprehensive approach to risk assessment is usually adopted. Quantitative assessment can be carried out when sufficient data are available, and qualitative assessment can be carried out otherwise. It should be noted that quantitative risk assessment with poor data quality or rigorous methods is far less scientific and convincing than qualitative assessment with a reasonable design and well-organized methods.

10.4 Case Analysis

10.4.1 Risk Assessment of the Impact of American Trypanosomiasis Transmission in China

10.4.1.1 Proposing the Issue

On May 29, 2012, an article entitled *Chagas Disease: The New HIV/AIDS of the Americas*, stating that American trypanosomiasis was spreading in the American continents, was published in PLoS Neglected Tropical Diseases magazine. This disease is called the "new AIDS" because it is similar to the early stage of the AIDS epidemic in terms of clinical symptoms, such as hidden onset, blood and mother-to-child transmission, high cost of treatment, and heavy burden of disease. The article attracted wide attention from domestic and foreign media. An evaluation was carried out specifically to evaluate the risk of importing the American trypanosomiasis epidemic into China and the possibility of further spread.

10.4.1.2 Assessment Methods

On the basis of a literature review, a qualitative analysis was carried out using the expert consultation method. The evaluation data and information sources included the WHO, CDC, Chinese Ministry of Health, CDC website information, and scientific and technological literature from the PubMed database.

10.4.1.3 Assessment Evidence

1. Event description. *Chagas Disease: The New HIV/AIDS of the Americas* compared American trypanosomiasis with AIDS in terms of the number of infected people, DALYs, the number of deaths per year, and the most affected countries (regions). American trypanosomiasis can be transmitted through blood or mother-to-child, and the incubation period is long. It also has the characteristics that patients cannot be cured if they cannot be identified in time, and treatment is 3 months of high-dose drugs. The article concluded that the epidemic situation of American trypanosomiasis was very similar to the early epidemic situation of AIDS, so it called this disease as the "new AIDS."

   After the article was published, some experts objected to its conclusions. For example, Tarleton, a famous professor at the University of California, USA, who specializes in American trypanosomiasis, believed that the author of this article was sensationalizing American trypanosomiasis by equating it with AIDS and refuted this comparison with examples: "an untreated HIV-infected person can basically be regarded as being sentenced to death, but most American trypanosomiasis patients (70%~80%) can survive for decades or even longer without treatment"; "American trypanosomiasis can be cured through scientific and complete treatment, but AIDS is currently incurable"; and "American trypanosomiasis is mainly transmitted through the bite of stinkbug, and almost all American cases are infected in Mexico and Latin America and returned to United States."

2. Disease features. American trypanosomiasis, also known as Chagas disease, is a zoonosis caused by *Trypanosoma cruzi*, with the vectors of *Panstrongylus megistus*, *Triatoma infestans*, and *Rhodnius prolixus* distributed mainly in Central and South America.
   (a) Epidemiology: According to WHO data and research reports, there are approximately ten million infected people worldwide, distributed mainly in Latin America. However, in recent years, cases have been reported in the United States, Canada, many European countries (Spain, Switzerland), and some Western Pacific countries (Australia, Japan), with the number of infected people in the United States exceeding 300,000. The WHO reported that in 2008, American trypanosomiasis caused approximately 10,000 deaths.
   (b) Sources of infection and host animals: Patients infected with *Trypanosoma cruzi* are the main source of infection, and dogs, cats, South American armadillos, bats, ferrets, foxes, opossums, anteaters, squirrels, and monkeys can be used as preservation hosts.
   (c) Transmission route: The disease is transmitted mainly by patients or preserved hosts → stinkbug → people, and more than 80% of human infections are transmitted by stinkbug. *Trypanosoma cruzi*-infected stinkbugs bite humans at night and excrete feces containing *Trypanosoma cruzi* at the same time as or after sucking blood. *Trypanosoma cruzi* invades the human body through bite wounds or skin wounds, eyes, mouth, and nose mucosa (it is not clear whether it can invade the human body through intact skin). In addition, the disease can be transmitted through ingestion of food contaminated by stinkbug feces, blood transfusion, mother-to-child vertical transmission, organ transplantation, and laboratory accidents.
   (d) Population susceptibility: The population is generally susceptible. People living in poor sanitation conditions are more susceptible to triatomine exposure due to more exposure opportunities. The epidemic has occurred mainly among rural, suburban, and urban residents living in poor conditions.
   (e) Clinical manifestations: The incubation period of the disease is 1–3 weeks. The acute phase after infection lasts approximately 2 months, and *Trypanosoma cruzi* propagates in large numbers in the blood. Most cases are asymptomatic or mild, with fever, headache, lymphadenopathy, muscle pain, dyspnea, edema, stomach discomfort, or chest pain. Approximately 50% of patients have broken skin and a swollen eyelid in one eye. Approximately 20–30% of patients will progress to the chronic stage.

Parasites in patients' bodies parasitize the smooth muscle of the heart and digestive tract. More than 30% of patients suffer from heart damage, and more than 10% suffer from digestive tract damage, such as esophagus and colon, nervous system, and multiple organ damage. Patients with advanced myocardial progressive damage experience heart failure or sudden death.

(f) Prevention and control measures: In the acute stage, patients should be treated with antitrypanosomiasis drugs as early as possible. Benznidazole or nifurtimox is effective.

There is no vaccine. Preventing triatomine bites and cutting off the transmission route are the key steps in preventing and controlling the disease. Prevention and control measures recommended by the WHO include spraying insecticides near homes to kill as many triatomines as possible; retrofitting housing to eliminate triatomine breeding sites in homes; strengthening personal protection, such as the use of mosquito nets; ensuring hygiene in food preparation, transportation, storage, and consumption; screening the blood of blood, organ, tissue, and cell donors; and screening newborns of positive pregnant women and siblings of infected children to provide early diagnosis and treatment.

10.4.1.4 Assessment Conclusion

1. The population is generally susceptible. Chinese citizens in the Americas are at risk of infection if they are bitten by triatomine bugs, eat unclean food, or undergo unsafe blood transfusions.
2. There is a risk that American trypanosomiasis will be imported into China due to the international movements of people. The risk level needs to be further evaluated according to the specific situation of people entering the Americas (such as the number of people traveling to countries where the disease exists and the chance of contact with stinkbugs infected with trypanosomiasis). There has not yet been a case of American trypanosomiasis in China, and the disease has no specific typical clinical manifestations, so doctors have insufficient ability to detect and diagnose it in a timely manner.
3. The possibility of transmission of American trypanosomiasis caused by imported cases is low. Although dermatitis caused by triatomine bugs has been reported in Southern China, such as in Hainan Province, China is not a natural focus of trypanosomiasis. Therefore, the likelihood of further spread of the disease from imported cases is low.

10.4.1.5 Measures and Recommendations

The China CDC will continue to pay attention to the development trend of American trypanosomiasis and conduct further risk assessments if necessary. It is suggested that tourism, businesses, and other relevant organizations and large-scale employers carry out effective risk communication for people traveling to the Americas and carry out key education on how to prevent insect bites; in addition, travelers should be advised to seek medical advice promptly when they have symptoms such as fever, headache, and unilateral eyelid swelling. For those returning from the

Americas, if the above symptoms occur within 1–3 weeks, they should seek medical advice promptly and provide information about their travel history. Relevant agencies should properly reserve benznidazole and nifurtimox. When a patient has a history of having been in the Americas and experiences symptoms such as fever, headache, swollen lymph nodes, and swollen unilateral eyelids, the medical institution should attach great importance to the case, report it to the local CDC or health administrative department in a timely manner, and send specimens to the NIPD China CDC for testing. The China CDC will report relevant information on its website for medical and health institutions at all levels.

10.4.1.6 Recommendations

With the increase in public health emergencies, sources of information and intelligence are increasingly extensive. However, in view of the unreliability of personal knowledge, experience, alertness, attention, and judgment, information and intelligence that have not been filtered, screened, preliminarily verified, and evaluated cannot be directly provided to relevant decision-makers. Therefore, it is necessary to establish a relevant risk assessment system and carry out risk assessment. Through risk assessment, public health events can be closely monitored so that the relevant departments can remain alert at all times, and information can be screened and filtered to promptly identify risks. On this basis, suggestions for early warning and emergency response mechanisms can be put forward, and the priority of response measures determined. By reducing uncertainty and information insufficiency, disease risk assessment can also help decision-makers in making coping decisions.

10.4.2 Rapid Risk Assessment of Schistosomiasis After an Earthquake in Mianzhu City, Sichuan Province

10.4.2.1 Proposing the Issue

On May 12, 2008, a devastating earthquake struck Sichuan Province. Mianzhu City, located in the northwestern Sichuan Basin and upstream of the Tuojiang River in the middle of the Longmen Mountains, was particularly hard hit. Historically, this city was one of the severe schistosomiasis-endemic areas in Sichuan Province. After 28 years of persistent control after initiation of national schistosomiasis control program in 1950s, it reached the national standard for schistosomiasis transmission control in 1984. At present, the schistosomiasis situation in the city is basically stable. No acute cases of schistosomiasis, new cases of schistosomiasis in children under 12 years old, or infected snails have been found in the city for many years. However, there are still a small number of patients with chronic and advanced schistosomiasis, and the *Oncomelania* snail is still distributed in a series of hilly regions with features of wide distribution, large area, numerous sites, and complex environment. Mianzhu suffered serious losses in the “5.12” earthquake. The natural ecological environment, including snail breeding areas, as well as residents’ living environment and living conditions and basic social health facilities, such as those providing disease prevention and control, was seriously damaged and affected. At the same time, to provide earthquake relief, a large number of rescue workers and volunteers from all over the country entered the local area. These changes in natural and social factors increased the potential risk of a schistosomiasis transmission in the region. To understand the impact of the snail breeding environment and changes in the production modes and lifestyle of residents on schistosomiasis prevalence in Mianzhu City after the earthquake, the former National Bureau of Disease Prevention and Control issued the *Notice on Carrying out Rapid Assessment of Schistosomiasis Transmission Risk after Earthquake in Key Endemic Areas of Sichuan Province* and *Implementation Plan of Rapid Assessment of Schistosomiasis Risk after Earthquake in Mianzhu and Three Other Cities of Sichuan Province*, requiring rapid assessment of schistosomiasis transmission risk in earthquake-stricken areas.

10.4.2.2 Assessment Methods

1. Organization of assessment. From July 19 to July 25, 2008, the rapid assessment team of schistosomiasis risk in Mianzhu City, Sichuan Province, was formed by schistosomiasis prevention professionals from the NIPD at China CDC, Mianzhu Medical and Health Corps of Jiangsu Province, the Institute of Parasitic Diseases of Sichuan CDC, Mianzhu Institute of Schistosomiasis Control, and other units.
2. Scope of assessment. According to the historical epidemic data of schistosomiasis in Mianzhu City and the preliminary field investigation, resettlement sites of Ganquan village in Qitian Town, Tongma village in Qifu Town, Shuiyue village in Qingdao Town, and the No. 2 overpass were selected as the evaluation survey sites.
3. Components and Methods of Assessment
   (a) Snail survey. All historical snail environments around the three evaluation villages and one residential settlement were investigated by systematic sampling combined with environmental sampling, and the distribution and infection rates of snails were investigated. Every village survey contained no fewer than 1000 frames that is $0.11 m^2$ in size, and all snails in the frames were captured. All captured snails, living and dead, were identified and then examined for schistosome infection. For environments with snails, it was necessary to count the density of snails and infected snails, measure the snail distribution area, record the latitude and longitude, and draw a snail distribution map. At the same time, the scour area under the dammed lake was investigated to determine whether there were signs of snail spread.
   (b) Livestock feces survey. This survey was carried out around three evaluation villages and one residential settlement. During the snail survey, all livestock feces in the environment were collected, the types of feces were determined, and the average density of feces was calculated. All fresh livestock feces were collected and examined by the egg collection hatching method to determine the positive status of different kinds of feces.
   (c) Questionnaire survey. Questionnaires were used to investigate the history of schistosomiasis, water contact with epidemic diseases, and related health behaviors of local residents and outsiders in the three assessment villages and one residential settlement. The number of people at each assessment point was at least 50.

(d) Survey of other relevant factors. Losses of schistosomiasis control institutions and teams: An investigation was conducted to evaluate changes in professional schistosomiasis control institutions and teams in cities, townships (towns), and other places and damage to all kinds of data collection sources for schistosomiasis control at all levels and to evaluate their influence on schistosomiasis control after the earthquake. Damage to schistosomiasis control projects: An investigation was conducted to evaluate damage to original schistosomiasis control projects, such as agriculture and water conservancy in the selected evaluation villages, and to record the names, scope, and damage degree of each schistosomiasis control project as well as whether the damage would cause snail diffusion.

(e) Quality control and data processing. The NIPD at China CDC was responsible for organizing an expert group to train technical personnel participating in the evaluation and investigation of the snail distribution, disease prevalence, questionnaire, and other investigation techniques, unify the investigation methods, clarify the quality requirements, and appoint a person to be responsible for collecting, checking, and collating the investigation statistics and evaluation data. All investigation and evaluation data were entered into the computer in a unified way. A database was established, and relevant statistical analysis was carried out.

10.4.2.3 Assessment Results

1. Basic Situation. Mianzhu has jurisdiction over 21 townships, 263 administrative villages, and 46 community resident committees, with a total population of more than 500,000. With the exception of Qingping and Tianchi mountain towns, the remaining 19 townships are schistosomiasis-endemic areas. According to data from before 1964, the average infection rate of residents in the county was as high as 26.12%, with the highest township reaching a level of more than 50%, and the average infection rate of cattle in key areas was as high as 47.5%. To date, the city has a total area of 12.89 million square meters of snails, 10,300 people with schistosomiasis, 831 people with advanced schistosomiasis, and more than 40,000 infected cattle.

   The city reached the national standard for schistosomiasis transmission control in 1984. According to the data from 2006, living snails were found in 214 administrative villages, 1246 villager groups, and 6413 environments in the city's 19 endemic towns, and the area of snails was found to be 4,097,200 m^2. Snail villages accounted for 90.7% of the historical endemic villages, and the actual snail area accounted for 31.8% of the historical snail area. In 2007, the city still had a snail area of 2,894,200 m^2.

   From 1997 to 2006, more than two million live snails were dissected in the city, but no infected snails were found. More than 1.6 million people were monitored for schistosomiasis. No any case of acute schistosomiasis or no any case in children under 12 years of age were found. One acute case imported from Hubei Province was identified in 2007.

At the evaluation sites, Ganquan Village of Qitian Town, Tongma Village of Qifu Town, Shuiyue Village of Qingdao Town, and No. 2 overpass residents' settlement site of Shuangsheng Village, the historical snail areas were 47,700 m^2, 62,600 m^2, 201,900 m^2, and 34,900 m^2, respectively. In 2007, the snail areas were found to be 21,300 m^2, 37,400 m^2, 62,300 m^2, and 1700 m^2, respectively, and the positive rate of blood tests was 0.86–2.56% before the earthquake in 2008. No any fecal test was positive (Table 10.1).

2. Snail Survey

A total of 178 blocks with 79,200 m^2 of suspicious environment were randomly examined in the three villages and one residential settlement. A total of 165 plots with snails were found, accounting for 92.70% of the total environment. Snails were found in 73,300 m^2, accounting for 93.49% of the total area investigated. A total of 3341 frames were set up, and snails were found in 1279. Thus, the total occurrence rate of live snail frames was 38.28% (the highest was 56.36% in Shuiyue Village of Qingdao Town, and the lowest was 9.56% in the No. 2 overpass residential settlement of Shuangsheng Village). A total of 12,339 live snails were captured, and the average density of live snails was 3.69 snails/0.1 m^2 (the highest was 4.99 snails/0.1 m^2 in Tongma Village, Qifu Town, and the lowest was 0.50/0.1 m^2 in the No. 2 overpass residential settlement of Shuangsheng Village). All snails were examined for schistosome infection by shell crushing under microscopy, and no infected snails were found (Table 10.2).

Among the 178 blocks investigated, 56 had a snail with frame occurrence rate of 100%, accounting for 31.46% of all the investigated environments. There were 43 plots with an average density of more than 10 snails/0.1 m^2, accounting for 24.15% of all the investigated environments.

The highest density of *Oncomelania* snails was found in a ditch behind the entrance to Tongma village, Qifu town, with a density of 556 snails/0.1 m^2, followed by a hardened ditch in Shuiyue village, Qingdao town, with a density of 380 snails/0.1 m^2.

In the survey, 140 of the 178 plots were in mud ditches, 131 snails (93.57%) were found, and the average density of living snails was 3.49/0.1 m^2. In 13 hardened cement trenches, 11 snails (84.62%) were found, and the average density of living snails was 6.29/0.1 m^2. In addition, in 20 plots in fields, snails were found in all 20 (100%), and the average density of living snails was 4.81/0.1 m^2.

3. Livestock Feces Survey

In addition to the snail survey in the three evaluation villages and one settlement site, the distribution of livestock feces in the surrounding environment was investigated. Ultimately, no livestock feces was collected.

4. Questionnaire

A total of 217 residents were interviewed in the three villages and one resettlement site. Among them, 171 were local residents, 13 were settled residents, 23 were construction workers from outside, and 10 were other people from outside.

The survey found that the total awareness rate of schistosomiasis among the respondents was 82.49%, with a lower rate of 73.91% among the nonlocal construction workers. Those interviewed in schistosomiasis-endemic areas

Table 10.1 Basic information on schistosomiasis risk assessment villages (resettlement sites) after the earthquake in Mianzhu city

Name of township (town)	Name of village (settlement)	Historical epidemic situation			Snails last detected			At last check		
		Cumulative number of snails (pieces)	Cumulative area with snails (m^2)	Cumulative number of patients	Year	Number of snails (pieces)	Area with snails (m^2)	Year	Method	Positive rate (%)
Qitian	Ganquan	65	47,700	604	2007	29	21,290	2008	ELISA	2.56
Qingdao	Shuiyue	105	62,577	855	2007	42	37,352	2008	ELISA	1.35
Qifu	Tongma	193	201,900	587	2007	76	62,291	2008	ELISA	1.20
Dongbei	Shuangsheng	56	34,944	390	2007	9	1720	2008	ELISA	0.86

accounted for 67.74%, with schistosomiasis cases accounting for 36.41%. The proportion of residents with sanitary toilets was 26.27%, with the ownership rate of local residents being the lowest, at only 14.04%, and that of residents in resettlement sites being the highest, accounting for 84.62%. The ownership rate of tap water was 17.97%, with the lowest being only 7.61% for local residents and the highest being 84.62% for resettlement residents. However, the proportion of those using well water was high, reaching 97.70%. The proportion of schistosomiasis prevention knowledge was 64.52%, and there was no significant difference among the different groups. Among the survey participants, 81.11% had contact with water, mainly concentrated in the spring and summer seasons, through washing, such as articles for daily use, hands and feet, as well as planting, weeding, and ditch cleaning. Moreover, most places of contact with infested water were in the snail-infested environment, such as areas around residences or ditches around fields (Table 10.4).

5. Other Risk Factor Investigation
   (a) Loss of schistosomiasis control institutions, etc. The evaluation team investigated the losses of professional schistosomiasis control organizations at all levels after the earthquake in Mianzhu city. Before the earthquake, schistosomiasis control in one of the municipal schistosomiasis control institutions and 19 towns of the whole city was undertaken by the prevention and protection departments of the township hospitals. Although there were no casualties among the 18 employees of the schistosomiasis control institution during the earthquake, the 260 m^2 workroom was damaged, 9 large-scale testing instruments and pieces of equipment were damaged, and most historical data of schistosomiasis control were damaged. Among the 24 hospitals in 9 towns, all but one were damaged to varying degrees. In particular, the historical data of schistosomiasis control in each town were seriously damaged: all data of 9 hospitals were damaged and lost, some data of 14 hospitals were lost, and the data of only 1 hospital were not damaged (Table 10.5). The earthquake greatly affected all levels of schistosomiasis control in Mianzhu city and had a serious impact on the normal process of routine schistosomiasis control in the future.
   (b) Damage of schistosomiasis control projects. Because the impact of this earthquake was greater on mountainous areas than on plains and the endemic areas of schistosomiasis in Mianzhu city were concentrated mainly in plains, the original agricultural water conservancy schistosomiasis control projects, such as snail-hardened ditches, were less damaged, and the risk of snail diffusion in these areas was less affected by earthquake damage to schistosomiasis control projects.

      At the same time, the assessment team performed a preliminary investigation of the diffusion of snails downstream of the dammed lake and upstream of Hanwang town. The results indicated that snails in the lower reaches of the dammed lake were distributed mainly in the ditches of its tributaries, and the density of snails in this area was relatively low due to continuous prevention and control for many years. Before the earthquake, the basin had also experienced floods every year. Therefore, the preliminary conclusion was that the flood discharge from the dammed lake caused by this earthquake did not increase the risk of snail diffusion in the area.

10.4.2.4 Assessment Conclusion

1. After the 5.12 earthquake, all medical and epidemic prevention personnel in Mianzhu were engaged in earthquake relief work, forcing them to stop the daily work of local schistosomiasis control.
2. A large number of relief workers frequently entered and left the Mianzhu disaster area, making it difficult to classify and count them as potential risk population.
3. Due to the emotional instability of residents after the disaster, the originally planned snail control could not be carried out normally.
4. Although there were no infectious snails and no sources of human and animal infection in Mianzhu city, people were in close contact with the snail-infested environment because of its wide distribution, high density, and complexity.
5. The living environment and living conditions of residents changed considerably. The harmless treatment of human and animal feces was low, with an accelerated flow of people and increased risks of infection source inputs. Based on these factors, the assessment team believed that potential factors and high-risk conditions of a schistosomiasis transmission existed in this area. In future postdisaster reconstruction work, if transmission is not monitored and prevented, there is a possibility of outbreaks.

10.4.2.5 Measures and Recommendations

According to the prevalence and risk factors for schistosomiasis in Mianzhu city, the assessment team put forward the following suggestions for schistosomiasis control work in Mianzhu postdisaster reconstruction:

1. Governments at all levels should further strengthen the leadership of schistosomiasis control and ensure that schistosomiasis control is an important component of disaster relief to prevent a major epidemic after the disaster. The routine control of schistosomiasis should be resumed and strengthened as soon as possible to improve the capacity to control schistosomiasis at all levels.
2. The monitoring of mobile persons of all types should be strengthened. All nonlocal rescue workers, returning workers, and other mobile personnel should be monitored, especially those from schistosomiasis-endemic provinces and cities (including endemic areas of Sichuan Province) and those returning from endemic areas. Follow-up monitoring and investigation should be carried out, and when necessary, disease detection should be carried out to search for potential infection sources among such personnel.
3. The surveillance of suspected schistosomiasis patients should be strengthened. To strengthen the diagnostic capacity of medical institutions at all levels for schistosomiasis, officials should be highly alert to patients with fever of unknown origin, pay attention to the detection of suspected schistosomiasis, identify schistosomiasis patients in the population as early as possible, and take corresponding countermeasures as soon as possible to avoid the spread of the disease.

4. The surveillance of nonlocal domestic animals should be strengthened, and all kinds of domestic animals from epidemic areas should be monitored to prevent the importation of schistosomiasis infection.
5. The monitoring of snails and infected snails should be strengthened, and the distribution of snails in key environments such as residential settlements should be determined. In view of the current high density of snails, snail investigations and the detection of infected snails should be carried out regularly, especially in residential areas and around villages.
6. In view of the snail environment and the key environment with high snail density around residential areas and resettlement sites, appropriate methods of emergency snail control should be adopted. All localities should combine postdisaster reconstruction projects to carry out a thorough transformation of the original snail environment to eliminate its potential harm.
7. Health education should be strengthened. Various forms of health education on schistosomiasis control should be carried out in endemic areas and resettlement sites for victims, and the self-protection awareness of people in endemic areas should be improved.
8. Feces should be treated to render them harmless, and safe water use should be promoted. All localities should gradually improve the treatment of human and animal feces and reduce the environmental pollution caused by various infection sources. Safe water use guidance should be strengthened in endemic areas to avoid or reduce the chance of residents' contact with infested water.
9. The basic data of schistosomiasis control at all levels that were damaged in the earthquake should be recovered as soon as possible, and data that cannot be recovered should be restored, improved, and rebuilt to the greatest extent possible.

10.4.2.6 Recommendations

Mianzhu city was once one of the areas of Sichuan Province where schistosomiasis was most prevalent. After decades of arduous prevention and control efforts by local governments and professional institutions, it reached the standard of schistosomiasis transmission control in 1984 and kept the transmission at low level. The prevalence of schistosomiasis is closely related to social, economic, and natural factors, and natural disasters such as floods can aggravate the prevalence and spread of schistosomiasis. Schistosomiasis control is a long-term and arduous task that requires unremitting efforts. Once those efforts are relaxed, endless troubles can ensue.

Major natural disasters, such as earthquakes, often cause serious damage and influence the natural ecological environment, residents' living environment and living conditions, disease prevention and control facilities, and other basic social health facilities in a region. At the same time, to provide disaster relief, a large number of rescuers and volunteers enter the affected areas. These changes in natural and social factors are likely to increase the potential danger of disease epidemics and spread in disaster areas. Therefore, it is of great importance to promptly carry out the necessary risk assessment to control and prevent disease spread and epidemics in disaster areas.

Response Procedures

11

Shi-Zhu Li and Qiang Wang

The process of emergency response to an outbreak is a dynamic test of local capacity building in terms of daily prediction and early warning, resource reserves, technology implementation, operation management, and access to external support to achieve quick and efficient responses to outbreaks. In ordinary times, it is necessary to strengthen capacity building to ensure dynamic coordination and effective action in emergency response. The specific process of emergency response to parasitic disease outbreaks has the following main aspects.

11.1 Reserves and Start-up

11.1.1 Reserves

Emergency reserves are an important basis of emergency response to epidemics. Primary-level parasitic disease prevention and control institutions should establish scientific, economic, and effective emergency material reserves and operation mechanisms based on the unified deployment of health administrative personnel, the working principle of "prevention first, preparedness first," the characteristics of parasitic diseases in the region, and possible emergency response needs. Emergency material planning, procurement, reserve creation, and routine replenishment should be carried out scientifically and smoothly, all kinds of emergency response technologies should be in good condition, and personnel should on standby for emergency response.

S.-Z. Li (✉) · Q. Wang
National Institute of Parasitic Diseases, Chinese Center for Disease Control and Prevention (Chinese Center for Tropical Diseases Research), Shanghai, China
e-mail: lisz@nipd.chinacdc.cn; wangqiang@nipd.chinacdc.cn

X.-N. Zhou (ed.), *Surveillance and Management of Parasitic Diseases in China*,
https://doi.org/10.1007/978-981-97-1241-0_11

11.1.1.1 Material Reserves

Emergency supplies are the most important material support for the emergency treatment of sudden epidemic events. According to the epidemic characteristics of local parasitic diseases, necessary reserves should be created for detection technology, treatment drugs, vector control, emergency equipment, and other aspects of controlling locally common parasitic diseases. At the same time, the management of emergency materials should be strengthened, and the capacity of unified allocation and support of materials should be improved. It is necessary to manage emergency materials in accordance with the principles of unified reserve, centralized management, classified storage, rational allocation, scientific use, and benefits. Dual considerations of normal time and emergency time are essential to achieve standardized use, careful storage, safety maintenance, and timely maintenance to ensure that emergency materials are in good technical condition and to meet the needs of emergency response tasks.

11.1.1.2 Technical Reserves

According to the epidemic characteristics of local parasitic diseases, key points should be highlighted, the construction of disease prevention and control institutions and grassroots prevention and health organizations should be accelerated, and the responsibility of medical and healthcare institutions and CDCs for emergency response to outbreaks should be strengthened. Additionally, a public health emergency response mechanism with complete functions, rapid response, and coordinated operation should be established; local epidemic emergency response plans based on the actual situation of the region should be formulated, and the planning mechanism should be updated as needed. A sensitive, efficient, fast, and unimpeded epidemic information reporting network should be established to enhance sensitivity to epidemic information and ensure timely detection of various outbreaks. Where conditions allow, rapid diagnostic methods, diagnostic reagents, and related vaccine research for emerging and rare parasitic diseases should be strengthened to create technical reserves.

11.1.1.3 Workforce Reserves

An emergency professional team should be established to improve the capability of epidemiological investigation, field treatment, and laboratory testing inspection. The workforce team should take both a regular and irregular form; organize emergency disposal technology training and drills for epidemic outbreaks; familiarize emergency personnel with epidemic emergency procedures, laws and regulations, and emergency response plans; and continuously enhance their awareness of emergency treatment to ensure that they can respond promptly to an emergency at any time. The training should include management awareness and public safety awareness, laws, regulations, plans and systems, technical norms and standards, professional knowledge and skills, processing procedures, and safety protection. Exercises can be conducted by organizing health emergency-related departments and health emergency teams through actual combat exercises, graded exercises, and joint exercises. Drills should be carried out according to the steps of planning, drawing up schemes, organizing the drill, and postdrill evaluation.

11.1.1.4 Social Public Science Knowledge Reserves

At ordinary times, radio, film and television, newspaper, Internet, manuals, and other media should be used to popularize scientific knowledge of disease prevention and healthcare among the public. When an outbreak occurs, it is necessary to carry out extensive education on emergency knowledge, publicize health science knowledge, and guide the public to treat the outbreak scientifically.

11.1.2 Initiation

According to the national public health emergencies contingency plans, public health emergencies can be divided into special major (I), major (II), large (III), and general (IV) public health emergencies, corresponding to four colors, red, orange, yellow, and blue, respectively, based on their nature including amount of damage caused and range. For special major emergencies (I), the state council, provincial, city, and county governments jointly initiate the response; for major emergencies (II), there is a joint response at the province, city, and county levels; for large emergencies (III), the response is initiated jointly at the city and county levels, while for general emergencies (IV), the response is initiated by the county government. The hierarchical response program is shown in Fig. 11.1.

After receiving the relevant information about an epidemic incident, county-level CDCs should immediately verify the information, report to the health

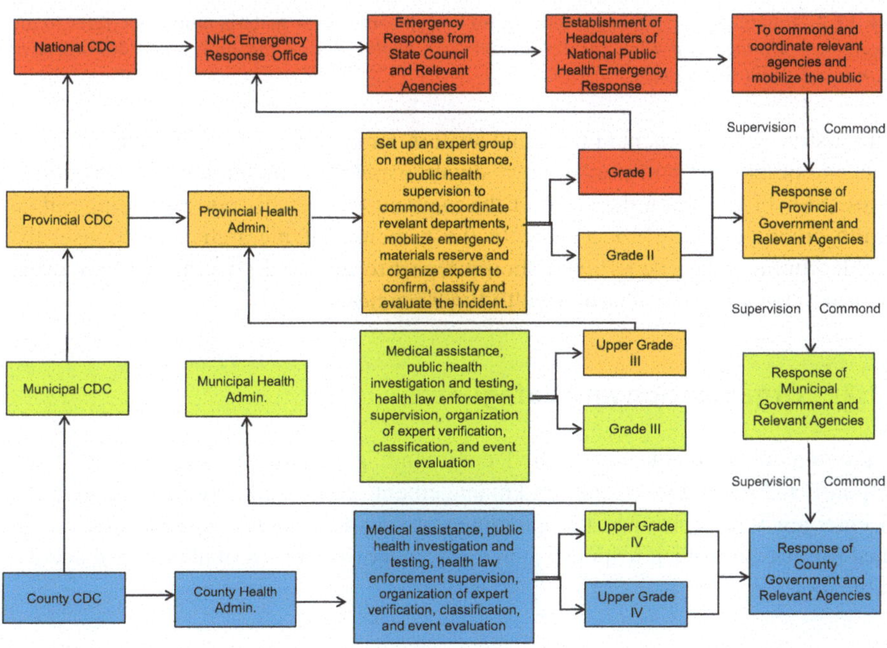

Fig. 11.1 Emergency response process for schistosomiasis or malaria outbreaks or events

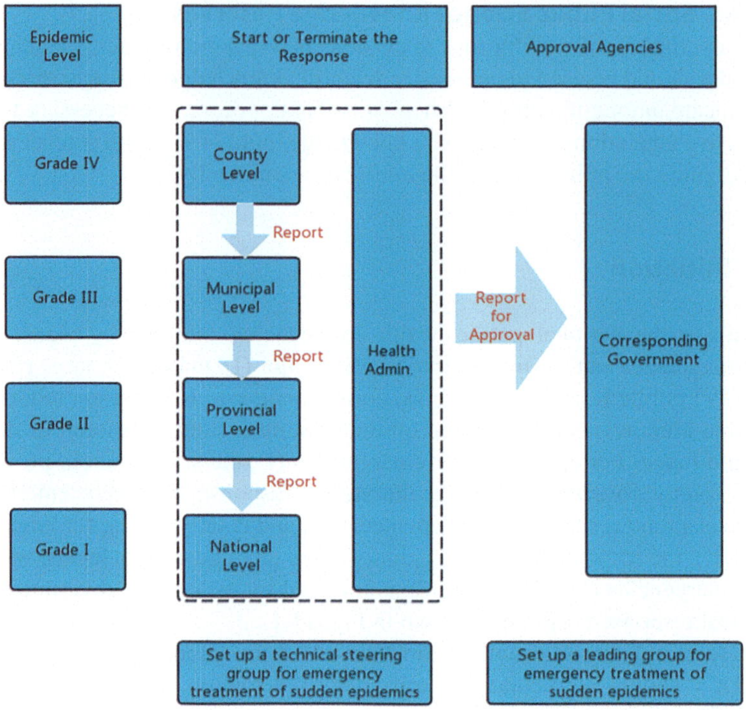

Fig. 11.2 Emergency response procedures for schistosomiasis or malaria outbreaks (refer to the *Emergency Management Plan for Schistosomiasis Outbreak and the Emergency Management Plan for Malaria Outbreaks*)

administrative department at the same level immediately after preliminary verification, and quickly organize on-site investigation and implementation of control measures (Fig. 11.2). When the scale of the incident reaches the corresponding public health emergency level, according to the principle of hierarchical response, the health administrative department should report to the local government to activate the Emergency Plan for Public Health Emergencies.

11.2 Organization and Safeguards

Following the principles of unified command, graded responsibility, and quick response and relying on science and management according to law, the organization of emergency response and the guarantee of on-site response should be strengthened for effective quick response to and emergency treatment of all kinds of sudden outbreaks.

11.2.1 Organizational Systems of Emergency Response

After an outbreak, all localities should set up a leading group, expert technical guidance group, and on-site emergency disposal working group for emergency response according to the local needs to achieve a unified command structure and scientific response.

11.2.1.1 Leading Group for Emergency Response

A leading group should be established for emergency response to epidemic emergencies led by government leaders. According to the needs, this group can organize the heads of various departments such as agriculture, water conservancy, forestry, public security exit and entry administration, and education under the heads of health administrative departments and business institutions.

Responsibilities: (1) Organizing, implementing, leading, and coordinating emergency response to sudden epidemic situations, (2) setting up a professional emergency response team, and (3) supervising and inspecting the implemented emergency response work of each unit.

11.2.1.2 Committee of Experts

An expert committee on emergencies should be set up, with the specific composition based on the relevant local plans or designated by the leading group.

Responsibilities: (1) making suggestions on determining the level of the outbreak and the response measures to be taken, (2) proposing suggestions for emergency preparedness for a sudden epidemic, (3) participating in the formulation and revision of the technical scheme for emergency treatment of a sudden epidemic situation, (4) providing technical guidance for emergency treatment of a sudden epidemic, and (5) putting forward opinions on the termination and evaluation of emergency response to a sudden outbreak.

11.2.1.3 Emergency Response Teams

According to the characteristics of epidemic events and the emergency response needs, professional response teams, such as medical rescue, epidemiological investigation, vector control, laboratory testing, and health supervision, should be set up. Their main responsibilities include patient diagnosis, reporting and treatment, on-site epidemiological investigation and data collection reporting, epidemic response implementation, collection and inspection of specimens, on-site health education, and on-site emergency response.

11.2.2 On-Site Emergency Disposal Preparation

On-site emergency response working groups should learn about the nature of the outbreak, place (unit), time, and numbers of cases, deaths, and people threatened before rushing to the scene. Then, the working group should analyze the existing data, form a preliminary hypothesis, draft a site work plan based on the hypothesis, and provide necessary support for site disposal in terms of technology, personnel, materials, and logistics.

11.2.2.1 Technical Preparation

According to the clues that have been obtained, the working group should carry out literature retrieval or review, consult relevant experts, and contact laboratories for on-site sampling and testing preparations.

11.2.2.2 Personnel Organization

According to the nature of the incident, relevant professionals should be organized to participate in the site disposal working group, with clear responsibilities and division of labor before departure.

11.2.2.3 Commodity Preparation

Response teams should be equipped with personal protective equipment, equipment and tools for collecting and transporting samples and specimens, on-site rapid detection equipment and reagents, preventive drugs or vaccines, disinfection equipment, investigation and evidence collection equipment (cameras, voice recorders, etc.), questionnaires, reference materials (professional, legal, etc.), publicity materials, communication equipment, computers, and on-site contact information (contact person and telephone number).

11.2.2.4 Logistics Support

This type of support includes vehicles, transportation, food, and accommodations.

11.2.2.5 Other Matters

The emergency response teams should make contact with the place where the incident occurred, organize a preparatory meeting, exchange information, and jointly discuss the on-site work plan and implementation plan.

11.3 Implementation and Evaluation

11.3.1 Implementation

After the response to the outbreak is initiated, on-site investigation and disposal must be carried out in an orderly manner according to the provisions of the plan (or technical plan), which can be divided into the following ten steps.

11.3.1.1 On-Site Visits and Discussion

Once the on-site working group arrives at the site, it should immediately hold a discussion with relevant local personnel, visit any locations that will help team members understand the situation, exchange opinions, and arrange relevant work.

1. The on-site visit includes (1) understanding the occurrence, development process, and latest situation of the incident; (2) determining whether there have been similar incidents in the local area in the past; (3) learning about the recent production modes, lifestyle, entertainment, and related activities of people in the affected areas; (4) determining the common means of exposure or contact; (5) understanding the measures taken and their effects; and (6) determining whether there are similar cases in surrounding areas or units.
2. Development of the on-site work plan. On-site work ideas should be discussed and tasks that must be completed on-site listed; additionally, the system of liaison personnel and regular on-site work meetings should be determined.
3. Consultation on the plan for implementing preliminary preventive and control measures. Based on the analysis of existing data and the effects of the measures taken, a preliminary prevention and control plan should be formed, and the implementation plan should be discussed.
4. Reporting. According to the preliminary situation, the relevant situation report to the health administration and higher-level guidance institutions should be prepared.

11.3.1.2 Verifying the Epidemic

The team should conduct interviews with clinicians involved in diagnosis and treatment, consult medical records, verify laboratory results, and collect relevant clinical data; visit some cases and conduct physical examination and sampling tests of patients in person if necessary. According to the clinical manifestations, laboratory results, and epidemiological data, a preliminary judgment on the nature of the epidemic should be made.

11.3.1.3 Questionnaire Design and Case Definition

Depending on the nature of the incident, existing questionnaires should be used or revised on the ground. In cases where the etiology or epidemic factors are not known, the questionnaire should be comprehensive and detailed.

The epidemiological case questionnaire should include the following components: (1) basic information: name, sex, age, occupation, address, work organization, contact information, etc.; (2) clinically relevant information: date of onset, date of doctor visit, clinical symptoms, signs, visit and treatment, clinical examination and laboratory results, and the progression or outcome of the disease; (3) epidemiological information: infection of patients, source and route of exposure, etc.; and (4) sampling, testing, and results.

On the basis of the preliminary investigation, a case definition is established, which should include the time, place, boundary of the population, clinical manifestations, laboratory results, and other basic components. Case definitions can be divided into confirmed cases, clinically diagnosed cases, and suspected cases. The most sensitive case definition should be used as much as possible in the early stage of field investigation or in the case search phase; a highly specific case definition should be used in the etiological study phase.

11.3.1.4 Search for Cases

Case searches are conducted according to established case definitions, with a list (or checklist) of case information, and epidemiological studies are conducted on the cases.

In addition to case searches through hospital, community, and contact tracking in the place where the event occurred, it is necessary to understand whether similar cases occurred in the surrounding area or unit. A temporary monitoring system can be established to dynamically collect information on new cases.

11.3.1.5 Epidemiological Investigation

1. Case investigation. A detailed epidemiological investigation should be conducted promptly for each case found and verified, and the case questionnaire should be completed. During the case investigation, attention should be paid to asking about and obtaining detailed descriptions of the components (clues) that are not included in the questionnaire, especially emphasizing collecting data for indicated cases and special cases.
2. Special investigation. On the basis of the case investigation, special investigations should be carried out according to the needs. Special investigations should be targeted and in-depth to meet the needs of a certain situation. Before the investigation, a special questionnaire and outline should be designed, and specimens should be collected during the investigation.

11.3.1.6 Sample Collection, Transportation, and Testing

During the investigation, samples such as patients, host animals, and vectors should be collected, and laboratory serological and pathogenic tests should be carried out in a timely manner to clarify the cause of an outbreak or to provide etiological clues:

1. Sample collection. In accordance with the principles of timeliness, accuracy, representativeness, and safety, sufficient samples should be collected separately. All samples should be stored for different testing purposes.
2. Transportation and testing. After a sample is collected, it should be sent to the laboratory as soon as possible in accordance with the provisions of the sample transportation management. The laboratory should test the sample immediately after receiving it. The patient's clinical symptoms and epidemiological investigation results should be synthesized so that the test report can be issued as soon as possible. Samples that can be tested on-site should be tested quickly.
3. Precautions. During the collection and testing of samples, relevant guidelines such as the *Regulations on the Management of Biosafety in Pathogenic Microbiology Laboratories* should be followed. Technical operating procedures should be strictly observed, and effective protective measures should be taken to prevent cross-infection and environmental pollution. Waste generated in the process of sample collection and testing should be disposed of in accordance with relevant regulations.

11.3.1.7 Descriptive Analysis and Hypothesis Formulation

On the basis of comprehensive investigation, the survey data should be sorted out, summarized, and analyzed. Appropriate statistical charts should be created to visually, intuitively, and clearly show the distribution of the diseases. If necessary, a hypothetical etiology should be established and proposed. Etiological assumptions should be reasonable, explain the characteristics of the various distributions, and be validated by the facts under investigation.

11.3.1.8 Recommendations for Prevention and Control Measures

In the early stages of an event, specific public health measures should be proposed and taken, even in the absence of clear epidemiological evidence related to the cause. First, the risk of the incident should be preliminarily assessed, and on this basis, emergency measures should be taken according to the corresponding principles for the purpose of reducing morbidity and mortality.

Control and preventive measures are determined according to the cause of the incident, the way it occurred and developed, and its characteristics. On-site control measures include controlling or eliminating sources of infection (hazard sources), reducing contact with exposure factors, protecting susceptible (high-risk) populations, and carrying out health rescue to control the further development of the situation. In the process of on-site interventions, staff should carefully follow personal protection measures.

11.3.1.9 Carrying Out Further Epidemiological Investigations

For possible risk factors, transmission routes, and exposed populations, epidemiological research methods such as case-control studies and cohort studies (mostly retrospective cohort studies) should be used to investigate etiological hypotheses and transmission patterns.

11.3.1.10 Writing On-Site Work Reports

On-site investigation reports can be divided into the initial report, progress report, stage report, and closing report. In the emergency response process, the corresponding on-site reports should be completed in a timely manner:

1. Preliminary Report. The preliminary report refers to the investigation report on the occurrence of the incident and preliminary investigation results after the incident or the initial on-scene verification of the incident. This report should be timely, quick, and concise. It should have five main parts: (1) a brief description of the occurrence, discovery process, diagnosis, or characteristics of the incident, a simple analysis of the nature and spread range of the incident, and an assessment of the degree of harm; (2) a brief introduction to the relevant characteristic data of the events, such as the time, population, and regional distribution of cases; (3) a brief analysis of the possible development trend of the event, such as the possible trend of an epidemic situation; (4) a simple analysis of the cause

of the incident (suspicious factors); and (5) a description of measures to be taken next or related work to be carried out.

2. Progress Report. This report is used mainly to dynamically reflect the progress of prevention and control effects and development of the investigation and handling of an incident as well as the evaluation of the preliminary work and arrangements or suggestions for follow-up work, with an emphasis on continuity. It should have three main aspects: (1) the report should be completed in a timely manner after new information is obtained so that it does not lose significance, as a report should be completed every 1–2 days after starting the investigation; (2) the time intervals between progress reports can be extended according to the situation in the middle and late stages of on-site investigation and treatment if the incident tends to be stable and there are no new changes as the investigation develops and on-site control measures are implemented; and (3) the report should be continuous and can be repeated.
3. Stage Report. A stage report is a periodic summary of the investigation events at regular intervals when the investigation and handling of an event last for a long time. It is used mainly to comprehensively summarize and review the preliminary investigation and research, evaluate the handling of the events in stages, and forecast the development of the event and later work.
4. Case Closing Report. A case closing report is a comprehensive review and summary of the investigation of and response to the whole incident after the investigation, including the discovery of the event, the treatment of patients, the development and results of the investigation and research, preventive and control measures and their effects, the occurrence of the event and problems exposed in the investigation and response, the experience and lessons worth summarizing, and suggestions for performing similar work or preventing similar events.

When the collection and preliminary analysis of on-site investigation data are basically completed and the event has been effectively controlled, the on-site work can be finished. Before the response team leaves the site, they should hold a meeting with the local authorities to formally summarize the on-site epidemiological investigation and disposal and provide feedback on the investigation results and work suggestions. After the fieldwork is finished, the summary of the fieldwork should be completed in a timely manner, and the field investigation data should be filed.

11.3.2 Assessment

After the treatment of all kinds of parasitic diseases, the epidemic situation should be scientifically and objectively evaluated. Assessment components include the categories and nature of the epidemic; the impact on society, the economy, and public psychology; the emergency response process; investigation steps and methods; treatment measures taken; and the conclusion of the investigation. This assessment summarizes the experience and lessons and puts forward suggestions for improvement.

11.4 Skills for Responding to the Media

To keep pace with the times, skills for responding to the media in the new media environment should be developed. On the one hand, the awareness of the role of media in a crisis should be strengthened; on the other hand, past modes of thinking about crisis management should be changed, including the media coping style, and necessary media coping skills should be mastered. Only in this way can emergency managers improve their ability to control the media, calmly face emergencies, and even turn a crisis into an opportunity.

11.4.1 Principles of Response

When dealing with the media, the following principles should be observed: first, a frank attitude; second, openness and transparency, including the timely disclosure of information; and third, prompt requests for the support of third parties and the avoidance of boastfulness and self-justification. Fourth, crises should be dealt with humanely, with concern shown for people to build trust. Fifth, one voice should be selected to speak to the outside world to avoid confusion. Sixth, there should be room for expression to avoid passivity. After an outbreak, it is necessary to skillfully apply these principles, the most important of which is to disclose information in a timely and transparent manner.

11.4.2 Tips and Precautions

11.4.2.1 Striving for the Leading Power of Public Opinion

To control a crisis situation, stabilize social order, and avoid panic, crisis managers must respond quickly, purposefully choose information sources and channels, choose authoritative people to disseminate news, choose mainstream media to report on the event, and effectively control the direction of public opinion, usually by holding press conferences or news briefings. It is necessary to prevent some media from "irregularly" expanding their discourse space to serve their commercial interests by attracting an audience, thereby intensifying the crisis situation. At the same time, it is necessary to prevent the media from spreading incorrect or incomplete news or even rumors, thereby misleading the audience and aggravating their fear.

11.4.2.2 Providing Clear and Authoritative Information

During a crisis, rumors spread quickly, and the order becomes unstable. The public, especially the victims, have a low psychological tolerance. Speculation about the situation could easily reduce their trust and support at the decision-making level. Therefore, the words and deeds of crisis managers must be consistent, unified, clear, and authoritative in order to reveal the truth, reassure people, and resolve conflicts. When authoritative information is lacking, rumors will spread widely.

11.4.2.3 News Release Without Interruption

Dynamic crisis development requires that crisis managers constantly provide new information, announce progress in a timely manner, and guide the media and the public to positive and feasible measures, actions, and effects.

11.4.2.4 Selecting a Suitable Press Spokesperson

The spokesperson plays an important professional role in the crisis management process, so he or she must participate in decision-making; make the progress of control measures clear; understand the attitude of senior managers toward crisis information dissemination; determine the scope, degree, and frequency of information disclosure; and finally establish the authority of the information released. The more serious the crisis, the higher the position of the spokesperson. In a more important situation, it is better for the spokesperson to be older.

11.4.2.5 Maintain Good Communication with the Media

When a crisis breaks out, the main crisis managers may confront the media or use diplomatic rhetoric such as "No comment." Regardless of whether the main crisis managers have expressed their position, the media will report this attitude, so it is necessary to maintain a good communication relationship with the media.

11.4.2.6 Determining the Information to be Released to the Public

When preparing a press release, the title should show its position and stand out. The components should be complete and clear with a focus on facts. To ensure that the measures and actions to be taken are clear, guesses and inaccurate information should not be included, jargon or trade terms should be avoided, and the information should be expressed clearly. At the same time, the contact information of the spokesperson should be indicated.

11.4.2.7 Preparing for a Press Conference

Preparation, especially ideological preparation, should be made for press conferences. The issues of public concern should be thoroughly considered, and reasonable explanations and satisfactory answers should be provided. The press conference should be rehearsed so that the spokesperson can clearly state the position of the government or unit. The events should be described accurately and clearly, and journalists should be resourceful in their pursuit of information. When there are serious problems, it is important to avoid the development of an overwhelming situation. The atmosphere at the scene of the crisis should be guided with the help of the media, and the crisis should be addressed in a manner that is beneficial to the government and response units.

11.4.2.8 Cooperation with Reporters

Journalists will definitely pursue information on a crisis. Officials should be careful not to avoid journalists and should speak with a unified voice. Nondesignated spokespersons should not be allowed to be interviewed but should politely provide contact information for the spokesperson. The receptionist or switchboard operator should not actively provide information to the outside world but after understanding the nature of a call should transfer it to the relevant department or staff member. All team members should have a positive attitude toward reporters' calls and inquiries. Information should not be provided exclusively, as this will arouse feelings of unfair treatment among other journalists. The publicity office should implement a 24-h working system to prevent the vacuum caused by out-of-control communication in a crisis. It is necessary to ensure that news agencies can always obtain the latest information and the measures being taken by the government and relevant units to control the crisis. Bad behavior should be admitted and positive corrective measures taken. Complete and accurate information and background details should be provided to avoid media speculation or snooping in other channels.

11.5 Case Studies

11.5.1 Response to the Adverse Event of Antimalaria Drug Reactions in Neijiang City, Sichuan Province

11.5.1.1 The Event

Mr. Lu, a villager from Group 5, Shashi village, Channan township, Dongxing District, Neijiang city, traveled for work to Uganda, Africa, in November 2008. He was infected with malaria twice in Uganda, in October 2009 and April 2010. After returning to China on April 22, 2010, he became ill on May 11 and June 6 and was treated with artemether injection and chloroquinine. On July 8, he was examined by blood smear because of recurrent fever, and *Plasmodium* parasites were found. To prevent the secondary transmission of imported cases in the local area, the CDC of Neijiang Dongxing District organized preventive medication in Group 2, Mawan village, Yinan township, according to the requirements of relevant departments of the local government at approximately 17:40 p.m. on July 12. By approximately 19:20 p.m., a total of 142 people had taken the medication. At 9:10, Meng, a 3-year-old girl, developed symptoms of headache, paleness, and weakness approximately 20 min after taking the medicine. After providing symptomatic treatment, the medical staff of Yinan Township Health Center immediately sent the child to the Yumu Town Central Health Center for emergency treatment. At 20:00 on the same day, Meng died due to ineffective rescue efforts. After that, many children and adults suffered from dizziness, vomiting, and abdominal pain to varying degrees. By 4:00 a.m. on June 13, there were 87 inpatients in the First People's Hospital of the city, including 40 pediatric patients, of whom 2 were critically ill and 7 were seriously ill; 20 internal medicine patients, of whom 3 were critically ill and 16 were seriously ill; and 27 patients in respiratory medicine. At 8:30 on July 13, the CDC of Dongxing District, Neijiang city, reported this major public health emergency, "a child died of preventive medication in a malaria group," through the "National Public Health Emergency Report Management Information System."

11.5.1.2 Emergency Response and Disposal

1. Emergency response. After receiving the report, the Emergency Office of the Ministry of Health assigned poisoning experts and pediatric experts from the China CDC and malaria experts from the Parasitic Disease Expert Advisory Committee of the Ministry of Health to form a joint expert group and travel to the site to guide the investigation and treatment. The expert group arrived in Neijiang city, Sichuan Province, at 15:20 pm on July 14 and carried out relevant investigations in coordination with Sichuan Province and Neijiang city.
2. Investigation. After the expert group arrived at the scene, they quickly organized a symposium to learn more about the occurrence, development, and treatment process of the incident and to gather information about the drugs, specifications, batch numbers, and sources used in the preventive medication. On-site visits were made to hospitalized children, and the process and dosage of preventive medication were investigated. According to the investigation results, after discussion and consultation, the expert group put forward suggestions for the next step.
3. Survey results. The preventive drug taken on this occasion was "chloroquine phosphate tablets/primaquine phosphate tablets," which met the relevant specifications. However, the dosage data showed that some children had been given therapeutic doses intended for adults. Therefore, the conclusions and recommendations of the expert group were as follows: the targets of this event should be limited to those who experienced discomfort after taking the "chloroquine phosphate tablets/primaquine phosphate tablets" issued by the local CDC on July 12, 2010. "Adverse drug reaction" and "overdose" were identified as health effects on "adults" (≥16 years) and "children" (<16 years), respectively. The cause of death in the child Meng was poisoning caused by excessive administration of chloroquine phosphate.
4. Treatment recommendations. Disposal recommendations were made for different groups of people.
   (a) For patients ≥16 years old, the dosage of the drug was within the prescribed range, and the reaction was caused by an adverse reaction to the drug. There were no serious adverse reactions, and patients ≥16 years old could be discharged from the hospital.
   (b) For patients <16 years old, older children had less severe symptoms due to mild overdose and could be discharged for home observation, with the local community health service center instructed to take the initiative to observe the children at home every day until they recovered. Younger children with obvious symptoms remained in the hospital for observation, but the treatment scheme was simplified (fluid replacement was mainly used to promote drug excretion, and other treatment drugs not related to the adverse drug reaction were stopped).
   (c) Publicity, explanation, and follow-up: The expert group recommended that the local government actively provide publicity, education, and explanation to relevant personnel and properly handle the follow-up work on the children who died.

11.5.1.3 Release of Investigation Results

At 10:30 a.m. on July 15, the expert group informed the provincial health department, the municipal health bureau, the provincial and municipal food and drug regulatory bureaus, and other units of its investigation conclusions. With the consent of the leaders of the emergency management office of the Ministry of Health, the expert group attended a communication meeting organized by the local government at noon to answer questions from patients and their families. Subsequently, the expert group answered the questions of other meeting participants.

11.5.1.4 Event Analysis

The analysis of the development and disposal process of the whole event indicated that the main problems were in the following six links (Fig. 11.3).

1. Malaria prevention and control of laborers going abroad. In this case, a clear diagnosis of imported malaria cases ("*P.f.*", "*P.v.*" or mixed infection) should have been made, and standardized treatment should have been provided to cure the cases and help the patients avoid recurrence after treatment.
2. Response to epidemic foci of imported malaria cases. Before this epidemic foci treatment, epidemiological investigation and risk assessment should have been conducted. Whether to implement epidemic foci treatment with suitable treatment measures should have been decided according to the transmission conditions and risk assessment.

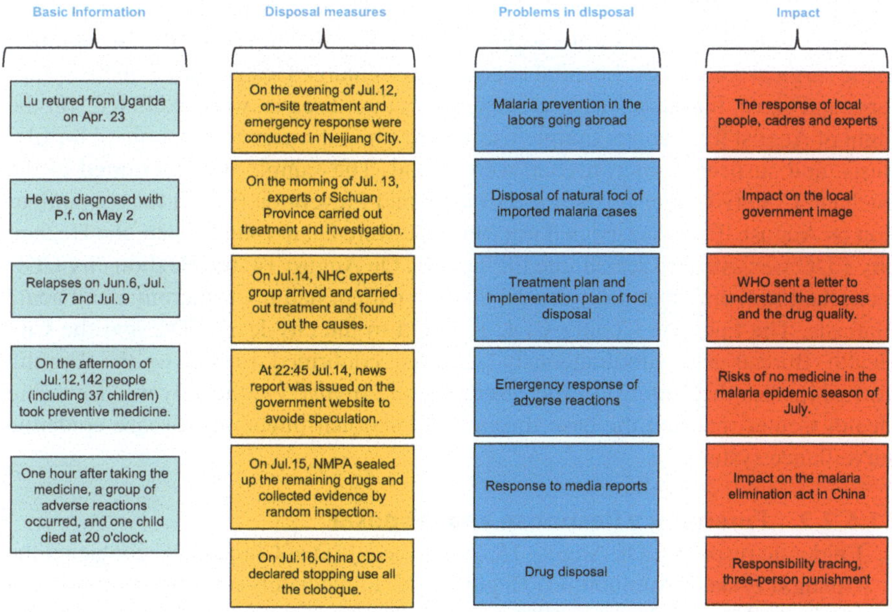

Fig. 11.3 Process of development and disposal of mass events of preventive medication in Neijiang city, Sichuan Province

3. Planning of epidemic disposal and implementation. Before the disposal of the epidemic site, a specific implementation plan and emergency response plan should have been formulated according to the relevant technical plan, especially for the drug subjects, the type of drug to be taken, and the dosage.
4. Emergency treatment of adverse reactions. After the occurrence of adverse reactions, the emergency response should have been promptly activated.
5. Responding to media reports. Once an emergency happens, a lack of correct media information guidance may result in misleading media hype. In this case, the government first reported the incident on the evening of July 14 and mentioned that the provincial health administration had organized experts to deal with the incident but did not further expand on the incident. The local government conducted on-site communication with local people, which ensured the disposal of the whole incident.
6. Drug disposal. The State Drug Administration sealed the remaining drugs of the same batch and sent samples to Neijiang city and the Shanghai Drug Administration for testing. No drug quality problems were found.

11.5.2 Focal Outbreak of Lymphatic Filariasis in Fuchuan Yao Autonomous County, Guangxi Zhuang Autonomous Region

11.5.2.1 Outbreak of the Epidemic

On August 8, 2007, patient Li, a 19-year-old male from Chaodong town, visited Fuyang Township Health Center in Fuchuan County due to fever. He was admitted to the surgical ward of the hospital because of an enlarged testis. Based on his symptoms and signs, filariasis was suspected after other diseases were excluded. On the night of August 9, a venous sample was taken, a blood smear was examined for filariasis, and suspected microfilariae were found. The sample was submitted to the Fuchuan County Center for Disease Control and Prevention (CDC) for reexamination on August 10, and suspected microfilariae were again found. Then, the Hezhou city CDC was asked to reexamine the sample. On August 14, the Hezhou city CDC rechecked the sample and confirmed that the patient had microfilaremia and acute filariasis. The Fuchuan County CDC reported to the city-level CDC and the City Health Bureau and simultaneously sent the confirmed results to the Fuyang Township Health Center. On the afternoon of August 14, the Fuyang Township Health Center reported the case through the national infectious disease epidemic surveillance system.

11.5.2.2 Emergency Response and Disposal

1. Epidemic response. On August 15, 2007, the NIPD at China CDC contacted the Guangxi Zhuang Autonomous Region CDC immediately after being notified of filariasis through the surveillance and early-warning system of the national infectious disease epidemic surveillance system, demanding that the case be verified immediately and that an expert group be sent to Guangxi on the same

day. The next day, experts and professionals from the Guangxi Autonomous Region CDC traveled to Fuchuan County to perform on-site blood collection and review. The case was confirmed as Bancroftian filariasis on August 16. Blood samples were collected from 16 family members and relatives on the evening of August 17 to provide further information about the infection location and transmission range.

After China officially declared that filariasis had been eliminated nationwide in 2007, the China CDC included the disease in the surveillance of focal outbreaks of infectious diseases and the early-warning system for single cases. A single case is treated as a sudden outbreak. Immediately after the confirmation of the outbreak, the Ministry of Health dispatched a team of experts on August 17 to carry out on-site investigation and emergency response. Based on the patient's clinical symptoms of acute filariasis (spermatic cord inflammation, spermatic cord nodule, epididymitis, and hydrocele) and the preliminary investigation results of two cases of microfilaremia found in his relatives, the expert group believed that the patient's original residence may have been the site of his infection and requested that Fuchuan County immediately organize an epidemiological investigation of the outbreak.

2. Epidemiological investigation. The survey subjects were residents over 2 years old of Changtang administrative village, Chaodong town, where the patients were located, and the examination rate was over 98%. Blood samples were collected in the evening, with thick blood film used in one ear and two pieces. After routine staining, microscopic examination was performed to look for microfilaria under an optical microscope and count the density. Of the negative blood films examined by microscopy, 30% were rechecked. In addition, at 06:00 am, the whole-catch method was adopted in the microfilariasis patient's home in the natural village, and all mosquito vectors were caught by the mouth suction mosquito trap. The captured *Culex pipiens quinquefasciatus* were examined by individual anatomy.

   At the same time as the on-site investigation, the experts consulted the historical epidemic and control data of local filariasis and had a discussion with the filariasis control personnel of that year to gain a better understanding of the filariasis control work in the county. They also visited villagers, village cadres, and village doctors in Changtang village and neighboring Huangsha, Chashan, and Youmu villages, especially people over 60 years old, to learn about local filariasis investigation and treatment and whether there were patients with chronic filariasis.

11.5.2.3 Survey Results

1. Epidemiological investigation. After the confirmation of the first case, as of 12:00 a.m. on August 22, blood samples were collected at night from residents of six natural villages belonging to Changtang administrative village. A total of 648 blood samples were collected, and a total of 1293 blood smears were made. In total, 14 cases with microfilaremia (including the first case with microfilaremia) were identified, and the microfilaremia detection rate was 2.16%. For

patients with microfilaremia, the average microfilaria density was 21.86/60 μl, and the median was 24/60 μl. Among them, the maximum was 54/60 μl and the minimum was 2/60 μl. No patients with chronic filariasis were found. A total of 140 mosquitoes were caught in Ganshang natural village, of which 54 were *Culex pipiens quinquefasciatus*. All were dissected and found to be negative. Fourteen patients with microfilaremia were distributed in Ganshang natural village (12 cases) and Yinshan natural village (2 cases), among whom 10 were over 50 years old, with the youngest being 19 years old and the oldest 70 years old. The male to female ratio was 11:3. The 14 patients with microfilaremia were distributed in 12 families, including ten families with one case each, and two families with two cases each. Among them, 12 patients with microfilaremia were related to one another.

2. Investigation of the history and control of filariasis. Since 1951, filariasis investigation has been carried out in Fuchuan County, and the county has been determined to be an endemic area of Bancroftian filariasis, with the main vector being *Culex pipiens quinquefasciatus*. In 1971, the microfilaria infection rate in Changtang village was 4.34%, making it a low epidemic area. Fuchuan County began to implement large-scale filariasis control in the 1970s, and it was confirmed in October 1985 that it had reached the standard of basically eliminating filariasis.
3. Investigation conclusion. According to the field investigation results, among the 14 cases with microfilaremia, 3 individuals had a history of being migrant workers, and the remaining 11 had no long-term history of being outside the village, showing spatial aggregation and family aggregation and indicating that the patients with microfilaremia had a high possibility of local infection. In addition, 71% of the patients with microfilaremia were over 50 years old, which was in accordance with the characteristics of filariasis foci. Therefore, the current filariasis epidemic situation was determined as the remaining epidemic foci.
4. On-site disposal. After the outbreak, the government of Fuchuan County attached great importance to the situation and set up an epidemic response leading group, a technical group, and a logistics group on August 14. According to the recommendations of the national, provincial, and municipal joint expert group, the county government launched an emergency investigation and treatment of microfilaremia patients in Changtang village as well as a focus investigation and environmental sanitation remediation. Health education on filariasis was also conducted to prevent unnecessary panic. To further prevent the spread of the epidemic, environmental sanitation campaigns were carried out, such as garbage cleaning, pothole filling, mosquito breeding ground clearing, and indoor cleaning.

11.5.2.4 Transmission Analysis

Filariasis was historically prevalent in the Guangxi Zhuang Autonomous Region. After comprehensive investigation, control, and monitoring, the region reached the criteria for the elimination of filariasis in 1995. Although filariasis surveillance has been carried out continuously for many years since reaching the standard of basic elimination in Fuchuan County, some remote villages with serious epidemic situations have not been covered. This was the case in Changtang administrative village of Fuchuan County. The detection rate and density of microfilaria in this village were high, and new infection cases existed. At that time, it was the filariasis transmission season. As the microfilaremia cases found in Changtang village were concentrated in the area, there was still a risk of local transmission. Considering that the main local vector, *Culex pipiens quinquefasciatus*, is a domestic mosquito species with a small activity range, the probability of long-distance transmission to surrounding villages through the vector was limited, and the threat to the surrounding population was not large. On this basis, the remaining filariasis foci in Fuchuan County could be considered local transmission only.

Surveillance and Reporting System 12

Shi-Zhu Li and Ying-Jun Qian

Timely, complete, and accurate disease surveillance information is an important basis for analyzing disease epidemiological trends and formulating preventive and control strategy and is an important part of disease prevention and control. A well-designed and effective surveillance reporting system is the basis for taking appropriate and timely action in responding to the necessary information.

Thus, disease surveillance and reporting systems are essential for parasitic disease prevention and control and are a key tool for decision-making in national public health. Surveillance data provide information that can be used for priority strategy adjustment, policy decisions, program development and implementation, resource mobilization and allocation, epidemic forecasting, and early detection.

12.1 Categories and Their Functions

12.1.1 Categories

At present, the largest platform for information management of disease prevention and control in China is the "China Disease Prevention and Control Information System," which is an important part of the monitoring information system for public health emergencies of the national emergency response mechanism. The information on infectious diseases is monitored in real time and dynamically by the central, provincial (autonomous region), municipal (regional), and county (district) CDCs, thus implementing integrated management and sharing of disease surveillance information.

S.-Z. Li (✉) · Y.-J. Qian
National Institute of Parasitic Diseases, Chinese Center for Disease Control and Prevention (Chinese Center for Tropical Diseases Research), Shanghai, China
e-mail: lisz@nipd.chinacdc.cn; qianyj@nipd.chinacdc.cn

X.-N. Zhou (ed.), *Surveillance and Management of Parasitic Diseases in China*,
https://doi.org/10.1007/978-981-97-1241-0_12

Three types of reporting systems are currently involved in parasitic disease surveillance in the China Disease Prevention and Control Information System: the notifiable infectious disease surveillance and reporting system, the emergency public health event reporting system, and the Parasitic Disease Control Information Management System.

12.1.1.1 Surveillance and Reporting System for Notifiable Infectious Disease

In 2004, a web-based national reporting network system was established and formally launched based on case reporting of notifiable infectious diseases, which greatly improved the coverage, timeliness, sensitivity, and accuracy of epidemic reporting and basically achieved the overall goal of "horizontal to the edge and vertical to the bottom" for epidemic monitoring and reporting. Subsequently, China's nationwide disease prevention and control information system, with the Infectious Disease and Public Health Emergencies Monitoring and Reporting Information System as its core, was gradually improved, pioneering the implementation of "case-by-case, real-time, online" reporting in China's public health sector. This fundamentally changed the way that the previous reporting system reported monthly and aggregated data by county.

Medical institutions and CDCs at all levels can submit surveillance data to the data center of the China CDC through the IE browser. Relevant departments of the Ministry of Health, China CDC, CDCs at all levels, and health administrative departments can make relevant inquiries and obtain statistics, analysis, and release of the data.

1. Reporting procedures. At present, surveillance information on infectious disease in China is collected mainly through the "Infectious Disease Report Card of the People's Republic of China" (hereinafter referred to as the Infectious Disease Report Card) and reported directly on the web-based China Disease Prevention and Control Information System. For legally reported infectious diseases, the reporter is responsible for filling out the Infectious Disease Report Card, the reporting unit is responsible for organizing the entry of the report card information, and CDCs at the county (district) level and above are responsible for auditing the information. Organizations with infectious disease reporting responsibilities are mainly medical institutions, CDCs, blood collection agencies, and health and quarantine agencies; the personnel responsible for reporting are mainly medical and quarantine personnel, disease control personnel, rural doctors, and individual practitioners. The infectious disease surveillance reporting workflow and information flow are shown in Fig. 12.1.
2. Reporting components. The following categories of infectious diseases are reported by law in China.
   (a) Notifiable communicable diseases
      - Category A: plague and cholera
      - Category B: severe acute respiratory syndrome (SARS), AIDS, viral hepatitis, poliomyelitis, human avian influenza (H5N1), measles, epidemic

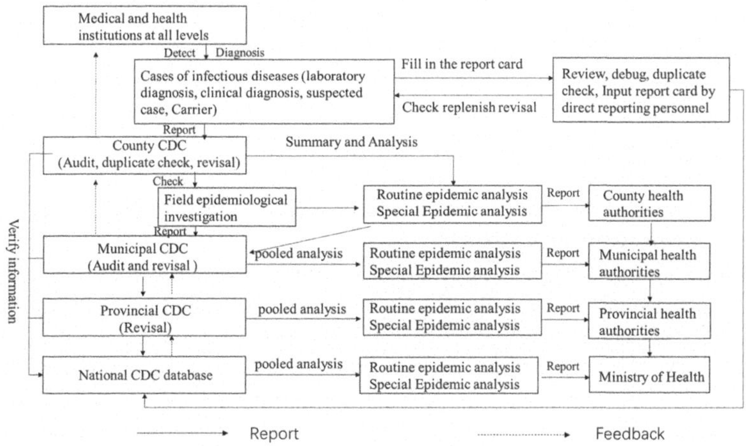

Fig. 12.1 Infectious disease surveillance reporting workflow and information flow

hemorrhagic fever, rabies, epidemic B encephalitis, dengue fever, anthrax, bacterial and amoebic dysentery, tuberculosis, typhoid and paratyphoid fever, epidemic encephalomyelitis, pertussis, diphtheria, neonatal tetanus, scarlet fever, brucellosis, gonorrhea syphilis, syphilis, leptospirosis, schistosomiasis, and malaria

- Category C: influenza, mumps, rubella, acute hemorrhagic conjunctivitis, leprosy, epidemic and endemic typhus, black fever, echinococcosis, filariasis, infectious diarrheal diseases other than cholera, bacterial and amoebic dysentery, and typhoid and paratyphoid

(b) Pneumonia of unknown causes stipulated by the NHC
(c) Other infectious diseases listed under the administration of Categories B and C infectious diseases determined by the NHC
(d) Other endemic infectious diseases under the administration of Categories B and C as determined by the government at the provincial level
(e) Other outbreaks, epidemics, or infectious diseases of unknown origin

3. Reporting methods. Infectious disease surveillance information is collected mainly through Infectious Disease Report Cards and reported directly on the web-based China Disease Prevention and Control Information System. The responsible reporters of medical institutions, CDCs, and health and quarantine institutions fill out the Infectious Disease Report Card after identifying cases of notifiable disease. Cases of infectious diseases found to be underreported should be reported in a timely manner. For notifiable diseases, the appointed reporter is responsible for filling out the Infectious Disease Report Card, the appointed reporting unit is responsible for organizing the entry of the report card information, and CDCs at the county (district) level and above are responsible for auditing the information. The appointed reporting units with network reporting conditions should enter the information on the Infectious Disease Report Card into the Disease Surveillance Information Reporting Management System in

real time (as soon as possible within the specified time limit); the reporting units without web-based reporting conditions should submit the Infectious Disease Report Card to local township health centers, community health service centers, or county CDCs with web-based reporting capabilities in a timely manner so that these centers can report directly on their behalf.

12.1.1.2 The Public Health Emergencies Reporting Information Management System

The Public Health Emergencies Reporting Management Information System is one of the subsystems of the Chinese Disease Prevention and Control Information System and has been in operation since 2003. The system is set up with one code per event, and surveillance agencies can continuously and systematically report information related to an event, including the source of information, time and location of the event, population affected, event category, event level, epidemiological distribution, laboratory surveillance results, hazard level, hazard factors, and control measures taken. Figure 12.2 depicts the reporting process of public health emergencies.

Incident management is the main focus of the public health emergency reporting system, and all reports are dependent on this component. The dissemination of early-warning messages, the management of cases, and the processing of events are all based on the event as the basic operational unit. Event management is mainly for users at each regional level, who can intuitively enter or query the cases, test results, and approval opinions at all levels corresponding to the event. Data from other monitoring systems can be used to consolidate the cases corresponding to an event to avoid duplicate data in two systems. The system can directly display data as a list of events from all levels within the jurisdiction arranged by region, last report time, or event category. The default arrangement can be modified individually, and a geographic information map can be used to display all events within the jurisdiction.

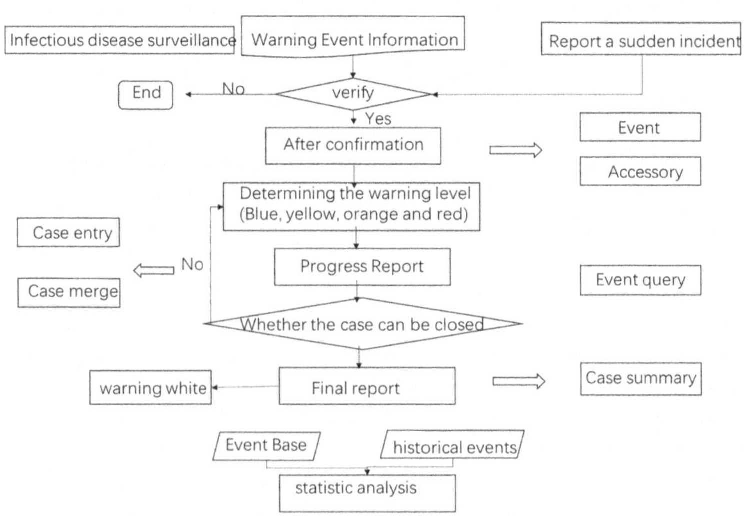

Fig. 12.2 Diagram of reporting process for outbreak or major events of parasitic diseases

The system's statistical analysis of the characteristic indicators of events (i.e., epidemiological three-dimension distribution) can be classified by the variables and information of each reporting system according to specified warning indicators and issue warning messages to notify relevant personnel in a timely manner.

12.1.1.3 Parasitic Disease Control Information Management System

The Parasitic Disease Control Information Management System is a unified national parasitic disease information management platform and data center that realizes dynamic statistical analysis, resource sharing, and data exchange of parasitic disease control information by developing a database structure and information exchange standards. Additionally, it realizes vertical management through authority settings at the national, provincial, municipal, and county levels.

According to the characteristics of different parasitic diseases, the main work components, and system maintenance needs, the system is divided into the schistosomiasis prevention and control system, malaria prevention and control system, echinococcosis prevention and control system, filariasis control system, soil-borne nematode disease control system, and five other subsystems as well as a "system management" auxiliary system. The report includes investigation of schistosomiasis prevention and control, investigation of acute schistosomiasis cases, registration of medical treatment of advanced schistosomiasis, malaria prevention and control and surveillance at all levels, soil-born helminthiasis prevention and control by CDCs at all levels, investigation of population monitoring and natural infection of filarial larvae since the elimination of filariasis, monitoring of laborers returning from overseas, case investigations of people with microfilaremia, mosquito vector classification investigation, and investigation of echinococcosis prevention and control.

12.1.2 Functions

12.1.2.1 Notifiable Disease Surveillance and Reporting System

The notifiable disease surveillance and reporting system has four major functions: information reporting, data management, data utilization, and information dissemination. It extends information collection to the grassroots level through the rational use of virtual special network technology and Internet resources. Furthermore, the system integrates the resources of various existing infectious disease reporting systems in China; strengthens and improves the management of infectious disease reporting; enhances the quality of infectious disease reporting in China; eliminates nonreporting, concealment, and misreporting; and improves the completeness, accuracy, and consistency of "data sources" of infectious disease epidemics, thus providing a rapid response to outbreaks of infectious diseases. The system includes report card management, data source management, and reporting of outbreaks of infectious diseases. It includes report card management, real-time statistics, statistical reports, data analysis, quality control, and other service functions and provides data dictionary maintenance and code maintenance. The system sets up a dashboard so that users can publish, reply to, and modify the information on the dashboard.

12.1.2.2 Public Health Emergencies Reporting Management Information System

This system is an important part of the emergency response mechanism and an important information channel for early prevention, detection, reporting, and response to emergencies. It is a national statutory information system for reporting management of public health emergencies based on reporting of multiple types of public health emergencies. The system is used mainly for comprehensive monitoring of 11 kinds of emergencies, such as major infectious diseases, mass diseases of unknown cause, food poisoning, occupational poisoning, pesticide poisoning, other chemical poisoning events, environmental health events, immunization events, nosocomial infections, radiological health events, and other public health events. Online reporting, confirmation, examination and approval, early warning, and other functions of public health emergencies can be realized, and dynamic monitoring of public health emergencies can be achieved by the four levels of central, provincial (autonomous region, municipality directly under the central government), municipal (region), and county (district) CDCs.

12.1.2.3 Parasitic Disease Control Information Management System

This system carries out management and maintenance for users at all levels according to the principles of management at each level. It is for users of CDCs or institutions of parasitic disease control at all levels. The functions of each subsystem are separated according to the disease categories under the jurisdiction of each unit. The system is divided into four modules by function: data collection, statistical analysis, data management, and system management. Through these four functional modules, information and data collection (entry, reporting, review, revision, and deletion) of parasitic disease control can be realized. Then, statistical analysis (query, statistics, reporting, analysis summary) and management (printing, import, export, and backup) of collected data, as well as system maintenance and management (set reporting process, system user, and organization management), and other functions, can be achieved. The main functional modules of the system are shown in Fig. 12.3.

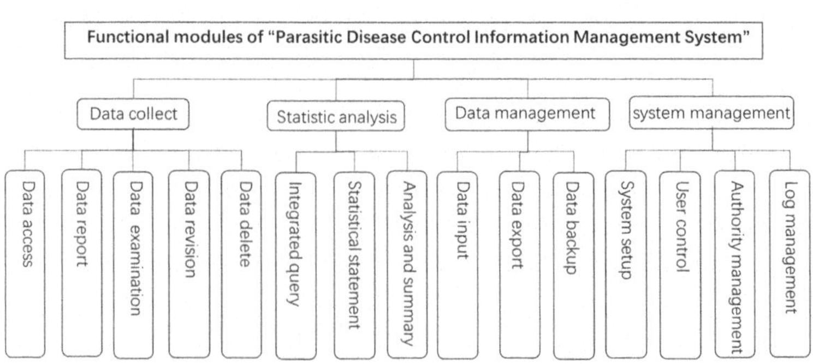

Fig. 12.3 Functional modules of the Parasitic Disease Control Information Management System

12.2 Report Data Verification and Correction

CDCs at all levels nationwide verify and audit the reported data to ensure data quality. Unconfirmed suspected cases may be reported by medical institutions to prevent possible underreporting and for possible early warning, but these unconfirmed cases cannot be input into a central database. CDCs at all levels must verify the cases reported by medical institutions, and only data verified by CDCs at all levels can be finally entered into the central database for analysis. Each CDC not only verifies and revises the data but also investigates, processes, and tracks suspected cases as needed. The data validation diagram of the infectious disease surveillance system is shown in Fig. 12.4.

12.2.1 Information Review

12.2.1.1 Internal Audit of Information Provided by the Reporting Unit

The responsible reporting unit's online reporting personnel are required to review the Infectious Disease Report Card filled out by the unit before reporting directly on the network or to the local county (district) CDC.

1. When incorrect items, omissions, or logical errors are found, the reporter is promptly notified to recheck. Cards with duplicate reports are marked, and no further online reporting is performed.
2. Further verification of the reported information is recommended immediately when the following are found:
   (a) Patients, suspected patients, and pathogen carriers of Category A infectious diseases and Category B infectious diseases managed according to Category A, and infectious diseases with unknown sudden causes stipulated by the NHC.
   (b) Discovery of cases of unexplained pneumonia and unexplained deaths.

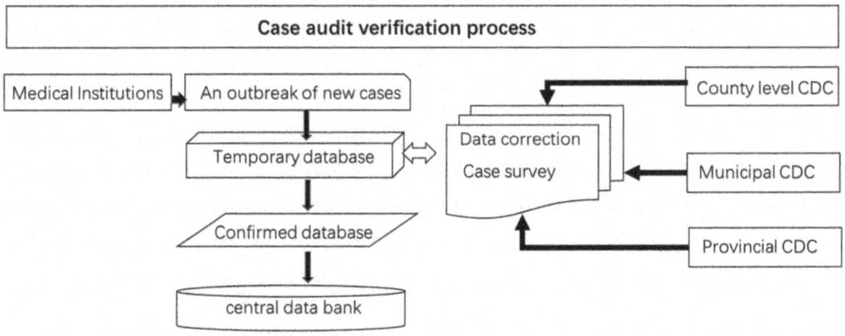

Fig. 12.4 Schematic diagram of data validation for the infectious disease surveillance system

(c) The same acute infectious disease is found in the same natural village, street, or collective unit (place) in a day with three cases or more or in a week with five cases or more.

If any of the above is true, the information should be reported directly to the network within 2 h, while the director of local county (district)-level CDC should be notified by phone.

12.2.1.2 Review of Information on Infectious Disease Report Cards by CDCs at All Levels

1. The state requires the epidemic managers of county-level CDCs to go online at least four times a day to review information on infectious diseases reported in their jurisdictions. Reported information with doubts is promptly fed back to the reporting unit or verified by the reporter. The epidemic managers of county CDCs should log on to the system daily in the morning and afternoon to review and confirm Infectious Disease Report Cards reported within their administrative areas.
2. When auditing the following types of cases, it is recommended that authenticity of the information in the report be further verified in combination with the actual local situation.
   (a) Patients, suspected patients, and pathogen carriers of Category A infectious diseases and Category B infectious diseases managed according to Category A, and infectious diseases with unknown sudden causes stipulated by the NHC.
   (b) Discovery of cases of unexplained pneumonia or unexplained deaths.
   (c) Discovery of clusters of unexplained cases (more than three cases).
   (d) Discovery of possible outbreaks of infectious diseases; e.g., the same acute infectious disease is found in the same natural village, street, collective unit (place) in a day with three cases or more.
   (e) Taking the county (district) as a unit, the incidence of an acute infectious disease in a short period (1 week or 1 incubation period) is found to be significantly higher than in the same period in the history.
   (f) Taking the county (district) as a unit, infectious diseases are found that are rare in the local area or have not happened within the 3 previous years.
   (g) Taking the county (district) as a unit, infectious diseases with extremely low incidence (or that have previously been eliminated) are found.

   If any of the above situations are true, the reported information should be confirmed through the network within 2 h or as soon as possible, the on-site personnel of the unit should be informed to conduct an investigation, and the CDC at the higher level and the health administrative department at the same level should be notified as soon as possible (such as by telephone).
3. To verify suspected duplicate report cards prompted by the system, the reasons for a false alarm and duplicate report cards should be briefly explained in the

remarks column and marked for deletion in a timely manner. It is recommended that the patient's name, sex, occupation, disease type, date of birth, and present detailed address be checked for duplication.

The web-based reporting system requires no network re-reporting card and regular network rechecking. When a duplicate card is found, which Infectious Disease Report Card to keep is determined as follows: when the same reporting unit reports many times or different reporting units jointly submit Infectious Disease Report Cards for the same case, the card with the earliest initial report time should be kept, and the remaining cards should be marked for deletion. If the diagnosis level of the reserved card is lower than that of other cards, the card information with the earliest initial report time will be revised according to the card information with the highest diagnosis level (case classification, diagnosis time, etc.). If the card pushed by the special disease management system needs to be revised again, the special disease management department should be notified, and the special disease management department should revise it. Duplicate cards caused by repeated reports by the same reporting unit should be deleted by this unit or local county (district) CDC; duplicate cards caused by the joint reporting of different reporting units should be deleted by the county (district)-level CDC where the patient lives.

4. For other infectious disease reporting information, the reporting information should be confirmed through the network within 24 h after verifying that there are no errors.

12.2.2 Information Correction

When a reported case is found to have a change, such as diagnosis change, the patient died, or the card filled out incorrectly, the responsible reporter should promptly create a revised report (complete a new Infectious Disease Report Card, select the revised report for the card type, and indicate the name of the disease reported before the revision) and complete the network correction operation according to the regulations. If a pathogen carrier or suspected case is diagnosed as a confirmed case (clinically diagnosed case or laboratory-confirmed case), the original diagnosis date should be changed to the confirmed date. If the revision is made by the responsible reporting unit, the local county (district)-level CDC should review the revised information reported. For cases where the investigation and verification address is incorrect, the local county (district)-level CDC should revise it to an unknown address. For infectious diseases that are subject to special disease reporting management, the corresponding special disease management agencies or departments should follow up on the reported cases and revise the wrong information or excluded cases found in the infectious disease reporting card in a timely manner.

12.3 Network Data Analysis and Expression

Epidemic analysis is the process of compiling and summarizing the collected case data of infectious diseases; using appropriate epidemiological and statistical analysis methods; describing the distribution characteristics of infectious diseases in the population, their development, and influencing factors; evaluating the effectiveness of disease prevention and control measures; and sending and feeding back the analysis results to relevant institutions and personnel in a timely manner for the formulation of infectious disease prevention and control strategies and measures as well as adjustment and evaluation.

Parasitic disease data analysis is based mainly on the infectious disease reporting system and the special report system, which regularly analyzes, compiles, and releases epidemic information to provide a basis for understanding the dynamics of a parasitic disease epidemic and evaluating and adjusting prevention and control strategies.

12.3.1 Regular Analysis

The NIPD at China CDC has dedicated staff who dynamically monitor the information reported on parasitic diseases in China on a daily basis and analyze the changing epidemiological situation; analyze and report the epidemic situation of Category A and Category B infectious diseases managed according to Category A in a timely manner; pay close attention to abnormal situations such as aggregated cases, possible outbreaks, unexplained cases, and deaths; analyze distribution characteristics, epidemiological history, and possible epidemics; and report weekly, monthly, and annually.

12.3.1.1 Weekly Analysis of Acute Schistosomiasis and Malaria

The epidemic profile and epidemiological characteristics of acute schistosomiasis and malaria are summarized and analyzed weekly, including the number of reported, clinically diagnosed, and confirmed cases and the reported incidence rates by region. Longitudinal comparisons and analyses of the epidemic are performed and represented with statistical charts. Imported epidemic analysis is also performed to identify outbreaks of acute schistosomiasis infection and to track the latest progress.

12.3.1.2 Monthly Analysis of the Surveillance of Echinococcosis and Kala-azar

Each month, a comprehensive overview of the epidemic surveillance of echinococcosis and kala-azar is provided, the morbidity and mortality of infectious diseases for the month are analyzed and summarized, statistical summaries of morbidity during the month are provided, and the quality of reported information, especially imported diseases, is comprehensively evaluated and archived as historical information for epidemic surveillance.

The number of cases and deaths are reported, as are the morbidity and mortality, and the monitoring level is analyzed and compared with that of the previous month and the same period of the previous year, and these data are then expressed and described with appropriate statistical charts. Monthly analysis of key epidemic situations is based mainly on daily and weekly analysis of infectious disease surveillance, summarizing and analyzing the key infectious diseases reported during the month, describing their epidemiological distribution characteristics and influencing factors, and applying appropriate statistical methods to establish a statistical model for infectious disease prediction.

12.3.1.3 Phase Analysis of Parasitic Disease Surveillance Data

Phase analysis includes quarterly analysis, semiannual analysis, and annual analysis. Phase analysis, especially annual analysis, is used to summarize and describe the epidemic monitoring data of parasitic diseases in a certain period, summarize outbreaks and epidemic events, and summarize their characteristics in detail; these data can be expressed by combining a large number of statistical charts. In addition, the annual analysis report should include the quality evaluation components of the annual epidemic surveillance information.

The annual analysis report of parasitic disease surveillance should be filed as important historical data of epidemic surveillance for future inquiry and use by disease control, information, scientific research, and other departments. In contrast to the weekly and monthly monitoring analysis, the annual analysis provides statistics according to the onset dates of cases, reflecting the actual level of new cases in the year.

12.3.2 Thematic Analysis

In special historical periods (such as earthquakes, debris flows, and floods) when a sudden epidemic situation of parasitic diseases or a public health emergency occurs, special analysis should be performed at any time according to the epidemic characteristics and the need for disease control. A daily zero reporting system of epidemic situations should be established to provide a reference for effective preventive and control measures; at the same time, the disease control effect can be evaluated.

The analysis components should include a description of the historical epidemic level and recent epidemic characteristics, combine the results of field epidemiological investigations to analyze and describe the whole epidemic situation, and propose reasonable prevention and control measures.

12.4 Evaluation of the Surveillance and Reporting System

Regular surveillance system evaluation is conducive to standardizing system operation, improving efficiency, and optimizing resource utilization. Surveillance and reporting system evaluation will clarify the evaluation objectives, specify the evaluation objects through evaluation purposes, and scientifically determine the functions, characteristics, and effects and other attributes of the evaluation. Generally, three indicators including technical, economic, and social, are selected for evaluation, and the evaluation should focus on comprehensively assessing direct, indirect, and immediate and long-term benefits and impacts as well as quantitative and qualitative indicators.

To evaluate the quality of disease reporting systems, different evaluation programs have been developed in various countries based on a series of indicators. In 1988, the US CDC published the Disease Surveillance Evaluation Guidelines Program, which first described the disease surveillance evaluation system centered on nine attributes reflecting the effectiveness of the system, namely, simplicity, flexibility, acceptability, sensitivity, positive predictive value, representativeness, completeness, timeliness, and stability. The evaluation framework should consist of four parts: first, system description, including the purpose of the system, data users, and operational processes; second, evaluation of disease outbreak surveillance in terms of the timeliness and authenticity of surveillance data; third, description of the system in terms of its usefulness, flexibility, acceptability, portability, stability, and operational costs; and fourth, recommendations for improvement based on the evaluation results. The WHO published the Guidelines for Surveillance and Evaluation of Response to Infectious Diseases in 2001 and revised them in 2004 and 2006. The guidelines divide the assessment into five parts, public health value, system structure, system core functions, system support functions, and system attributes, with 95 specific indicators, including input, process, output, outcome, and effect indicators; system attributes include completeness, timeliness, effectiveness, simplicity, acceptability, flexibility, sensitivity, specificity, positive predictive value, and representativeness.

According to the definition of the WHO Surveillance and Evaluation Framework for Communicable Disease Surveillance and Response Systems, the assessment of a surveillance reporting system should include both supervision and evaluation. Supervision refers to a routine evaluation based on the daily operation of the system. Its purpose is to identify problems in the operation of the system at an early stage, as it is necessary to find them early through regular evaluation before the usefulness of the system is reduced. Evaluation refers to a phased evaluation aimed at changes in the predetermined results (objectives) related to the monitoring system and involves attribution analysis of changes in the input, implementation process, results (positive and negative), and effects (expected and unexpected) of the monitoring system.

After the disease prevention and control information system was established in China, regular quality evaluation modules were set up within the system, including indicators such as the unaudited rate of cards, replacement card rate, direct reporting, proxy reporting statistics, timeliness (untimely reporting and auditing), and zero missing report rate at counties and districts levels. In addition, system attributes such as positive predictive value, specificity, sensitivity, accuracy of reported information, and user satisfaction (acceptability) were evaluated in some thematic

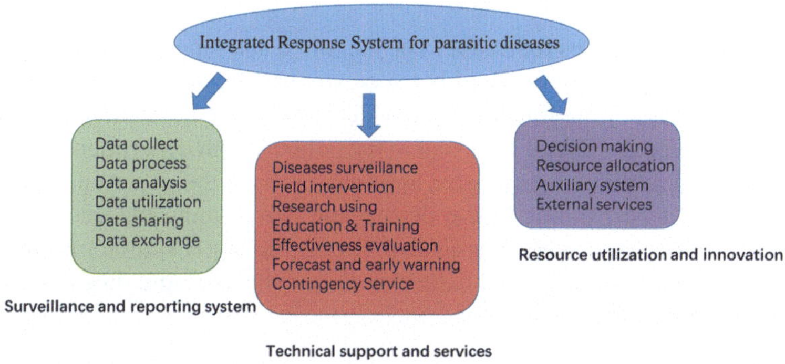

Fig. 12.5 Conceptual framework for an integrated response system of parasitic diseases in China

surveys, and regular supervisory evaluation of the system was performed to continuously improve the system functions. The evaluation of the parasitic disease surveillance reporting system in China is shown in Fig. 12.5.

12.4.1 Evaluation Indicators

Surveillance and reporting system evaluation indicators include the sensitivity, positive predictive value, timeliness, simplicity, acceptability, representativeness, stability, flexibility, and completeness of the system information reporting.

12.4.1.1 Sensitivity

Sensitivity refers to both the percentage of true-positive cases detected by the surveillance system as a percentage of all cases and the ability to detect outbreaks, including the ability to track changes in the number of cases over time. The sensitivity of a surveillance system can be influenced by (1) whether people with certain diseases or health conditions actively seek health services, (2) whether the disease or health condition is correctly diagnosed (this factor reflects the skills of health service personnel and the accuracy of diagnosis and testing), and (3) whether cases with definite diagnoses are reported to the monitoring system. Sensitivity questions are usually raised when a disease changes. The sensitivity of the surveillance system can be improved by increasing the knowledge of certain diseases, introducing new diagnostic tools, and improving the monitoring methods.

12.4.1.2 Positive Predictive Value

The proportion of people who are actually sick among the people identified as sick by the monitoring system is an important index that reflects the sensitivity and specificity of case definition and the occurrence of diseases in people. When evaluating the positive predictive value, the confirmation of cases reported by the monitoring system should be emphasized.

12.4.1.3 Timeliness

The number of days it takes for a case to be identified and reported to the monitoring system reflects the processing speed between any two steps in the monitoring system. Usually, the first time interval to consider is from the beginning of a negative health event to the reporting of the event to the public health department responsible for formulating control and preventive measures or from the occurrence of a negative health event to the public health department determining the trend, outbreak, and effective control measures. For acute diseases, the initial symptoms are usually used as the beginning of negative health events, and the exposure date is sometimes used as the starting time.

12.4.1.4 Simplicity

The structure of the system is simple, and it is easy to operate, that is, the structure of the system itself is simple, and the whole system and each of its components are easy to operate. The design of the monitoring system should be as simple as possible on the basis of fulfilling its purpose. The information flow chart describing the monitoring system is helpful in evaluating whether the system is simple. The indicators include (1) the quantity and types of information needed to diagnose diseases, (2) data and types of reporting sources, (3) methods of transmitting case information or data, (4) number of institutions receiving case reports, (5) personnel training requirements, (6) method and depth of data analysis, (7) number and occupation categories of personnel using case information, (8) method of distributing report or case information to users, and (9) time spent on system maintenance, case information collection, case information transmission, case information analysis, and preparation and sending of the monitoring report. The simplicity of the surveillance system is reflected mainly in the design and scale of the system. The surveillance system, which is simple and clear in design, should be a system in which cases can be easily diagnosed, and the individuals who identify cases should also be the ones who analyze and use this information. For more complex surveillance systems, special laboratory tests should be provided to confirm cases, and multilevel reports should be made on the information obtained.

12.4.1.5 Acceptability

Acceptability reflects the subjective willingness of individuals and organizations to participate in a surveillance system and refers to the spontaneous motivation of those who operate the system internally and those who use it externally (e.g., those who are asked to report data). Quantitative indicators of acceptability include individual or institutional participation rates; visit completion rates and refusal to answer questions; completion rates of reporting forms; physician, laboratory, or hospital reporting rates; and timeliness of reporting. Acceptability is a subjective attribute that reflects the willingness of the individuals on whom the surveillance system is based to provide accurate continuous, complete, and timely information. Factors that may have an impact on the acceptability of a surveillance system are the public health importance of the health event, the surveillance system's perception of the individual's contribution, the surveillance system's response to recommendations and comments, and whether the legislature places restrictions on data collection and confidentiality.

12.4.1.6 Representation

The surveillance system can accurately describe the distribution of diseases across people, times, and places. The representativeness of the surveillance system can be evaluated by analyzing and comparing the characteristics of the reported cases with those of the disease. The representativeness of the surveillance system can be more precisely evaluated by examining all cases in one sample. Data quality is an important part of representativeness and is affected by the clarity of the surveillance forms, the training and instruction of those completing them, and the seriousness of the people's attitudes during data management.

12.4.1.7 Stability

Stability refers to the reliability (e.g., the capacity to collect, manage, and deliver data accurately without errors) and availability of the surveillance system. The guideline also provides the indicators and methods when evaluating each attribute.

12.4.1.8 Flexibility

Flexibility is the ability to adapt to changes in information needs or operating environment with little consumption of time, personnel, and financial resources. A flexible monitoring system should be able to change the type of information collected at any time or change its operation with little additional time, staffing, or cost.

12.4.1.9 Integrity

Integrity refers to the completion rate of required items in the infectious disease reporting system.

12.4.2 Evaluation Methods

According to the actual situation at all levels, an objective evaluation of the quality of the epidemic report of infectious diseases should be made. The evaluation methods include daily evaluation, supervision evaluation, and special investigation evaluation.

12.4.2.1 Daily Evaluation

The current indicators for evaluating infectious disease surveillance information are the number of unaudited cards, the number of replaced cards, the timeliness of the process from responsible reporting to the audit of the cards by the county (district) CDC, and the county (district) zero and missing report rate:

1. Number of unaudited cards. The number of cards that have not been audited by the local county (district) level CDC in the administrative area being queried from 0:00 to 24:00 of the queried time period.

2. Number of duplicate cards. The number of duplicate cards reported in the queried administrative area from 0:00 to 24:00 of the queried time period.
3. Timeliness of diagnosis to audit. The sum of the two time intervals of "diagnosis to report" and "report to audit" is used to evaluate the timeliness of reporting by direct reporting units and the timeliness of report card audits by county and district CDCs, respectively.
4. County (district) zero and missing report rate. The zero and missing report (zero infectious disease reports) rates of the county and district within the queried administrative area from 0:00 to 24:00 of the queried time period.
5. Average index. The report quality of the epidemic situation information in the queried administrative region is comprehensively evaluated through the comprehensive evaluation index calculated by the above indicators. At present, the query level "comprehensive evaluation of information quality of infectious disease report" in the system reaches the direct reporting unit. The average index is the average value of each individual evaluation index. The lower the value, the better the report quality of the evaluated county (district) or direct reporting unit is.

12.4.2.2 Supervision Evaluation

Regular supervision and evaluation of direct network reporting to CDCs and medical institutions at all levels to grasp the current problems of direct network reporting will help improve the quality of data collected by the web-based reporting system and ensure the normal operation of the system, thus ensuring the effective monitoring of infectious disease epidemics and providing valuable suggestions and opinions for the continuous improvement of the system. On-site supervision includes the following:

1. Investigating the financial investment, network construction, and online reporting capabilities of CDCs or medical institutions in a jurisdiction in terms of online reporting and verifying whether special funds and equipment are provided in a timely manner
2. Understanding the online staffing and training, assessing the professional knowledge and field operations of the online reporting staff, and providing timely guidance on any identified problems
3. Investigating the management mode of CDCs and medical institutions at all levels in online reporting, such as the management of infectious disease registers, online reporting accounts, and their own supervisory and inspection mechanisms
4. Investigating the use of information and evaluating the quality of information analysis
5. Collecting suggestions and comments from all levels in a timely manner

The selection of supervisory evaluation indicators should reflect the comprehensiveness of online reporting as much as possible. The main indicators include network online reporting funds (including budget, expenditure, and network communication costs), network connection methods, network online reporting rate, number of training sessions and people, online reporting staff questionnaire scores, and information utilization and analysis.

12.4.2.3 Thematic Survey Evaluation

The thematic survey includes the organization system establishment of infectious disease information reporting management; the quality of registration of various types of basic information; the work carried out; the implementation of the system, such as self-inspection of units; the rate of reporting infectious diseases and timely reporting; the accuracy and timeliness of Infectious Disease Report Card entry; the infection reporting and supervision of responsible epidemic reporting units; the investigation of underreporting of notifiable infectious diseases in residential areas; and archival epidemic information management.

Indicators of special investigation and evaluation include the following components: (1) the completion rate, accuracy rate, and timely reporting rate of Infectious Disease Report Cards; (2) the completeness, accuracy, and timeliness of outpatient record and inpatient registration reporting; (3) the completion rate, timely correction rate, and timely reporting rate of each link in infectious disease reports; (4) the missing report rate for infectious diseases; (5) the awareness rate of relevant knowledge in infectious disease information reports; (6) assessment of the quality of the epidemic reporting organization, system, and personnel management; and (7) relevant monitoring indicators for other special diseases.

After the monitoring system is evaluated, a summary should be created to show whether the surveillance system has monitored an important public health problem and whether it has achieved its goal, and suggestions should be made regarding whether it needs further improvement. When making suggestions for the improvement of a surveillance system, it should also be noted that the attributes and costs are interdependent, and increasing the sensitivity or specificity, timeliness, or representativeness requires increasing the costs of the surveillance system. Before the surveillance system is changed, the interaction among the attributes should not be ignored, and the function of one attribute should not be improved without determining whether it affects the function of another. Evaluation is not an easy task, and there is no perfect system. In practice, all aspects should be weighed. Each system has its own characteristics. To achieve the expected goal, it is necessary to maintain a balance between resources and effects. There is no absolute authority over the evaluation of the surveillance system, and the evaluation process should be clear and objective.

Compared with the evaluation system of the WHO, China's evaluation indicators include both real-time monitoring of the system and periodic evaluation. They basically cover the evaluation indicators of the system's ability to detect events and comprehensive capability but lack evaluation indicators for decision support and action. Compared with the US CDC's evaluation guidelines, China's guidelines concentrate on the evaluation of system attributes but lack the evaluation of simplicity and flexibility. As China's disease prevention and control system continues to improve, for example, through the integration of surveillance systems and other health information systems, the establishment of data standards, changes in data exchange methods, and changes to accommodate the surveillance of new/recurring infectious diseases, the corresponding evaluation system will surely become increasingly better.

It is important to note that since different systems vary from design to operation to specific purposes, the evaluation of a system must take full account of its own characteristics to select the optimal indicators and methods. In addition, each monitoring system encompasses the broad and long-term purpose of the system builder, and therefore, a roadmap of the system cannot be established with any one evaluation method.

12.5 Case Studies

12.5.1 Role of the Epidemic Reporting System After the Elimination of Lymphatic Filariasis in China

Lymphatic filariasis (filariasis) was formerly one of the major parasitic diseases in China, with a large population at risk. Filariasis not only posed a serious public health risk but also became one of the major causes of poverty. After unremitting efforts, the WHO confirmed that China had achieved the goal of eliminating filariasis in May 2007, making China the first country in the world to eliminate this disease. The following is a case study of two cases of filariasis reported directly on the Internet respectively, showing the importance, timeliness, and necessity of the disease surveillance system.

12.5.1.1 Case One

1. The problem. On August 8, 2007, a patient with swollen testicles was seen at Fuyang Town Health Center in Fuchuan County and was initially diagnosed with filariasis after a medical history was obtained and symptoms were identified. On the evening of August 9, microfilariae were detected by the double-slice method of thick blood film, and the results were reported to the Fuchuan County CDC. The local CDCs in Fuchuan County, Hezhou city, and Guangxi Zhuang Autonomous Region successively conducted etiological examination and case investigation of the first suspicious case to determine the time and place of onset, the process of diagnosis and treatment, and contact history and traveling history; they finally confirmed the diagnosis of Bancroftian filariasis. The filariasis case was reported on the afternoon of August 14 through the national infectious disease surveillance network system. The administrative village of Changtang was designated as the epidemic focus village.

   Guangxi took the lead in reaching the national standard of filariasis elimination in 1995, and in the same year, its filariasis control program was transferred to surveillance system. After more than 10 years of surveillance, no new infections were found. The discovery of the remaining foci of Bancroftian filariasis in Fuchuan County occurred after the NHC submitted the national report on lymphatic filariasis elimination in China to the WHO in 2006, so this discovery has attracted extensive attention from the international community. To assess the situation and consolidate the elimination results, the NHC deployed the relevant health departments to conduct a survey of filariasis epidemic foci in the surrounding suspected areas from August 2007 to March 2008.

2. Investigation objects and methods. The investigation was conducted through pathogenology, serological testing, mosquito vector survey, review of historical data, and visits. The target population included all residents of Changtang administrative village (including Ganshang and Yinshan natural villages), where the first case was located, residents aged 10 years or older of the four administrative villages adjacent to the focus area, and students of one secondary school and one primary school in Chaodong township.
3. Investigation results. Ganshang and Yinshan natural villages in Fuchuan County, Guangxi, were the remaining foci of Bancroftian filariasis. These villages were in low-prevalence areas, and the epidemic scope was limited, with the majority of patients being middle-aged and elderly.
4. Recommendations. This incident fully reflected the role of the network web-based reporting system in infectious disease reporting and emergency response, which enabled the remaining foci of filariasis to be promptly identified and cleaned up in the past-elimination stage. Therefore, in the future, it is still necessary to carry out surveillance after filariasis elimination and to continue to include filariasis in the network direct reporting system of infectious diseases.

12.5.1.2 Case Two

1. The problem. On August 17, 2007, one case of filariasis was reported directly through the network by the CDC of Jiangyou City, Sichuan Province, after having been diagnosed by the Institute of Occupational Disease Control of the Jiangyou CDC.
2. Investigation process. On September 4, professionals from the Institute of Parasitic Diseases of the Sichuan CDC traveled to Jiangyou City to investigate the suspected case in accordance with the "Diagnostic Criteria and Treatment Principles of Filariasis of the Ministry of Health" and "Diagnostic Criteria of Filariasis" and to review the blood film collected by the Institute of Occupational Disease Control of the Jiangyou CDC.
3. Results. The results revealed that the worms within the blood film were significantly more elongated than filarial microfilariae and had no internal structure. The conclusion was that the blood film had been contaminated by mycobacterial filaments. The antimycobacterial treatment was effective, thus confirming the misdiagnosis.
4. Recommendations. This incident exposed the fact that the diagnostic capacity for filariasis had been weakened since the elimination of filariasis in China. However, it also showed the key role played by the disease reporting system in China, which enabled the false-positive case to be verified and corrected in a timely manner. The reporting system is not only a platform for health departments to promptly grasp an epidemic situation and the basis for health policy formulation but also an important means of early warning after disease elimination.

12.5.2 Surveillance and Monitoring of Imported Malaria Cases in TC County

12.5.2.1 Proposing the Issue

The results of malaria surveillance in a county in 2013 are analyzed, and the importance of surveillance leading to malaria elimination is discussed.

12.5.2.2 Investigation Process

The infectious disease surveillance and reporting system and the Parasitic Disease Control Information Management System were used to collect malaria epidemic data and surveillance information in TC County in 2013 and to analyze the prevalence and surveillance of malaria in the county.

12.5.2.3 Results

Malaria Prevalence

A total of 138 malaria cases were reported in the county in 2013, all of which were imported cases, with no local infections; all cases were confirmed by laboratory testing. The cases were mainly *P. vivax* (118 cases), which accounted for 85.51% of the total number of cases; the remaining 20 cases were *P. falciparum*, which accounted for 14.49%.

1. Time distribution: Cases were reported every month of the year, and the peak season was from April to June, during which the number of cases was 88, accounting for 63.77% of the total cases in the year. The peak incidence was consistent with the peak period of round trips of the mobile population in this county.
2. Spatial distribution: The cases were distributed in 16 townships of the county, and the incidence of individual townships was relatively concentrated.
3. Distribution of population: The infected people were mainly young migrant workers, accounting for 97.83% (135/138). Of the cases; 128 were male, accounting for 92.75%, and 10 were female, accounting for 7.25%; the sex ratio was 13:1. The patients ranged in age from 4 to 58 years old, with the main cases in the range of 20 to 40, accounting for 82.61% (114/138).

Malaria Surveillance Status

1. Blood tests for patients with unexplained fever: In 2013, there were 8135 blood tests for patients with unexplained fever in the county, including 4572 tests with rapid diagnostic reagents (RDTs). A total of 138 malaria cases were found, for a positive blood test rate of 1.7%.
2. Case reporting and case investigation: A total of 138 malaria cases were reported by medical institutions in the county, for a 100% reporting rate within 24 h and a 100% timely case investigation rate within 3 days.
3. Active case detection: According to the requirements of the "Technical Program for Malaria Elimination (2011 Version)," the TC County CDC organized 57

active investigations and found no positive cases. Residual spraying was carried out for 124 foci where imported cases were found, for a total of 248 kg of pesticides covering an area of 335,414 m^2 and 9463 people.

4. Vector population and density monitoring: From May to October 2013, one village was selected in each of three townships. Mosquitoes were trapped by lamps twice a month, and a total of 31 investigations were conducted. The monitoring results showed that the mosquito density in each village was higher in cattle sheds than in human houses. The results of the vector population survey showed that *An. sinensis* and *An. liangshanensis* were the main malaria vectors in this county, followed by *An. minimus*, and *An. maculatus*. Preliminary statistics indicated that the distribution of malaria vectors was related to altitude.
5. Screening of returned travelers: Screening of malaria cases imported from abroad (the same group of people returning from infected places overseas) was carried out. All individuals with fever symptoms and 10% without fever symptoms were examined by earlobe blood microscopy or RDT, and other people were followed up. A total of 528 people were screened, with no positive results.

12.5.2.4 Recommendations

The malaria transmission in TC County was basically stable in 2013, since the local malaria cases were all imported cases. Imported cases become key to monitoring when local malaria program moving toward elimination. Comprehensive use of web-based reporting system has demonstrated that it can improve the quality of infectious disease monitoring and reporting systems, parasitic disease prevention and control information management systems and other monitoring systems, therefore, malaria case information and vector monitoring data should be collected and reported promptly, monitoring data should be analyzed and utilized in a timely manner, local epidemic characteristics and evolution trends should be determined precisely, and additional guidance for malaria prevention and elimination should be provided scientifically.

Establishment and Implementation of the Surveillance Network 13

Xiao-Nong Zhou

The parasitic diseases in the routine reporting system for notifiable infectious diseases include only schistosomiasis, malaria, hydatid disease, Kala-azar, filariasis, and amoebic dysentery. Related disease surveillance reports can quickly reflect the epidemic situation and development of these parasitic diseases to a certain extent. However, as hospitals at all levels still have certain limitations in reporting notifiable diseases, such as incomplete reports and unrepresentative reports, and these reports are limited to cases requiring medical treatment after the onset of disease, it is difficult for conventional reporting systems to truly and comprehensively reflect the transmission characteristics and epidemic trends of these notifiable parasitic diseases. Therefore, it is necessary for the parasitic disease monitoring network system to combine the objectives of disease prevention and control planning in different periods and establish different monitoring systems to support the whole parasitic disease monitoring network to achieve the purposes of timely detection of epidemic situations, effective evaluation of prevention and control effects, and scientific guidance of prevention and control planning. These monitoring systems include laboratory monitoring based on laboratory tests, symptom monitoring based on hospital cases, active monitoring based on specific populations, sentinel monitoring based on specific sites, pest-resistant host and vector monitoring based on animals or vectors, parasite population distribution monitoring in pathogen-based environments, intervention input monitoring based on database areas for disease prevention and control, public opinion monitoring based on media information, epidemic factor monitoring based on transmission risk factors, and parasite resistance and vector insecticide resistance monitoring based on drug use.

X.-N. Zhou (✉)
National Institute for Parasitic Disease at Chinese Center for Disease Control and Prevention (Chinese Center for Tropical Diseases Research), Shanghai, China
e-mail: zhouxn@nipd.chinacdc.cn

X.-N. Zhou (ed.), *Surveillance and Management of Parasitic Diseases in China*,
https://doi.org/10.1007/978-981-97-1241-0_13

However, the implementation of parasitic disease surveillance varies from place to place due to different species of parasitic diseases and different prevention and control efforts. Some use only one monitoring system to monitor parasitic diseases, and others use a combined parasitic disease monitoring network composed of various monitoring systems to achieve the purpose of effectively controlling or even eliminating parasitic diseases. The reasons for the imbalance of monitoring efforts in different places are as follows: unclear prioritization of monitoring needs, improper application of monitoring data, insufficient training of monitoring staff, weak systematization of monitoring and evaluation, and new monitoring tools not yet applied. The above reasons are also the challenges faced by monitoring in the new period.

13.1 Configuration and Establishment

As a grassroots disease prevention and control institution, the very important duty of the surveillance network is to monitor parasitic diseases correctly and scientifically to provide a scientific basis for local effective control and even elimination of these diseases.

13.1.1 Configuration

The monitoring network of a certain area should be configured to comprehensively plan for the parasitic diseases that need to be monitored in the area and determine what monitoring system to adopt for parasitic disease monitoring. Diseases to be surveilled should be determined according to the disease burden of various locally existing parasitic diseases and their impacts on local public health. For example, in southern China, in addition to malaria and schistosomiasis as the main problems affecting local public health, liver fluke disease has a high disease burden, and some areas are even being affected by new parasitic diseases, such as angiostrongyliasis. Therefore, local monitored diseases should be determined according to the disease burden or the impact on public health on the basis of mastering the local disease spectrum.

The selection of the surveillance system should be determined by the characteristics of disease species and local resources available for surveillance. In general, in addition to conventional surveillance and reporting systems, zoonotic parasitic disease surveillance requires laboratory surveillance, host and vector surveillance, active surveillance, sentinel surveillance, pathogen surveillance, and interventional input surveillance. Insect-borne parasitic disease surveillance requires laboratory surveillance, host and vector surveillance, active surveillance, sentinel surveillance, epidemiological factor surveillance, and interventional input surveillance. Soil-transmitted parasitic disease surveillance requires active surveillance, sentinel surveillance, pathogen surveillance, and epidemiological factor surveillance. Foodborne parasitic disease surveillance requires laboratory surveillance, symptom surveillance, pathogen surveillance, epidemiological factor surveillance, and media information surveillance. Opportunistic parasitic disease surveillance also requires laboratory surveillance, symptom surveillance, pathogen surveillance, epidemiological factor surveillance, and media information surveillance (Table 13.1).

13.1.2 Construction

The construction of the surveillance network is based on the scientific configuration. According to different surveilled species and surveillance systems that have been identified, specific monitoring methods should be selected and combined to surveil parasitic diseases. The appropriateness of the selection is fully reflected in the sensitivity, specificity, and complete performance of the whole surveillance network as well as its operability, acceptability, and representativeness.

13.1.2.1 Principles for the Selection of the Combination of Surveillance Systems

In the construction of a parasitic disease surveillance network, the issue of completeness should be considered first. A complete surveillance network should reflect the current situation and development of all parasitic disease epidemics so that it can guide the prevention and control of parasitic diseases. In terms of completeness, first, a surveillance system should be organically integrated, and the data and information should complement each other so that human resources can be dedicated to surveillance. Second, the selection of populations and areas for surveillance should be representative, involving the selection of surveillance areas, samples of the population, and specific time periods. If only one surveillance or data collection method is used, it will not be sufficient to explain a problem or a category of problems. Local prevention and control efforts must be combined with the appropriate surveillance method to compensate for the shortcomings of different methods. For example, in endemic areas of Kala-azar disease, if only the reported cases of Kala-azar in various hospitals were analyzed, due to the small number of cases and scattered distribution, it would be difficult to sensitively detect the trend of group onset. Therefore, it is necessary to conduct a comprehensive analysis of the spread and incidence of Kala-azar in various regions based on death statistics, on-site investigations of the population distribution of sandflies, and local livestock feeding data to determine whether intervention measures should be taken to prevent the occurrence of a local mass Kala-azar epidemic.

13.1.2.2 Selection Principles of Surveillance Network Coverage Area

When selecting surveillance areas, the following principles should be observed: first, the surveillance area should be representative of the overall level, so random sampling methods should be chosen as much as possible. Second, the disease spectrum and the incidence rate of the main diseases in the surveillance area should be considered. The size of the surveillance area varies with changes in the disease spectrum and the incidence rate of the surveilled diseases. Third, the special

Table 13.1 Selected parasitic disease surveillance systems in China

Monitoring system	Zoonotic parasitic diseases	Insect-borne parasitic diseases	Soil-transmitted parasitic diseases	Foodborne parasitic diseases	Opportunistic parasitic diseases
Routing reporting surveillance system based on notifiable infectious diseases	Schistosomiasis Echinococcosis	Malaria Filariasis Kala-azar			Amoebic dysentery
Laboratory monitoring system based on laboratory tests	Schistosomiasis			Clonorchiasis, angiostrongyliasis	*Cryptosporidium* infection, *Giardia lamblia* infection
Symptom monitoring based on hospital visits	Schistosomiasis Echinococcosis	Malaria Filariasis Kala-azar Babesiosis		Clonorchiasis, angiostrongyliasis	*Cryptosporidium* infection, *Giardia lamblia* infection
Proactive monitoring based on specific population	Schistosomiasis Echinococcosis	Malaria Kala-azar	Pinworm infections	Clonorchiasis, angiostrongyliasis	Protozoa infections
Area-specific sentinel monitoring	Schistosomiasis	Malaria	Roundworm infections, Hookworm infections	Clonorchiasis, angiostrongyliasis	
Animal or vector-based hosts and vector monitoring	Schistosomiasis Echinococcosis	Malaria Kala-azar		Clonorchiasis, angiostrongyliasis	
Pathogen-based monitoring of the distribution of parasite populations in the environment	Schistosomiasis	Malaria	Roundworm infections, Hookworm infections Whipworm infections	Clonorchiasis, angiostrongyliasis	
Intervention input monitoring based on a database of specific disease control	Schistosomiasis Echinococcosis	Malaria	Roundworm infections, Hookworm infections Whipworm infections	Clonorchiasis, angiostrongyliasis	
Monitoring of media information	Schistosomiasis Echinococcosis	Malaria		Trichinellosis Clonorchiasis Cysticercosis, angiostrongyliasis	
Prevalence factor surveillance based on transmission risk factors	Schistosomiasis Echinococcosis	Malaria		Clonorchiasis, angiostrongyliasis	
Insect resistance and vector pesticide resistance monitoring based on drug use	Schistosomiasis	Malaria			

characteristics of the surveillance area should be comprehensively considered, including the surveillance capacity, accessibility, and resource allocation. To ensure the representativeness of the surveillance area while reflecting its specificity and feasibility, stratified random sampling is often adopted for the selection of the surveillance area.

13.1.2.3 Case Analysis

To illustrate the significance of the configuration and construction of the parasitic disease monitoring network, the configuration and construction of schistosomiasis sentinel surveillance and symptom surveillance for *Angiostrongylus cantonensis* infection are taken as examples to illustrate the functions of different monitoring methods and the principles and ideas in the construction of the monitoring network.

1. Configuration and construction of sentinel surveillance for schistosomiasis in Jiangsu Province
   (a) The problem. Jiangsu Province is located in the lower reaches of the Yangtze River and the five provinces in the lake region where schistosomiasis is the most prevalent in China. Schistosomiasis occurs occasionally in some areas along the Yangtze River and rivers directly connected with the Yangtze River, and there is a high incidence of schistosomiasis in the annual flood season. How to accurately grasp the situation of schistosomiasis infection in the risk environments environments and water bodies along the Yangtze River in a timely manner, take effectively active and pertinent measures, eliminate the threat of schistosomiasis infection, and further improve the level of prevention and control of acute schistosomiasis is very important for realizing the long-term planning goal of schistosomiasis prevention and control. Since 2009, Jiangsu Province has begun to explore the establishment of monitoring and early-warning mechanisms for schistosomiasis infection in the high-risk environments and waters, aiming to determine environments and water areas that are susceptible to schistosomiasis transmission in advance, issue early warnings in time, take effective prevention and control measures, eliminate the hidden dangers of schistosomiasis infection, and prevent the occurrence of acute schistosomiasis epidemics. In this way, the emergency response capability and the level of acute schistosomiasis control in Jiangsu Province have been further improved, the process of schistosomiasis control in the whole province has been accelerated, and the achievements of schistosomiasis control have been continually consolidated.
   (b) Monitoring Methods
      - Scope of monitoring: The environment and water bodies along the banks of the Yangtze River and around marshland of the islands inside of the Yangtze River in Nanjing, Zhenjiang, Yangzhou, Taizhou, and Changzhou cities where *Oncomelania* snails are distributed are covered, including the north shore from Wujiang town, Pukou district of Nanjing city, to Yongan town, Gaogang district of Taizhou city; the south shore from

Tongjing town, Jiangning district of Nanjing city, to Xiaohe town, Xinbei district of Changzhou city; and the Jiangsu-Anhui-Shuiyangjiang river basin.

- Identification of monitoring sites: (a) Based on work documents for the prevention and control of schistosomiasis, monitoring sites were identified in accordance with the following factors, such as key susceptible environments and water bodies, selected environments and water bodies where infected snails were found in previous years or where acute schistosome infections had occurred. These selections of the monitoring sites were based mainly on the following five aspects: fishing villages with snail distribution, environments with snails around densely populated areas (enterprises along rivers, large construction sites, and wharves), environments where infested snails have been found in the previous 3 years, environments with snails on the borders of administrative regions, and snail-infested environments with frequent human and animal activities (including streams leading to the Yangtze River). (b) A surveillance database of high-risk susceptible areas and water areas was established. Early-warning monitoring sites were determined according to previous schistosomiasis epidemiological survey information and key environmental and water conditions in Nanjing, Changzhou, Yangzhou, Zhenjiang, and Taizhou. Each monitoring site was subjected to global satellite positioning, on-site photography, and video recording to assess the basic situation of the adjacent environment (riverbanks) and draw a distribution diagram. After on-site inspection and consultation between the provinces, cities, and countries (districts and cities), 50 monitoring sites for schistosomiasis infection in key environments and waters of the province were identified. Additional municipal mobile early-warning monitoring points can be set up in municipalities in relation to demand.
- Time of monitoring and surveillance: According to the epidemiological characteristics of schistosomiasis and the high-risk infection season of schistosomes along the Yangtze River, the monitoring time for schistosome infection in key environments and water areas is from May to September every year, with 2 days of high water levels every month.
- Early-warning and monitoring methods: Methods of placing sentinel mice on site to monitor cercariae on the water surface, manually placing sentinel snails to monitor a water body, and recording human and animal contact with infested water were used to monitor the high-risk factors for schistosome infection.

  The sentinel mouse assay was used to determine the infectivity of the water bodies by the mouse cage method, with 20 mice placed at each site and divided into 2 cages spaced 10–20 m apart. The assay was carried out once a month for 2 days (on the 18th and 19th or third and fourth days of the lunar month) for 8 h and daily for 4 h from 1000 to 1400 in May each year (starting from the 18th or 3rd day of the lunar month). Each sentinel mouse was dissected 30 to 35 days after having been in contact with water to check whether it was infected by schistosomes. If the results were positive, an emergency warning was issued, and measures against acute schistosomiasis were taken immediately.

For the sentinel snail assay, 500 snails that were not infected by schistosomes were collected from each site, equally divided into two pylon cages, and placed 5–10 cm below the water surface at intervals of 10–20 m. The sentinel snails were placed for the same duration as the sentinel mouse assay (except that the snails were not retrieved during the night when they were placed continuously). The infertility of the snails by miracidium in the water was measured. After each measurement, the snails were kept in a constant temperature chamber at 25 °C. After 2 months (the time it takes for the miracidium to infect the snails and develop into larvae), the sentinel snails were dissected and observed by crushing to check for infection and identify the type of larvae. An early warning of source infection was issued at sites with infectious snails, and relevant measures were taken to strengthen the control of the source of infection. A site with infectious snails was also classified as a key environment for snail inspection and eradication in autumn.

In addition, it is important to monitor relevant factors. While carrying out sentinel mouse and snail measurements, the exposure of people to infested water should be registered, and records should be kept of boats at anchor and their activities and the activities of people and animals on nearby riverbanks. At the same time, people who live on boats around the monitoring sites where infested snails have been found need to be checked for infection, and the source of infection should be investigated.

- Quality control: Technical training should be provided to the relevant personnel involved in monitoring so that they can be trained before they start work, and the staff should operate in strict accordance with the technical protocols. The on-site measurement time should be strictly 4 h per day for 2 consecutive days, and the total hours should not be fewer than 8 hours. The sentinel mice should be covered tightly after being placed in the measuring cage to prevent them from escaping. When the cage is placed in the water, the distance between the cage and the water surface should be adjusted to ensure that the limbs and tails of the sentinel rats are in contact with the water surface. The cage should be secured to prevent it from being washed away by the river and to prevent cats and snakes from coming close to it and damaging the sentinels. The sentinel mice should be brought back to the laboratory for feeding and watering after the measurement.

  Sentinels should be monitored in accordance with the requirements of "unified scheme, time, training and specification." Provincial and municipal technicians should supervise and inspect the whole process of early warning and monitoring, while professional technicians at the county level should be responsible for the on-site work and look after the monitoring sites in accordance with the requirements of personal responsibility.

(c) Monitoring results: In 2009, a total of 23 positive sentinel mice were detected in Jiangsu Province, for a total positive rate of 0.53%. A total of 53 schistosomes were detected, with an average worm load of 2.30/mouse and a maximum worm load of 12/mouse. Twelve positive sites were detected in 5 batches (of which 1 site was a repeated positive site). Eleven positive monitoring points were detected, of which 7 were located on the north bank of the Yangtze River, 3 on the central island, and 1 on the Shuiyang River system bordering Anhui. From May to September, there were 1, 2, 0, 1, and 8 positive sites, respectively, suggesting that the infectivity rate increased significantly after the Yangtze flooded.

Based on field observations and data from laboratory tests, four emergency warnings for positive pathogenic tests and one early warning for infection sources were issued in Jiangsu Province. The provincial schistosomiasis and endemic disease control office and relevant cities were provided with 11 high-risk early-warning maps of schistosomiasis infection in key water areas, maps of key distribution areas of infection sources, and relevant results. The municipalities concerned quickly activated their emergency response mechanisms based on the results of the warnings and immediately formulated an emergency work plan that focused on targeted prevention and control measures for the 11 sentinel mouse anatomically positive sites and the 5 sentinel snail monitoring positive sites, which effectively prevented the occurrence of an epidemic.

(d) Indications

- Overall improvement of prevention and control of acute schistosomiasis in riverside areas. Early warning of schistosomiasis infection was carried out in key environments and water bodies along the river from May to September, mainly through active monitoring and on-site measurement and observation, comprehensive detection of larvae and trichinae in water and high-risk groups, detection of areas with a high risk of schistosome infection, and people in contact with infested water. All these results were first released through Internet technology to schistosomiasis prevention and control agencies at all levels. The symptoms and areas of schistosomiasis infection along the Yangtze River were identified in advance, and prevention and control measures were taken in advance, which effectively prevented and controlled the occurrence of acute schistosomiasis, enabled all places to implement schistosomiasis prevention and control measures, and truly achieved the goal of emergency response and preparedness.
- Preliminary mapping of high-risk areas and times for schistosome infection in the Yangtze River in Jiangsu Province. Through continuous testing, the northern bank of the Yangtze River and the central island were initially mapped as high-risk areas for schistosome infection, and there was also a potential risk of acute schistosome infection in the Shuiyang

River on the border with Anhui Province. The results of detection in the different months indicated that snails infected in the previous year released their cercaria mainly in May to June and then died. Infected snails on the riverbanks began to mature in August, and a peak in the release of cercaria occurred after September, resulting in a very high level of schistosome infection in the river. Therefore, the period from August to October is a critical time of year for the prevention and control of acute schistosomiasis in the Yangtze River basin in Jiangsu Province.

- Work experience was initially accumulated to guide the direction of future work. The principles of early-warning point determination should be strictly followed, the scientific and overall configuration strengthened, the sensitivity of surveillance and early-warning methods ensured, and work efficiency improved. Research on early diagnosis methodology should be performed to improve the timeliness of surveillance and early warning, thus allowing more time for on-site prevention and control of schistosomiasis infection.

2. Configuration and construction of symptom surveillance of *Angiostrongylus cantonensis* infections in the southern region of China

   (a) The problem. *Angiostrongylus cantonensis* infection in human is a new parasitic disease with eosinophilic meningoencephalitis as the main clinical manifestation. It is caused by *Angiostrongylus cantonensis* larvae parasitizing the human body. In southern China, most cases are caused by eating raw or undercooked *Pomacea canaliculata* or *Achatina fulica*, the snail intermediate host of the parasite which contain the third-stage larvae of *Angiostrongylus cantonensis*. From 2006 to 2007, the national survey on the natural foci of *Angiostrongylus cantonensis* showed that *Pomacea canaliculata* was widely distributed in 10 southern provinces (autonomous regions and municipalities), among which 7 provinces (autonomous regions) presented the natural foci of *Angiostrongylus cantonensis*, and *Pomacea canaliculata* and *Achatina fulica* were sold in farmers markets or restaurants in more than 20 counties (cities), among which infection was detected in 9 counties (cities). To explore and establish a sensitive and rapid prevention and control mechanism for *Angiostrongylus cantonensis*, reduce the risk of transmission, and effectively reduce the incidence and control of the outbreak, a pilot program of symptom monitoring and early warning for *Angiostrongylus cantonensis* was carried out in southern China.

   (b) Monitoring methods

   - Pilot selection. Based on the distribution of natural epidemic sites and previous outbreaks across the country, pilot sites were selected in three categories: first, highly endemic areas, including Fuzhou and Xiamen cities in Fujian Province; second, areas that were susceptible to being natural epidemic sites and had reported outbreaks, including Kunming, Dali Autonomous Prefecture, and Jinghong in Yunnan Province; and third, areas that were not natural epidemic sites but had reported outbreaks, including Beijing. Sixteen health institutions (including CDCs and sentinel hospitals) were selected in the pilot areas to conduct symptom surveillance, behavioral surveillance, and vector surveillance.

- Symptom surveillance. According to the definition of surveillance cases, cases with clinical symptoms of meningoencephalitis and a recent history of eating mollusks such as edible snails (mainly freshwater snails and terrestrial snails) or slugs were monitored throughout the year. When the medical and health institutions in charge of surveillance found the cases, they made a definite diagnosis according to the epidemiological history, clinical symptoms, and laboratory examination results, referring to the diagnostic criteria for *Angiostrongylus cantonensis*. At the same time, they kept complete case-related data, stored serum, and reported the cases through the *Angiostrongylus cantonensis* Monitoring and Reporting System of the NIPD within 24 h after finding the symptom surveillance cases. When the CDCs in the pilot areas found *Angiostrongylus cantonensis*-infected snails in their vector monitoring, they reported these findings to the local symptom monitoring medical and health institutions within 24 h. After a local CDC found reported monitored cases or was notified of cases by the monitoring medical and health institutions, they verified the reported cases and conducted a case investigation according to the component requirements of the "Angiostrongyliasis Case Questionnaire." Medical and health institutions responsible for symptom monitoring assisted in case investigation. According to the case investigation information, food sources were traced, and markets and restaurants were checked.
- Edible snail monitoring. The monitoring targets included terrestrial or freshwater snails, such as *Ampullaria gigas* and *Achatina fulica*, which are sold in markets or restaurants. According to the local situation, one large aquatic product or farmers market and three large catering places (including food stalls) that sold edible snails such as *Pomacea canaliculata* and *Achatina fulica* were selected as fixed monitoring sites in cities, urban-rural fringe areas, and rural areas, respectively. They were then sampled twice a month in the snail sales season every year (May–October), with an interval of 2 weeks each time. When infected snails were found, samples were taken for 5 days continuously. When the CDCs in pilot areas found *Angiostrongylus cantonensis*-infected snails in the vector monitoring, they reported these findings to the provincial CDCs within 24 h. The provincial CDCs then reported the findings to the NIPD within 24 h.
- Behavioral monitoring. This included individual and group behavioral monitoring. In individual behavioral surveillance, the target population included parents of primary and secondary school students in the surveillance area. The main components of the survey were personal eating habits and knowledge of *Angiostrongylus cantonensis* control. The primary and secondary school students brought the "Behavioral Survey Form on Angiostrongyliasis cantonensis" to their parents for completion. Primary

and secondary schools were selected in areas where the vector surveillance pilot sites were located. For each pilot site, a cross-sectional survey covering 3 years was performed at the beginning of edible snail surveillance. Group behavioral monitoring mainly involved monitoring the volume of edible snails such as *Pomacea canaliculata* and *Achatina fulica* sold in markets and restaurants. The monitoring method was mainly to ask the owners of each market booth and restaurant to summarize the volume of edible snails sold.

- Epidemic early warning. In the pilot areas, an epidemic early-warning mechanism was established, especially for mass epidemics. Epidemic tracking and investigation were carried out, and epidemic trends were analyzed so that timely preventive and control measures could be implemented. When a case of angiostrongyliasis cantonensis was detected in Guangzhou (including local infection and imported cases) in a surveillance hospital in the pilot area, it was reported through the online reporting system in a timely manner. The local CDC reported to the local health administrative department within 24 h upon receipt of the reported information, and the provincial CDC and the NIPD at China CDC were also notified. If a hospital reported a case of local infection, the local CDC increased surveillance of edible snails in markets and catering establishments and expanded the scope and frequency of testing when appropriate. If the number of reported cases increased to five or more and if edible snails sold in markets or catering establishments were found to be infected with *Angiostrongylus cantonensis*, local authorities were advised to issue warnings as appropriate and to ban the sale of these edible snails in markets and catering establishments accordingly.

(c) Monitoring results. Between 2009 and December 2011, a total of 198 cases were reported, and 10 cases were excluded from the direct reporting through the network system. Fujian Province accumulated the highest number of reported cases, accounting for 60.6%, and Yunnan Province reported 35.4% of the cases. No case was reported in Beijing.

Except for Beijing, edible snails were found to be sold in all monitoring pilot sites. Among them, *Pomacea canaliculata* was sold in Fuzhou, Xiamen, Dali, Jinghong, and Kunming cities, and *Pomacea canaliculata* infected by *Angiostrongylus cantonensis* was detected in Fuzhou, Xiamen, Dali, and Jinghong cities. However, no infectious *Pomacea canaliculata* was found in Kunming. Only one batch of *Pomacea canaliculata* was found to be infected with *Angiostrongylus cantonensis* in Jinghong City. Although the other two species of freshwater snails (*Bellamya* spp. and *Cipangopaludina* spp.) were routinely monitored at all monitoring points, only Fujian had positive results. However, the infection rate and infection degree were very low.

The data of three monitoring pilot sites with high snail infection rates in Dali, Fuzhou, and Xiamen indicated that there was a significant fluctuation in the snail infection rate in Dali, due mainly to the different origins of the snails. *Pomacea canaliculata* comes mainly from Ruili, and there is a high infection rate around the Spring Festival every year. The main source of *Pomacea canaliculata* in summer and autumn is Eryuan County (a nonnatural epidemic focus of *Angiostrongylus cantonensis* according to the national survey). In Xiamen and Fuzhou cities of Fujian Province, *Pomacea canaliculata* was rarely sold in farmers' markets, but it was sold in small-scale restaurants with wide channels. Fujian Province was a serious epidemic area of *Angiostrongylus cantonensis* in China, and all the provinces were natural foci of *Angiostrongylus cantonensis*. The infection rate of *Ampullaria canaliculata* was maintained at approximately 12.3% to 12.4%, which was similar to the previous investigation results of natural foci and had no obvious seasonality.

Three consecutive outbreaks of angiostrongyliasis cantonensis occurred in the surveillance pilot sites of Dali and Kunming City during the surveillance process, and positive emergency responses were implemented. The first outbreak occurred in February–March in Dali, with 18 people being infected at a dinner party, 16 people by consuming snail meat, 14 patients with clinical symptoms, and 6 patients hospitalized. NIPD immediately dispatched professional staff to conduct a questionnaire survey and clinical diagnosis verification of the patients in collaboration with the local disease control authorities. According to the questionnaire survey, the patients had purchased the snails themselves at the Xiaguan market, and the snails originated from Ruili, Yunnan. Therefore, the response team traveled to Ruili and collaborated with technicians from the Yunnan Institute of Parasitic Diseases (YIPD) to sample and test *Pomacea canaliculata* from 9 to 12 April in Ruili, which was identified as the natural source of the outbreak. The outbreak of angiostrongyliasis cantonensis in Dali was closely related to food sources. Based on the monitoring of marketed *Ampullaria gigas*, the main sources of *Pomacea canaliculata* sold in Dali City were the local Eryuan County and Ruili in the southern region of Yunnan Province. There were seasonal variations in the sales of *Pomacea canaliculata* from the two different sources. For example, the snails came mainly from the local area in the second half of the year, while around the Chinese New Year and in the first half of the year, they came mainly from Ruili. The monitoring data indicated that the infection rate of *Pomacea canaliculata* was high in southern Ruili, and positive *Pomacea canaliculata* was rarely found in most sampling batches in the area. The results of two surveillance experiments (case surveillance and snail marketing surveillance) showed that the outbreak in Dali City in February 2011 had a significant relationship with the infection rate of *Pomacea canaliculata* in the market. Through publicity about the outbreak and food processing hygiene, the outbreak was effectively controlled.

During the monitoring process, training courses on monitoring technology and seminars on prevention and monitoring were held every year to discuss problems in the monitoring process and revise the monitoring pilot program. At the same time, problems can be found in time through the supervision and investigation of on-site monitoring every year, and the information exchange between the monitoring pilot sentinel hospitals and the disease control departments can be strengthened so that a "regular meeting system" and a "zero reporting system" can be established between the two departments to promote the exchange of monitoring experience.

(d) Recommendations

It is essential to strengthen cross-sectoral collaboration. When CDCs find positive *Pomacea canaliculata* through surveillance, especially during times of high incidence (e.g., during the Chinese New Year), there should be appropriate channels for communicating with the business administration and health supervision departments. The establishment of such channels requires coordination at the administrative level in order to control the occurrence of the disease.

13.2 Surveillance Scope and Procedures

13.2.1 Scope

The scope of parasitic disease surveillance includes all surveillance components. The components of parasitic disease surveillance are determined by the transmission characteristics and control objectives of different diseases (e.g., zoonotic parasitic diseases, vector-borne parasitic diseases, soil-transmitted parasitic diseases, foodborne parasitic diseases, and opportunistic parasitic diseases) in order to achieve prevention and control or elimination of parasitic diseases. The overall surveillance procedure also determines the process of implementing different disease surveillance components.

13.2.2 Procedures

The surveillance procedure can be summarized in three steps, namely, the early, middle, and late stages of monitoring, with different components in each stage, as shown in Fig. 13.1.

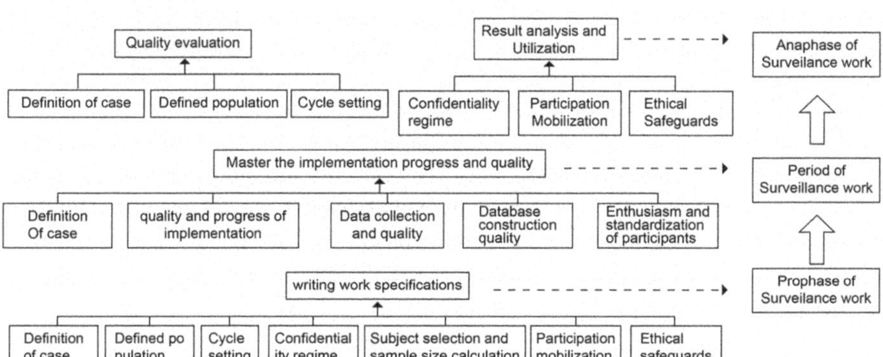

Fig. 13.1 Procedures and components of parasitic disease surveillance

13.2.2.1 Before Monitoring

First, a monitoring work plan should be written and the monitoring scope determined according to the parasitic diseases in different regions, including the definition of cases to be monitored, the determination of the population to be monitored, the setting of the monitoring period, the provisions of the monitoring confidentiality system, the selection of monitoring objects and sample size, the mobilization of monitored participants, and the protection of monitoring ethics.

13.2.2.2 During Monitoring

The implementation progress and quality of monitoring should be mastered, including the completion of monitoring work, the quality and progress of monitoring technology implementation, the quality of data collection and recording, the quality of database construction, and the enthusiasm and standardization of monitored participants.

13.2.2.3 The Later Stage of Monitoring

The quality of monitoring is evaluated after the surveillance finished, and the monitoring results are then analyzed and utilized. The evaluation includes checking the completion rate of monitoring, checking the compliance degree of monitored samples, and verifying the quality of the monitoring database. The analysis includes statistics of monitoring results, writing and auditing of monitoring reports, and feedback and transmission of monitoring information.

In short, the scope of parasitic disease monitoring includes the formulation of a monitoring plan, implementation of monitoring, quality control, writing and feedback of monitoring reports, and evaluation of monitoring effects.

13.3 Formulation of Surveillance Plan

The formulation of the monitoring program is conducive to monitoring different diseases, and different monitoring methods can achieve maximum results in the overall monitoring system. Different monitoring methods can be selected according to different needs and resources, and different monitoring methods can be applied at the same time. These methods include the notifiable disease reporting system, laboratory monitoring results based on laboratory testing, epidemiological surveys based on populations, monitoring results based on field networks, statistical results based on cause-of-death information, and sentinel monitoring results based on sentinel detection of unknown health factors. The above monitoring work is fixed in the form of a scheme, and the correlation analysis between the monitoring data of a certain area and other information resources can expand the scope of the monitoring data and lead to unexpected complementary effects. In the monitoring program, the following six aspects should be defined.

13.3.1 Case Definition

The definition of a surveillance case is the most basic task, as the goal and effects of the surveillance system must be evaluated. The definition requires a balance between the need for monitoring sensitivity, specificity, and feasibility. During this process, the accessibility of diagnostic tests, the method of use, the ability of personnel to interpret the results, and the preservation of diagnostic test evidence files are all important. Furthermore, the definition of a monitored case should be as simple as possible.

Some diseases may be classified according to information sources, such as suspected cases and confirmed cases. For behavior and exposure monitoring of health problems, the definition of monitoring can be determined according to self-reports, observation, and collection and testing of biological samples. It is reasonable to have no definition of individual diseases or health problems, as the definition can vary greatly in different environments according to the information needs, methods of obtaining reports or data, staff education, and resources. For example, the definition of hepatitis A surveillance is more successful. However, in practice, due to the lack of diagnostic tools, some asymptomatic infected people or groups of children are often excluded because their exposure history is ignored. In this case, the definition of surveillance cases needs to include these epidemiologically related cases in order to improve the sensitivity of surveillance and meet the needs of the actual situation, but it may be too complicated to implement.

For diseases with a long or delayed course, the definition of monitored cases should be determined according to the monitored course of disease, such as early symptoms, late symptoms, or death. For chronic parasitic diseases with a long cycle, the monitoring stage of the disease course should be determined according to different requirements, but the infection period is the most important. For example, the definition of symptom monitoring of *Angiostrongylus cantonensis* is set according to the diagnostic criteria. When monitoring the infection of *Pomacea canaliculata*, if the third stage of cercariae is found, then it is infectious. However, because infected cercariae are easily missed by enzymatic digestion, infected *Pomacea canaliculata* can be diagnosed by dissecting the nodules on the gills. This method is simple and sensitive. Therefore, the epidemic trend in an area should be judged when the definition of surveillance is determined, and the method should be simple and easy to implement.

In the clinical diagnosis of parasitic diseases, doctors can identify different clinical stages by diagnosing each individual. Some patients can receive the necessary medical subsidy or enjoy the national free medical policy according to the diagnosis of clinicians, which is often inconsistent with the definition of monitored cases. The definition of surveilled cases is determined according to the characteristics of group transmission. As long as there is a certain level of infectivity or potential infectivity, a case should be regarded as a surveillance case, with the aim of improving surveillance sensitivity and taking effective prevention and control measures as soon as possible to achieve the purpose of blocking transmission. The national policy of free medical care reflects the government's concern for the population in endemic areas under certain historical conditions. Sometimes, only patients requiring extensive medical treatment receive free treatment, such as the free treatment of advanced schistosomiasis cases and echinococcosis surgical treatment cases. Therefore, when determining whether a patient can benefit from the national policy of free medical care, the definition of a monitored case cannot be adopted. In addition, the specificity of monitoring cases should be emphasized, and the specificity of cases should not be ignored in order to improve sensitivity. The definition of surveillance cases is generally based on national or industrial diagnostic standards, and the monitoring objectives should be fully considered in formulating it.

Usually, it is not enough to monitor the disease only to implement disease prevention and control measures. It is necessary to monitor the exposure of people or the dangerous behaviors that can cause diseases, especially when public health resources allow relevant measures to avoid exposure factors or dangerous behaviors. For example, if the monitoring purpose is to eliminate malaria in a certain area, then the monitoring should include the monitoring of mosquito population and density and the environmental factors related to population distribution. It is also necessary to observe or master the exposure behaviors of people in general in this area, as well as the characteristics of particular people (i.e., the main exposed people), and measure the exposure frequency of such high-risk people, which is conducive to designing measures to change the exposure behaviors of high-risk people to promote malaria elimination in this area. This shows that when designing monitoring indicators in a certain area, the observed data indicators should be simple and easy to collect for large groups of people so that information can be obtained in a large range. For example, the exposure behaviors of different groups of people or in different areas can be separated. However, collecting data for small or high-risk groups calls for increasing accuracy, such as changes in the exposure frequency of high-risk groups in different times and spaces, to facilitate further analysis and formulation of accurate and effective prevention and control measures. It is very important to observe the principle of combining this with other aspects for the design of monitoring indicators with different monitoring purposes.

13.3.2 Population Identification

All surveillance systems must target specific populations, either in particular units (e.g., hospitals, schools, factories) or among residents of a specific community, region, or ethnic group, or even a group of people living in different countries. Some monitoring is intended to monitor all events, some to monitor representative events, and some to monitor health events among people in a certain area (based on a mass monitoring system). In addition, some monitoring is based on the results of a previous representative evaluation or on monitoring people at different points of participating in a monitoring system, until these points are finally connected to a monitoring network (convenience sampling).

Population-based surveillance systems are mainly using routine reporting systems for notifiable infectious diseases that require health staff to provide information about special cases to health authorities for statistical analysis. Other population-based surveillance systems are designed on the basis of representative sample populations or institutions, such as special cross-sectional surveys, and this information can be used for national-level surveillance of multiple diseases or health events; this approach can provide data with sufficient frequency and regional coverage. However, national surveys may provide better information for particular regions.

Although monitors would like monitoring systems to include all or statistically significant events, in many cases, this type of approach is more difficult to implement. Participants have different interests, aspirations, and abilities, and some monitoring systems are used primarily for certain monitoring sites that are nonrandomly selected, for example, when the target population is expected to be representative of a diverse composition of participants. In such cases, the actual population in the surveillance system may be influenced by certain outpatient primary care services, living in the chosen city or working in the chosen factory. For example, for the surveillance of *Angiostrongylus cantonensis*, China has established a surveillance network in Guangdong, Fujian, Yunnan, Guangxi, and Shanghai Provinces (cities). The main targets of this surveillance network are suspected patients who visited sentinel hospitals in endemic areas during the epidemic season. Through the analysis and follow-up investigation of information on the consultation and treatment of these cases, the location of *Angiostrongylus cantonensis*, the transmission route, and the source of the epidemic could be determined to guide prevention and control measures.

13.3.3 Cycle Setting

The surveillance system can be presented as a cycle of information. The exchange of information consists of two parts: collection and aggregation from the various collecting agencies and distribution to the various agencies that need to use the information. A typical surveillance cycle originates with the occurrence of a health event and includes reporting to health agencies, analysis and interpretation of the aggregated data, and finally distribution of the information. This process involves manual data collection, reporting, and data transmission, with some data transmission also involving telegraph, telephone and satellite technology, web-based technology, etc. Regardless of the technology used, it is critical that important information be communicated correctly and quickly to those who need to use the data. In parasitic disease surveillance programs, the surveillance cycle is often set at 5-year intervals, which are then partially adjusted to the situation and requirements of disease control program and changes in the epidemic in order to further advance surveillance. In particular, when some diseases (e.g., malaria) enter the elimination phase, surveillance and response become the main focus for this period. The surveillance components and indicators should then also be further adjusted.

The timing of surveillance is determined by the epidemiological season, the indicators monitored, and the purpose of the surveillance. For example, malaria surveys are conducted at the peak of the epidemic season or before or after the break in treatment (non-transmission season), schistosomiasis surveys are conducted 1 month after the epidemic season, and snail distribution and density surveys are conducted in spring.

13.3.4 Confidentiality

Personal information is an important basis for identifying whether a case has been repeatedly reported. Further investigation is required not only to provide individuals with a reporting service but also certain monitoring information can be used as a basis for further research. The physical protection of data security and the confidentiality of surveillance information are two issues that involve ethical responsibilities and the need to maintain the trust of those involved in the surveillance process. Regulations require that disease reports be kept in civil administration, that information be protected, and that there be penalties for the misuse or disclosure of confidential information. Measures to ensure data security include restricting access to sensitive data, keeping documentary data in a dedicated room, and setting passwords within computer systems or network systems. In addition, certain approval procedures or access requirements should be established for those who need to use the data (including employees of the unit or those from outside the unit who need to analyze the data). These approval procedures and access requirements should be part of the training of new employees and the evaluation of the quality of data storage.

To further protect data from leakage, some unnecessary personal information may not be collected and stored. Given that many data analysts do not require personal information, monitoring data should be stored electronically in different versions of files, some with and some without personal information, so that they can be made available to users with different needs. Personal information is sometimes needed only locally and does not need to be reported.

13.3.5 Target Selection and Sample Size Calculation

The target population for surveillance is a sensitive group of people determined by the purpose of the surveillance. For example, emerging cases of soil-transmitted parasitic diseases and echinococcosis are monitored mainly in school-age children, while adult males are monitored for clonorchiasis.

The sample size for surveillance is determined primarily by the level of morbidity of the disease being monitored and the quality requirements of the surveillance and is commonly calculated in two ways.

13.3.5.1 Calculation of Sample Size When Resources Are Adequate

When surveillance resources are sufficient, the formula for the minimum sample size (n) for a sample survey can be calculated as follows:

$$n = \frac{Z^2\sigma^2}{d^2}$$

where z is the confidence interval, n is the sample size, d is the sampling error range, and σ is the standard deviation, generally 0.5. The formula shows that the sample size is directly proportional to the confidence level. The sample size is proportional to the population variance, and the larger the overall difference, the larger the required sample size. The sample size is inversely proportional to the square of the allowable estimation error, that is, the larger the square of the estimation error is, the smaller the sample size is.

13.3.5.2 Calculation of Sample Size When Resources Are Inadequate

When surveillance resources are insufficient, a minimum sample size can be calculated according to the lot quality assurance sampling (LQAS) method. This method was first used for quality monitoring of industrial products and then began to be used in the early 1980s to monitor and evaluate health programs. The term "lot" refers to a defined group of units in a sampling study, e.g., the smallest unit commonly used in the surveillance of parasitic diseases is the administrative village. The advantage is that only a small number of samples need to be surveyed, and this is a low-cost, cost-effective, and rapid method of evaluating preventive measures that has been used in parasitic disease surveillance. The basic formula is

$$P_a = \frac{n!}{a!(n-a)!} P^{\alpha} \times q^{n-a}$$

where Pa is the probability of the number of infected cases drawn from sample n; p is the actual infection rate; q is the actual noninfection rate, $q = (1-p)$; and n-a is the number of noninfected cases in the sample, namely, the characteristic value (d).

For example, an administrative village with a population of 2000 people was selected as a monitoring site. If the schistosomiasis infection rate (P_0) in this village is 1%, the result (or calculation according to the above formula) indicates that the required monitoring sample is 530 people (Table 13.2).

13.3.6 Participant Mobilization

Surveillance participants include monitoring targets and personnel participating in monitoring data collection. Participant mobilization refers to the mobilization and training of personnel participating in the collection of monitoring data before monitoring and the mobilization of monitoring targets to ensure that monitored

Table 13.2 Sample sizes determined by LQAS under different schistosomiasis infection rates ($d = 2$, $a = 0.05$)

Administrative village (batch) population	Assumed prevalence of schistosomiasis infection in the population of the endemic area (P_0)					
	0.01	0.03	0.05	0.01	0.15	0.2
500	364	166	106	55	36	27
1000	459	181	112	56	37	27
1500	514	192	117	58	38	28
2000	530	192	117	58	38	28

participants can be involved in the whole process of monitoring. The motivation of monitored participants is key to ensuring the success of monitoring. Therefore, mobilizing monitored participants to actively participate in monitoring is important to ensure the implementation of monitoring as well as its quality.

Within the surveillance system, there is a need to establish long-term and effective cooperation between the various participants. On the one hand, surveillance participants should be mobilized to actively participate in all monitoring work; on the other hand, the results of the monitoring analysis should be fed back to the participants in a timely manner so that monitoring information can be effectively applied. The best way to ensure that the active participation of participants continues is to provide timely feedback to the participants after the analysis of the monitoring information they provide.

Feedback on monitoring information can take the form of reports, seminars, or even inviting participants to participate in the analysis of the data themselves. Often, doctors in clinics or hospitals are interested mainly in how local monitoring information is compared with information from elsewhere, so it is a good idea to report the results to all participants.

Other methods of mobilization include paying participants directly. Direct payments to health service providers are not desirable for agencies implementing monitoring, as they are too expensive and do not promote a spirit of cooperation among health personnel. However, in some cases, payment may be an effective and appropriate method, for example, to increase the response rate of those who participate in the survey as well as of questionnaire participants.

In addition, the investigation process may involve relevant legal requirements, such as the need for reports to be combined with the licensing or certification of health affiliates. Legal requirements also need to be enforced; otherwise, the quality of reporting and transmission of information may be affected, even in the context of infectious disease control program.

13.3.7 Ethical Consideration

Ethical issues in public health require efforts to achieve a balance between conflicting interests and dangers. Conflicts of interest include people's legitimate right to protect their privacy versus the disclosure of government information. For example, if health information is leaked, it may lead to people being under pressure or experiencing discrimination. Many surveillance data may not be published when the total number of cases has not reached a certain threshold; e.g., if the total number of cases is only five, publishing the data could result in the identification of individual cases. Conversely, the release of information on groups with a high prevalence of disease in surveillance data can lead to inequities in health services, especially in regard to equity in health services for certain groups with lower economic and social status.

To reduce such individual crises, surveillance data should be kept in a responsible manner. To avoid stigma for people at risk of disease, it is particularly important to stress that surveillance data alone cannot explain the causes of health inequities. If constructive action is taken to address the risk factors suggested by surveillance data, the risks to these individuals and groups may disappear.

There is a risk that the surveillance system will be formally objected to by the ethics committee. As disease surveillance is a statutory act, it is distinct from research and certain studies.

13.4 Surveillance Implementation and Quality Control

The implementation of parasitic disease surveillance in China is carried out mainly at county-level CDCs after the Bureau of Disease Prevention and Control of the NHC issues a surveillance program and CDCs at all levels specifically formulate the implementation protocols. The following is an example of the implementation process and considerations for surveillance at county-level CDCs.

13.4.1 Work Schedule

Under the guidance of the superior CDC, the county-level parasitic disease monitoring organization needs to formulate the disease monitoring work plan and time arrangement every year and provide the corresponding professionals with the equipment needed to complete annual monitoring.

13.4.2 Quality Control of Surveillance Techniques

Quality control is often carried out in the form of spot checks by experts from higher-level CDCs on CDCs at or below the county level. The quality control of surveillance includes interviews and questionnaire surveys of technicians, rechecks of preserved samples, assessment of quality inspection standards (samples), and timely correction of problems to ensure the quality of the surveillance implementation process. If the technical quality of each surveillance implementation process can be followed up or supervised at any time, this can ensure not only that the surveillance will be completed according to the original plan but also the surveillance quality.

13.4.3 Data Collection and Trace Management

The collection of surveillance data is carried out by professionals in grassroots units and is the original information used for the entire monitoring process. The raw information, whether obtained by survey or by testing, must be clearly registered, fully documented, and consistent with the requirements for completing the forms. All original information should be kept in a safe place and entered into the computer in a timely manner.

Trace management is the basis for keeping a complete record of the entire surveillance and management process, for keeping evidence of the work quality, and for traceability in parasitic disease surveillance. Therefore, all units involved in surveillance at all levels should develop appropriate data management methods or programs to ensure the quality of trace management and that it can withstand inspections and supervision by higher authorities.

13.4.4 Database Construction

A disease information system is a large, general, and extensive database that can be used in disease surveillance, and certain databases can sometimes be used for other purposes. Coroners' records are usually legal texts that provide official formal proof of origin and death. Hospital case admission and discharge data can provide the basis for monitoring hospital usage and costs. The consumption of diagnostic reagents is also good data for monitoring the number of disease cases. Private or government insurance bills can also provide information on hospital inpatient and outpatient cases. For example, the number of advanced schistosomiasis cases can be calculated from government bills for the treatment of schistosomiasis in different regions, and the number of surgical treatments and free ultrasound diagnostics for echinococcosis can be used to estimate the incidence of echinococcosis in a region.

Due to the large number of data types in these disease information systems, they require further analysis before they can be used for surveillance purposes. Given the quality of the large amount of data in the information systems, it is sometimes difficult to convince epidemiologists to use these data to address their concerns. In addition, the source of data depends on the possibilities of the collection site and is subject to changes in the system that artificially affect the trend of the data. For example, changes in the international classification criteria for diseases (e.g., echinococcosis) or changes in diagnostic criteria (e.g., malaria) can lead to artificial changes in the information system, causing the incidence artificially changed for the year when the change occurred. In addition, changes in diagnostic methods can lead to changes in incidence rates. For example, before the 1980s, most schistosomiasis-endemic areas used fecal incubation to detect disease. In the 1990s, the Kato-Katz method was used more often. In the 2000s, serological screening was used first, and the Kato-Katz method was then used for confirmation. Whenever the diagnostic methods alternate, the incidence rate will obviously fluctuate. In practice, it is difficult to explain this fluctuation through the effect changes of interventions.

When using disease information system data, we must be cautious. We should consult, analyze, and use it after approval according to the data management requirements and the responsibilities of the supervisor. These data may be published or used publicly, but the situation of the data being too complicated to be applied by users should be avoided.

13.4.5 Evaluation Indicators

Regular evaluation of surveillance, including process evaluation and result evaluation, is important in quality control. The discussion here focuses on process evaluation. Process evaluation is the continuous supervision, inspection, and feedback of the implementation process of the surveillance program. Through the process of surveillance, the problems existing in the surveillance program or in the implementation process are identified or predicted to provide effective information on how to improve the surveillance program for decision-makers.

The main process evaluation indicators are the completion rate of surveillance, implementation quality of surveillance techniques, completeness of surveillance data collection, quality of original records, quality of database construction, and work standardization of surveillance participants.

The process evaluation is generally carried out by a group of experts sent by higher-level professional organizations. It should be carried out throughout the whole process of surveillance. The evaluation results should be fed back to the implementers of the surveillance in evaluation reports in a timely manner to promote surveillance quality and provide a basis for improving the surveillance program.

13.5 Management of Surveillance Sites

Surveillance sites or laboratories are often set up depending on the surveillance situation to improve the efficiency as well as the collection and analysis of data. In practice, administrative departments at all levels attach great importance to the work of surveillance sites or surveillance laboratories and arrange special funding to implement regular work in this area. However, standardizing the management of surveillance or monitoring laboratories is also an important task, which includes the establishment of surveillance sites or laboratories, work manuals, management programs, and evaluation programs.

13.5.1 Establishment of Surveillance Sites

Representative pilot sites for surveillance are selected within an administrative area based on the division of the natural environment or economic conditions. The number of surveillance sites needs to be determined according to the state of the disease epidemic, the number and capacity of surveillance personnel, and other conditions. Sometimes surveillance sites are set up for specific surveillance purposes, and these are called sentinel sites. For example, there are diseases with fewer cases that are associated with specific exposures and therefore require separate surveillance. Sentinel surveillance is extremely important, as the detection of such cases allows higher authorities and the administration to make timely referrals, to predict or warn that people with the same exposure history are also at potential risk, and to intervene in a timely manner. Cases detected during sentinel surveillance are predictive of larger public health problems and can therefore be referred to as sentinel cases or sentinel events. Surveillance of sentinel cases or events can provide a scientific basis for conducting public health research or for implementing interventions in terms of target population and regional coverage. For example, in Jiangsu Province, sentinel sites of schistosomiasis are set up every 50 km along the Yangtze River. If a positive sentinel site was found during the transmission season, it can provide effective information for implementing the necessary preventive and control measures to prevent the occurrence of acute schistosomiasis. After schistosomiasis transmission control was achieved in Jiangsu Province, the sentinel sites provided much more effective preintervention monitoring information for control, and the province became a model place nationwide for early-warning surveillance in schistosomiasis transmission control areas.

13.5.2 Establishment of the Reference Laboratory

This is based mainly on the best service conditions and capacities of laboratories in a certain administrative area that can carry out the tests required for surveillance work in real time. The role of these tests is, first, to serve as a demonstration of the standardized operation of special techniques; second, to serve as a guide and trainer for similar laboratories in the service area; and third, to detect sentinel cases or sentinel events in a timely manner, thus driving laboratory testing in the whole region. Each region can set up a network of laboratories in its subregions according to the work needs and capacity to improve surveillance coverage. For example, in Zhejiang Province, after schistosomiasis transmission was interrupted, a schistosomiasis fecal examination reference laboratory was set up at the prefecture-level CDCs with unified quality tracking and evaluated by the provincial center for schistosomiasis control. Several schistosomiasis fecal test network laboratories were set up at county-level CDCs or sentinel hospitals below the municipal level so that the management of the schistosomiasis testing network could be formulated and standardized. This improved the surveillance capacity; for example, imported schistosomiasis cases were found every year among medical patients or medical examiners; these cases can receive timely treatment, and the further spread of infection sources can be effectively controlled, thus ensuring the consolidation of local schistosomiasis transmission blocking results.

13.5.3 Working Manual

Whether it is the work of surveillance sites or of reference laboratories, the operational steps and procedures, operational specifications and requirements, matters needing attention, etc. of each technology should be clearly defined after the work scope is determined in view of the surveillance technical work carried out. These specifications are compiled into a volume of manual, which becomes the specifications for each surveillance site or reference laboratory.

Surveillance sites or reference laboratories set up by higher-level technical bodies generally have their own specification manuals; thus, each surveillance site or reference laboratory can directly use the working manual developed by professionals at a higher level as its own working manual.

13.5.4 Management Program

The management program is based mainly on the approaches and requirements of the higher technical bodies for the standardized management of each surveillance site or reference laboratory, including the construction, organization, technical supervision, and inspection of the surveillance site or reference laboratory. At the same time, a special leading group and working group will be set up in combination with the management organization at the corresponding level to form a management plan for each surveillance site or reference laboratory.

13.5.5 Evaluation Program

The evaluation program addresses the requirements in the process evaluation. The evaluation can be carried out with the participation of experts from higher-level institutions to facilitate the timely identification and resolution of problems. The evaluation program includes the evaluation of each technology implementation and the evaluation at the technical and management level of the whole surveillance site or reference laboratory.

13.6 Surveillance Reporting and Feedback

13.6.1 Data Analysis

Data analysis refers to the analysis, interpretation, and presentation of monitoring data. The following sets out the relevant components.

13.6.1.1 Analysis and Interpretation

The analysis of surveillance data begins with a general direct description, primarily using standard epidemiological techniques. Analytical strategies used for other epidemiological studies can be applied to the analysis of surveillance data, including age-specific standardization rates in comparative analyses, other group contribution standardization rates, control of confounding factors, sampling response to surveys, and treatment of missing or unknown values. Particular attention should be paid to characteristics such as the time and place of occurrence of the data.

13.6.1.2 Time of Occurrence

In trend analysis, it is often necessary to decide whether to carry out a trend test for the time of occurrence (diagnosis or reporting). Using the reporting time is simpler, but it is limited by the irregularity of reporting. The time of diagnosis provides a better indication of trends in the occurrence of diseases. However, by applying the data of diagnosis to the analysis, the incidence in the most recent period will be underestimated if there is a long interval between diagnosis and reporting. Therefore, it is necessary to adjust the number of delayed case reports based on previous experience.

13.6.1.3 Place of Occurrence

In monitoring, it is often necessary to confirm the exact location of the health event or area of exposure and where people live or the area covered by healthcare, as some people may not have the same healthcare location as their place of residence. People's place of healthcare is more important for monitoring the role of healthcare. Where people live may be more important when it is necessary to understand the needs of preventive measures in different regions. For population censuses and disease incidence statistics, the population of the place of residence is generally used. In the case of a routine reporting system for notifiable infectious diseases, higher administrative levels need to screen or adjust for differences between the place of occurrence and the place of diagnosis or reporting.

13.6.1.4 Application of Geographic Information Systems (GIS)

Information on the location of health events (latitude and longitude) must be entered into information reporting systems. When this information is analyzed statistically along with information such as risk factors, areas of environmental exposure, or places of preventive measures, GIS allows for spatial correlation analysis of exposures or healthcare and health events. Monitoring rates can be increased when monitoring data are presented in maps.

13.6.1.5 Confirmation of Trend Changes

A variety of statistical methods can be applied to identify upward or downward trends in disease incidence from expected levels. The selection of a statistical method is based mainly on the nature of the disease trend, such as seasonal changes, long-term decline, whether there is an accumulation of long-term historical data, whether the analysis is for sudden abnormal trends (e.g., whether the trend is analyzed for a day, a week, a month or a year), and whether the purpose of monitoring is to analyze abnormal temporal changes or to analyze both temporal and spatial trends.

For example, when analyzing trend changes in surveillance data, the primary question is "Is the change real?" because such changes often occur due to a variety of types of human interference rather than representing a real change in disease incidence. These include changes in the number of staff in the surveillance system, changes in staff or reporting staff in the surveillance information system due to holidays, changes in screening or diagnostic criteria, or changes in screening, diagnosis, or access to healthcare. The second question is whether such changes are meaningful. If the increase in prevalence is considered to be an abnormal phenomenon or if the statistical threshold is simply exceeded, it is necessary to confirm whether the observed change has local public health implications and further research is needed, including a reanalysis of the original surveillance data or a full-scale epidemiological survey. This investigation is extremely important to determine whether the monitored phenomenon is a mild public health problem or a potentially serious early public health phenomenon. When threshold levels are set low, "pseudowarning" often occurs, as the likelihood of a warning is greatly increased by random changes. Such shortcomings suggest that surveillance practitioners must be familiar with the real sources of surveillance data and know how to analyze the basic information of local healthcare.

13.6.1.6 Evaluation of Surveillance Integrity

We suppose that there are two different independent monitoring systems or datasets, both of which record individual case information with correlation. Then, we can count the cases that can be found in both datasets or the cases that can be found in only one dataset and not in the other dataset. The capture-recapture method can be used to estimate the number of cases missed by both datasets at the same time so that all cases in a certain group can be completely identified, and the proportion of cases found in each dataset can be calculated. The correctness of this method is judged mainly by the correlation between cases found in one dataset and cases found in another, unrelated dataset, which is based on the assumption that it is difficult for cases to be included in two datasets. If this hypothesis is not true, it may lead to underestimation of the number of cases in a certain group and overestimation of the integrity of the surveillance system. The correctness of this method depends on the correctness of the record correlation and on the correctness and specificity of the related information. If a name is not provided, the individual is represented by a mark (such as a number combining birth date and sex). Even if the name information exists, it may be changed or incorrectly written, or an alias may be used. The software can convert names into numbers, thus avoiding mistakes in these related problems. However, errors in other records can also lead to matching errors or mismatches. Complete linear matching criteria, which should be set in advance, are required for matching, and whether to accept or reject similar matches should be evaluated. Although the computer system can complete most matches and can measure the probability of correct matching, it is still recommended that the simple matching and unmatched work be manually completed.

13.6.1.7 Smoothed Data

Bar charts using incidence or prevalence as the Y-axis and time or smaller areas as the X-axis may have statistical variations that obscure underlying trends or geographical features, resulting in irregular or uneven graphs. A variety of temporal and spatial smoothing techniques can be used to clarify these underlying trends and features.

13.6.1.8 Integrity Protection

Individual identification may occur when very few cases are reported. For this reason, in addition to not using these data, statistical methods can be used to introduce perturbed data in order to achieve the accurate and complete presentation of overall tables or maps without individual identification.

13.6.2 Reporting

13.6.2.1 Main Components of a Report

Because monitoring data have many uses, it is extremely important to spread the information widely and effectively, not only to those who participate in data collection but also to data users, including public health epidemiologists, project management, media, the public, and policy-makers. Therefore, the information presentation should be oriented to the expected audience. The presentation can provide data reviewers with comprehensive resources that change with time. The designed graphics or maps can directly convey the key points.

In addition to publishing monitoring reports, public health institutions have begun to gradually use the Internet or wall newspapers so that information can be updated frequently and widely used. In addition, network-based facilities allow users to obtain monitoring reports.

According to the monitoring results and the characteristics of diseases, the distribution of monitoring reports can attract media and public interest. This requires cooperation with the media, preparation to answer media questions, distinguishing core public health publicity information from monitoring data, and persuading project managers, policy-makers, and members of the public to care about and promote public health publicity.

13.6.2.2 The Role of Reporting

In principle, surveillance reports from different sources can be linked to help extend the service for monitoring because a single information source often suffers from information deficiency. For example, to monitor the development of malaria treatment resistance, it is necessary to monitor the use of local therapeutic drugs as well as the standard treatment of local hospitalized cases and the therapeutic effects of antimalarial drugs. At the same time, comprehensive analysis and statistical evaluation should be carried out in combination with vector distribution and local malaria incidence to fully and correctly understand the trend of local antimalarial drug resistance, put forward effective measures to prevent the spread of drug resistance, and achieve the original goal of surveillance.

In addition, by analyzing the correlation between the monitoring data and different independent data sources, new diseases that have not been discovered before can be confirmed, thus improving the integrity of surveillance. For example, an analysis of the correlation between the data of the second epidemiological survey of parasitic diseases and the cases of clonorchiasis diagnosed in hospitals all over China revealed that the border areas of Guangdong, Guangxi, and Hunan Provinces and the northeast area were the high-incidence areas of clonorchiasis, and the incidence of hepatic cholangiocarcinoma was also higher in the high-incidence areas than in other areas, suggesting that long-term clonorchiasis infection may lead to the occurrence of hepatic cholangiocarcinoma. Therefore, it is necessary to monitor clonorchiasis cases and cholangiocarcinoma cases simultaneously in such areas to effectively control these two diseases.

13.6.3 Information Feedback

Information feedback occurs mainly in the form of written reports or meetings with representatives of professional bodies at the same or lower levels on the surveillance results, the test results of reference laboratories, and the problems found in the process evaluation in order to facilitate the quality control of the surveillance and the work progress.

Professional institutions at all levels should pay attention to information feedback. If major problems are found in the process of monitoring or evaluation, it is necessary to provide timely feedback to the leaders of professional institutions and make a special report to higher-level institutions so that the identified problems can draw inferences from others and play a role in forecasting and early warning.

Before feedback, it is best to communicate effectively with local surveillance sites or reference laboratories, especially to put forward suggestions and opinions for improvement so that these suggestions and opinions can be effectively implemented in future work.

13.6.4 Examples of Surveillance Reports

13.6.4.1 Report on Surveillance of Soil-Transmitted Nematodiasis in 2012

1. Overview of surveillance in 2012
   (a) Overview of the transmission status nationwide: According to the information collected by the Information Management System for the Prevention and Control of Parasitic Diseases in 2012, 22 provinces (autonomous regions and municipalities directly under the central government), including Anhui, Beijing, and Fujian, carried out investigations and treatment of soil-transmitted nematodiasis in the population. A total of 436,338 people were examined, and 12,964 people were found to be infected with soil-transmitted nematodiasis. Ultimately, 5,477,270 people were dewormed.
   (b) Overview of the epidemic situation at surveillance sites: In 2012, surveillance of soil-transmitted nematodiasis was carried out in provinces (autonomous regions and municipalities directly under the central government). A total of 22,755 people were investigated, and 1569 people were infected with soil-transmitted nematodiasis, for an infection rate of 6.9% at 22 surveillance sites.

2. Overview of the Surveillance System in 2012
   (a) Overall: There were 22 surveillance sites of soil-transmitted nematodiasis nationwide, located in 22 provinces (autonomous regions and municipalities directly under the central government).

   Based on the results of a national survey on the status of important human parasitic diseases completed in 2004, the surveillance sites were divided into three categories according to the prevalence of infection:
   - Category I: Ten sites with infection rates >20% distributed in Hainan, Guizhou, Sichuan, Guangxi, Hunan, Chongqing, Fujian, Hubei, Jiangxi, and Yunnan.
   - Category II: Seven sites with infection rates ≥5% and ≤ 20% located in Anhui, Guangdong, Shandong, Gansu, Qinghai, Henan, and Jiangsu.
   - Category III: Five sites with infection rates <5% distributed in Heilongjiang, Hebei, Shanxi, Beijing, and Shanghai.

   (b) Operational analysis: After funding and equipment for soil-transmitted nematodiasis surveillance were released in May 2012, all surveillance sites attached greater importance to their work and carried out surveillance during that year according to the time and circumstances. Some provinces continued to expand the coverage of soil-transmitted nematodiasis surveillance on the basis of the previous 2 years. By the end of January 2013, all surveillance sites had completed the surveillance, but very few sites had not been tested to the required number. In addition, more than 90% of the surveillance sites had reported data to the "Information Management System for Parasitic Diseases Control" after standardized data collation.
   (c) Quality control: Based on the focused feedback from the surveillance sites, the professional staff of NIPD participated the tasks of quality control. After improved the equipment and methods for the isolation and identification of hookworm larvae in October and November, NIPD staff focused on guiding the staff of the surveillance sites with weaker technical capacity, such as Yunnan, to jointly complete the detection of hookworm larvae in the soil samples sent for testing. In addition, a total of 289 Kato smears (137 positive and 152 negative) were reviewed in November and December according to the requirements of the surveillance program, for a total compliance rate of 88.58%, including 94.16% for positive slides and 76.97% for negative slides.
3. Analysis of Surveillance Results in 2012
   (a) Surveillance Components and Methods
   - Surveillance of population infection: From September to November, permanent residents aged 3 years or older were examined for roundworm, hookworm, and whipworm eggs using the modified Kato thick smear method (one fecal triple test), and other detected species were also registered; children aged 3 to 12 years were additionally examined for pinworm eggs using transparent adhesive paper. The infection rate and degree of infection were calculated separately for roundworms, whipworms, and hookworms, and the infection rate was calculated for pinworms.

- Monitoring of roundworm egg contamination in soil: In 13 provinces (cities), Guizhou, Chongqing, Hunan, Hubei, Shandong, Gansu, Henan, Qinghai, Shanghai, Heilongjiang, Hebei, Beijing, and Shanxi, 10 households were randomly selected from each monitoring site, and soil samples were collected from each household in 4 types of locations (vegetable gardens, around toilets, courtyards, and kitchens). Fertilized and unfertilized ova of roundworms were examined microscopically, and soil culture was used to distinguish between dead and live fertilized ova of roundworms and to count them separately.
- Surveillance of hookworm contamination in soil: In 9 provinces (autonomous regions) including Yunnan, Jiangxi, Fujian, Sichuan, Guangdong, Guangxi, Hainan, Anhui, and Jiangsu, 15 households were randomly selected at each surveillance site. Soil samples from each household were collected from vegetable, fruit, or other crop cultivation sites for hookworm isolation. Hookworms were microscopically examined and counted.
- Investigation of relevant factors: the geographical environment, temperature, humidity, rainfall, crops, economic level, health status, control measures, and other natural and social factors were investigated at the surveillance sites.

(b) Analysis of Surveillance Results

- Basic Information on Surveillance Sites

  Population distribution: The 22 surveillance sites had a total of 15,102 households with a total population of 63,024 and a resident population of 49,864.

  Economic status: Except for the main economic resources of the Shandong surveillance sites from industry and commerce, the main economic resource of other surveillance sites was agriculture. The average annual net income of the 22 sites was RMB 6244 (according to the townships where the sites were located), with Shanghai being the highest at RMB 13,912, followed by Beijing (RMB 10,000) and Henan (RMB 2000).

  Drinking water and toilets: Except for the main drinking water at the Hainan monitoring station, which was mountain ditch water, among the remaining 21 surveillance sites, 9 had tap water, and 11 had well water. There were 6982 toilets without harmless treatment, 1386 toilets with biogas digesters, and 3812 toilets with other harmless treatments; thus, the harmless rate of toilets was 42.68%.

  Parasitic disease control: Intensive deworming operations were carried out at the surveillance sites in Guizhou, Guangdong, Henan, and Jiangsu provinces, either at the individuals' own expense or free of charge.

- Population infection

  Overall infection with soil-transmitted helminths: In 2012, out of 22,755 people surveyed at 22 soil-transmitted disease surveillance sites, 1569 were infected with soil-transmitted helminths, representing an infection rate of 6.9%, a decrease of 28.65% compared to 2011 (9.67%). The difference between the rates of infection with soil-transmitted helminths in the 2 years was statistically significant (chi-square test, $\chi^2 = 116.20$, $P < 0.05$).

  Regional distribution: The highest prevalence of soil-transmitted helminth infection was in Yunnan, with 34.05%, followed by Sichuan (20.69%) and Hainan (17.42%); among the Category II sites, Guangdong had the highest prevalence at 10.91%; soil-transmitted helminth infections were not detected in any of the Category III sites.

  Gender distribution: the number of infected males was 678, for an infection rate of 6.02%; the number of infected females was 891, for an infection rate of 7.75%. The difference in the infection rates between males and females was statistically significant (chi-square test, $\chi^2 = 116.20$, $P < 0.05$).

  Age distribution: The highest rate of infection with soil-transmitted helminths was found in the age group over 60 years, at 10.90%, followed by the 45–59 age group and the 0–6 age group. The infection rates in these four age groups were 7.35%, 5.73%, 4.86%, and 3.75%, respectively. The difference in infection rates between the age groups was statistically significant (chi-square test, $\chi^2 = 178.40$, $P < 0.05$) (Table 13.3).

  Occupational distribution: In 2012, farmers and students accounted for 90.09% of the total number of people surveyed at the national surveillance sites. Farmers had the highest rate of soil-transmitted helminth infection at 8.07%, followed by students, children living in the diaspora, retired workers, and children in childcare, with infection rates of 4.33%, 3.82%, 3.77%, and 3.16%, respectively. The rest of the occupational groups had infection rates of less than 1%.

  Distribution of education level: In 2012, the survey population had the highest number of people with primary and secondary education levels, accounting for 40.21% and 34.97%, respectively. Those with junior secondary education or below accounted for 90.75% of the total population. The highest prevalence of soil-transmitted helminth

Table 13.3 Soil-transmitted helminth infections by age group

Age group	No. of people inspected	No. of positives	No. of negatives	Infection rate (%)
0 to 6 years	1628	61	1567	3.75
7 to 14 years	2919	142	2777	4.86
15 to 44 years	8423	483	7940	5.73
45 to 59 years	5169	380	4789	7.35
60+ years	4616	503	4113	10.90
Total	22,755	1569	21,186	6.90

infection was among those with a high school education, at 8.62%, followed by those who were illiterate or semiliterate (8.60%), had completed primary school (8.41%), or had completed junior high school (4.26%). Those with technical school and college education were not examined for infection.

Ascaris lumbricoides infection: 260 of the 22,750 people examined were infected with *Ascaris lumbricoides*, for an infection rate of 1.14%, including 128 infected with fertilized *Ascaris lumbricoides*, for an infection rate of 0.56%. Three surveillance sites in Guizhou (8.59%), Gansu (5.93%), and Qinghai (5.93%) had the highest roundworm infection rates. The infection rate of ascariasis at ten surveillance sites in Beijing, Shanghai, Guangdong, Guangxi, Heilongjiang, Anhui, Hebei, Jiangxi, Shandong, and Shanxi provinces was 0, and the infection rates in the other nine surveillance sites were all below 1% (Table 13.4). The prevalence of *Ascaris lumbricoides* was 1.15% in males and 1.14% in females, with no statistically significant difference (chi-square test, $\chi^2 = 0.001$, $P > 0.05$). There was a statistically significant difference in the prevalence of *Ascaris lumbricoides* infection between the age groups (chi-square test, $\chi^2 = 135.60$, $P < 0.05$). The highest prevalence of *Ascaris lumbricoides* infection was 2.98% in the 7–14 year age group, and the lowest prevalence was 0.48% in the 45–59 year age group (Fig. 13.3).

Whipworm infection: 144 of the 22,755 people examined were infected with whipworm, for an infection rate of 0.63%. Most infections were found in Guizhou (79, 7.71%), followed by Sichuan (20, 1.98%), Hainan (13, 1.26%), and Guangxi (10, 0.99%). There were no infections at ten surveillance sites in Beijing, Gansu, Hebei, Henan, Heilongjiang, Jiangsu, Shandong, Shanxi, Shanghai, and Chongqing. The rest of the surveillance sites had fewer than ten infections each, and the infection rate was below 1% (Table 13.4). The prevalence of whipworm infection was 0.59% in males and 0.67% in females, with no statistically significant difference (chi-square test, $\chi^2 = 0.51$, $P > 0.05$). There was a statistically significant difference in the prevalence of whipworm infection between age groups (chi-square test, $\chi^2 = 19.65$, $P < 0.05$). The highest prevalence of whipworm infection was 1.13% in the 7–14 year age group, and the lowest was 0.39% in the 45–59 year age group (Fig. 13.3).

Hookworm infection: A total of 1202 hookworm infections were found in 22,755 people examined, with an infection rate of 5.28%. Among them, Yunnan had the highest infection rate of 33.85%, followed by Sichuan (18.91%), Hainan (16.55%), Guangxi (12.61%), Guangdong (10.58%), and Chongqing (10.55%). No infected persons were found at nine surveillance sites in Beijing, Gansu, Hubei, Henan, Heilongjiang, Qinghai, Shandong, Shanxi, and Shanghai provinces; the infection rate in the remaining surveillance sites was below 10% (Table 13.4). The prevalence of hookworm infection was significantly different in terms of both gender distribution (chi-square test, $\chi^2 = 424.70$, $P < 0.05$)

Table 13.4 *Ascaris lumbricoides*, whipworm, and hookworm infections at surveillance sites

Surveillance site categories	Provinces	No. of people infected	*Ascaris lumbricoides*		Whipworm		Hookworm	
			No. of infections	Infection rate (%)	No. of infections	Infection rate (%)	No. of infections	Infection rate (%)
Category I	Yunnan	1028	4	0.39	1	0.10	348	33.85
	Guizhou	1024	88	8.59	79	7.71	9	0.88
	Jiangxi	972	0	0	4	0.41	51	5.25
	Fujian	1052	2	0.19	1	0.10	65	6.18
	Hainan	1033	2	0.19	13	1.26	171	16.55
	Sichuan	1010	9	0.89	20	1.98	191	18.91
	Chongqing	1100	1	0.09	0	0	116	10.55
	Hubei	1071	1	0.09	1	0.09	3	0.28
	Hunan	1000	25	2.50	2	0.20	1	0.10
	Guangxi	1007	0	0	10	0.99	127	12.61
Category II	Guangdong	917	0	0	3	0.33	97	10.58
	Shandong	1052	0	0	8	0.76	0	0
	Gansu	1029	61	5.93	0	0	0	0
	Anhui	1067	0	0	2	0.19	20	1.87
	Henan	1003	4	0.40	0	0	0	0
	Qinghai	1046	62	5.93	0	0	0	0
	Jiangsu	1106	1	0.09	0	0	3	0.27
Category III	Shanghai	1051	0	0	0	0	0	0
	Heilongjiang	1005	0	0	0	0	0	0
	Hebei	1025	0	0	0	0	0	0
	Beijing	1026	0	0	0	0	0	0
	Shanxi	1131	0	0	0	0	0	0
Total		22,755	260	1.14	144	0.63	1202	5.28

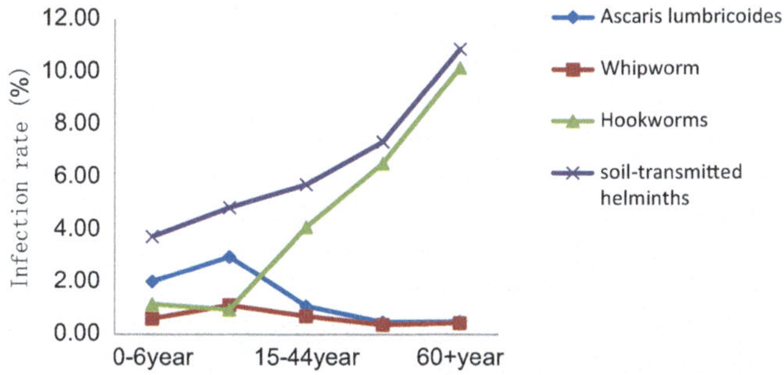

Fig. 13.2 Ascariasis, whipworm, and hookworm infections by age group

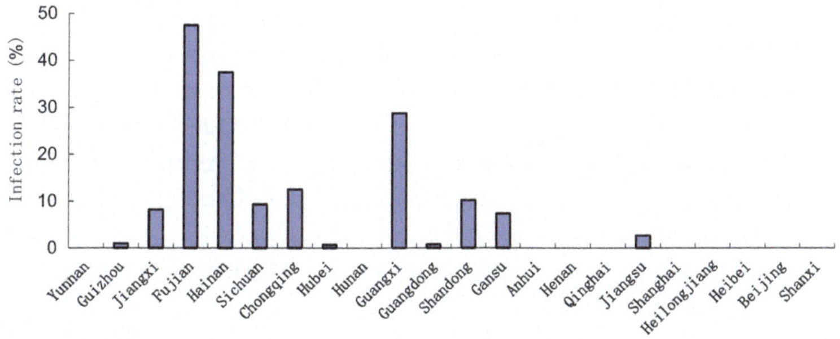

Fig. 13.3 Pinworm infection in children aged 3 to 12 years at the surveillance sites

and age group distribution (chi-square test, $\chi^2 = 33.05$, $P < 0.05$). The prevalence of hookworm infection was 4.42% in males and 6.13% in females. The highest prevalence of hookworm infection was 10.8% in the 60+ age group, and the lowest prevalence was in the 7–14 age group, at 0.96% (Fig. 13.2).

Mixed infections: There were 76 people with mixed infections of 2 or more species of soil-transmitted helminths in the surveyed population, with a mixed infection rate of 0.33%, and the proportion of mixed infections among all soil-transmitted helminth infections was 4.84%. Guizhou Province had the highest number of mixed infections, at 35, followed by Gansu and Sichuan Provinces, both at 11.

Pinworm infection in children aged 3 to 12 years: A total of 2841 children aged 3 to 12 years were examined for pinworms by anal swab method, and the number of infections was 224, with an infection rate of 7.88%, an increase rate of 24.29% increase from 2011 (6.34%). The difference in pinworm infection rates between the 2 years was statistically significant (chi-square test, $\chi^2 = 34.36$, $P < 0.05$). The highest infection rate was in Fujian (47.49%), followed by Hainan (37.50%) and Guangxi (28.75%) (Fig. 13.3).

- Monitoring of human roundworm egg contamination in soil: In 2012, monitoring of human roundworm eggs in soil was conducted at 13 surveillance sites nationwide, and a total of 130 soil samples were tested. A total of 25 soil samples were found to have human roundworm eggs, for an overall infection rate of 19.23%, and soil samples from sources around toilets, courtyards, kitchens, and vegetable garden plots were found to have human roundworm eggs. These positive soil samples were distributed in five surveillance sites in Guizhou, Qinghai, Shandong, Gansu, and Chongqing. All soil samples from ten villagers' homes in Guizhou showed human roundworm eggs, with an infection rate of 100%, while no human roundworm eggs were found in soil samples from the remaining eight surveillance sites.

- Monitoring of hookworm contamination in soil: In 2012, monitoring of hookworms in soil was carried out at seven surveillance sites across the country. A total of 121 soil samples were collected, and 59 samples were found to be contaminated with hookworms, with an infection rate of 48.76%. The surveillance site in Anhui had the most serious situation of contaminated hookworms in soil, with an infection rate of 93.33%, and hookworms were found in the soil at all surveillance sites except Jiangsu. The main infection at the surveillance site in Anhui was *Ancylostoma americanum*; four monitoring sites in Fujian, Guangdong, Guangxi, and Yunnan were infected with *Ancylostoma duodenale*; and a mixed infection of *Ancylostoma americanum* and *Ancylostoma duodenale* was present in Sichuan Province.

4. Conclusion
   (a) The overall prevalence of soil-transmitted helminth infection in the population at the 22 surveillance sites was lower than in 2011, but the areas where the Category I sites located were still at high infections with soil-transmitted helminths.
   (b) The characteristics of soil-transmitted helminths in the population had changed, highlighted by a significantly higher rate of hookworm infections than ascariasis and whipworm infections, mainly in the group over 60 years of age.
   (c) The level of pinworm infections among children aged 3 to 12 years remained high at some of the surveillance sites, and there was no downward trend.
   (d) The contamination by human roundworm eggs and hookworms of the living environment around rural residents was still serious, and different areas had different dominant species of hookworms.
5. Challenges and Recommendations
   (a) Challenges
      - As the surveillance was carried out on a "village rotation" basis from 2011 onward, the target populations differed, resulting in a more significant difference in the infection situation at some surveillance sites compared to previous years.
      - Some areas with a high incidence of soil-transmitted helminths had underutilized the Parasitic Disease Control Information Management System, resulting in the system not reflecting comprehensive information.
      - Frequent turnover of staff at the grassroots level and instability of the prevention team largely restricted the effective implementation of vari-

ous tasks, such as surveillance, which was reflected by the compliance rate of retesting of Kato smears this year. In addition, Nanchang city of Jiangxi Province was unable to carry out local soil hookworm monitoring for various reasons.

(b) Recommendations
- On the basis of continued surveillance, scientific analysis of the changing characteristics of the monitoring data collected by the "site rotation" method and comparison with the monitoring carried out by the "fixed point" method will provide a basis for future improvement of the surveillance program.
- Supervision of provinces with low utilization of the Parasitic Disease Control Information Management System should be strengthened so that the system can provide a comprehensive picture of soil-transmitted helminthiasis control.
- The training of young and middle-aged professionals should be strengthened to enhance their capacities in microscopic examination, identification of hookworm species, and data analysis which not only provide a reserve of personnel technical capacity for surveillance but also lay a foundation for analyzing the distribution of hookworm species in China and the relationship with population infection.
- To take the opportunity of the release of the midterm evaluation report on the implementation of the National Plan for the Prevention and Control of Key Parasitic Disease 2006–2015, localities should be encouraged to establish provincial (municipal) surveillance sites while carrying out prevention and control to lay the foundation for a comprehensive grasp of the epidemiological trends of soil-transmitted helminths.

13.7 Evaluation of Surveillance Effects

13.7.1 Roles

Effect evaluation of surveillance is the evaluation of results; i.e., the evaluation effectiveness of the surveillance is considered, interpreted, and judged. It is generally necessary to collect all kinds of information and data related to the results or effectiveness, relate the results to the surveillance objectives and information about the whole process of implementation, and analyze and evaluate the effectiveness and achievements of the surveillance.

The results of the evaluation are used mainly to improve the functionality and usefulness of the surveillance system. Surveillance systems have a variety of functions, among which are to evaluate existing systems or design new ones.

13.7.2 Indicators

As the resources are limited, strengthening the resources of one system will definitely weaken the resources of another system, so the use and budget of the surveillance system should be balanced according to various functional indicators to meet the minimum demand. Evaluation indicators mainly include the following.

13.7.2.1 Sensitivity

Sensitivity is the extent to which the system monitors all target events. If only epidemic trends are monitored, lower sensitivity monitoring is sufficient as long as the surveillance system operates continuously and the events detected are representative. If the purpose of monitoring is to evaluate the impact of One Health issue, higher sensitivity is required so that nondeterministic factors can be detected.

13.7.2.2 Timeliness

Timeliness refers to information flows in the surveillance cycle from data collection to information release. Timeliness is determined by the contingency of the public health event and the implementation of interventions.

13.7.2.3 Prediction Value

The prediction value is the accuracy of the reported cases as true cases.

13.7.2.4 Representation

Representativeness is the degree of events monitored by the surveillance system through representative groups in the target population. If the monitoring is not representative, it will mislead decision-makers regarding the use of health resources.

13.7.2.5 Data Quality

Data quality refers to the correctness and completeness of descriptive data in case reports, field investigations, and information systems.

13.7.2.6 Simplicity

The difficulties of surveillance implementation include the following issues: whether the monitoring procedure and implementation process are simple or too complicated to operate, whether the form is easy to fill out, whether the data collection can cover the most basic information, whether the software is easy to operate, whether the web page links are easy to open, and whether the surveillance report directly reflects the problem.

13.7.2.7 Flexibility

Flexibility is the extent to which the surveillance system can adapt to new situations or changing information needs.

13.7.2.8 Acceptability

Acceptability refers to participants' level of enthusiasm for the monitoring system, including whether the information generated by the participants' work is the information they want and whether public supporters can enable health agencies to search for personal information for monitoring purposes.

Some indicators may have similar functions or influence each other. For example, simplicity is likely to enhance acceptability and completeness, while improving timeliness may affect completeness and data quality. For example, efforts to complete the monitoring report may include those in the survey who have no diseases. The balance of these indicators is very important for the surveillance system to detect an epidemic situation early. For example, setting a low threshold enables a possible epidemic situation to be detected in a timely manner and completely, but there will be many "false warnings."

The evaluation results can also provide a basis for revising the new monitoring technical scheme, including its scientificity and feasibility. Therefore, when conditions permit, it is necessary to evaluate the monitoring costs, effects, or benefits of each technology in the surveillance program and analyze the impact of monitoring on local disease control program, which is conducive to improving the sustainability and operability of the surveillance program.

In summary, surveillance is the process of investigating and reporting special health problems in a defined population. Each part of the surveillance system can actually have different practices, including the selection of data sources, the application of statistical methods, and the method of information dissemination. When choosing the method, it is necessary to consider the special purpose of the surveillance system, the information needs of possible users, and the best performance of various functions of the surveillance system (such as timeliness and completeness). The most fundamental standard of the inspection system is whether the system succeeds or fails in preventing and controlling health events such as diseases, injuries, disability, and death.

13.8 Case Studies

13.8.1 Schistosomiasis Reference Laboratory Planning and Management Scheme

13.8.1.1 Proposing the Issue

To further strengthen the management of schistosomiasis diagnosis, improve the quality management system of schistosomiasis diagnosis in China, and make the on-site survey data reflect the epidemic situation of schistosomiasis more scientifically, the former Advisory Committee of Schistosomiasis Experts of the Ministry of Health organized the construction of provincial reference laboratories for schistosomiasis diagnosis at provincial CDCs and institutes of schistosomiasis control in 2009–2010 and gradually established a multilevel and multicenter network platform for schistosomiasis diagnosis laboratories at national, provincial, and municipal (county) levels. This scheme was specifically formulated for the construction of a provincial reference laboratory for schistosomiasis diagnosis.

13.8.1.2 Objectives

1. To establish a platform for diagnostic assessment of schistosomiasis, including the establishment of a national quality control product bank (standard serum bank) for schistosomiasis, a provincial sample (serum sampling) bank, and basic equipment and related support for the laboratory

2. To establish a base for exchange and training in schistosomiasis diagnosis, including staffing by an appropriate number of technicians, the establishment of a laboratory system, and annual technical guidance and training for reference laboratories at the same or lower levels

13.8.1.3 Working Principles

1. Grading responsibility. The Schistosomiasis Advisory Committee was responsible for the network construction of reference laboratories, the NIPD at China CDC was responsible for the construction of national reference laboratories, and Hunan, Jiangxi, Hubei, Sichuan, Anhui, and Jiangsu were responsible for the construction of provincial reference laboratories respectively.
2. Integrated management. The construction of participating laboratories at all levels was planned and managed in a unified manner.
3. Resource sharing. Information on diagnostic technology assessment at all levels of the reference laboratories was shared, and resources were shared at the same time.

13.8.1.4 Functions

1. National reference laboratories were responsible for interlaboratory quality control of relevant tests in schistosomiasis reference laboratories at all levels nationwide, preparing and issuing quality control samples and standards, and guiding the construction of provincial reference laboratories.
2. Provincial reference laboratories undertook testing of schistosomiasis laboratory diagnostic items; participated in laboratory comparisons and proficiency testing organized by higher authorities; were responsible for implementing the interlaboratory quality control of local, municipal, and county laboratories in the provinces; and organized technical training of schistosomiasis laboratory personnel in the province.

13.8.1.5 Access Criteria

1. Facility requirements. The schistosomiasis diagnosis reference laboratory should have a separate laboratory room, which is divided into a pathogen detection area, immunology detection area, diagnostic reagent, and test sample storage area.
2. Equipment
   (a) The laboratories should be equipped with equipment related to diagnostic experiments, including an enzyme marker reader, general refrigerator, low-temperature refrigerator (−20 °C), ultralow-temperature refrigerator (−80 °C), water bath, thermostat, centrifuge, shaker, pipette, light biochemical incubator, microscope, and dissecting microscope.
   (b) The laboratories should be equipped with a dedicated computer, the necessary camera equipment, and a dedicated filing cabinet.
   (c) The laboratories should be equipped with disinfection and dirt handling equipment, safety precautions, and laboratory thermostats.
3. Positions. The Schistosomiasis Diagnostic Reference Laboratory had at least seven technical positions, including one laboratory management position, three

Table 13.5 Rules for scoring provincial reference laboratories for schistosomiasis diagnosis

Indicators	Score (points)	Review components	Scoring criteria	Score
Organizations	40	The applicant unit laboratory has a clear mission, clear functions, and sound positions. (40 points)	1. Clear leadership, and the director of the declared laboratory is the legal representative of the organization (20 points)	
			2. With a reasonable mechanism (10 points)	
			3. With clear division of duties (10 points)	
Instruments and equipment	150	1. The type and number of instruments and equipment should meet the methodological requirements for testing. (80 points)	1. The equipment is fully configured to ensure that common diagnostic methods of testing are carried out properly. (70 points)	
			2. Equipped with a dedicated computer and data backup system. (10 points)	
		2. There is an instrument and equipment file, and each instrument should have records of unpacking, installation, commissioning, installation, use, and maintenance, with a dedicated person in charge. (30 points)	1. List of instruments and equipment and manuals is available. (10 points)	
			2. Instrument operating procedures are available. (10 points)	
			3. Various records are complete and standardized. (10 points)	
		3. The necessary instruments should be tested in accordance with the relevant requirements and have not exceeded the verification period. (10 points)	To check the calibration certificate, self-testing instruments must have self-testing protocols and records. (10 points)	
		4. In-use instrument integrity rate >90% (30 points)	Scoring of field tests, with deductions where appropriate. (30 points)	
Personnel qualifications	150	1. The number of relevant professional and technical staff engaged in testing should meet the needs of the testing work. (60 points)	1. At least four technical positions, including laboratory management positions, testing positions, and quality management positions (40 points)	
			2. A reasonable ratio of technical staff titles, with at least two staff with intermediate titles or above. (20 points)	
		2. The technical person in charge and the person in charge of quality assurance should be familiar with the business and have the corresponding titles. (30 points)	The technical person in charge and the person in charge of quality assurance should carry out work related to schistosomiasis testing and have senior professional titles in schistosomiasis or testing-related disciplines. (40 points)	
		3. The testing personnel should be familiar with the norms, standards, and methods of schistosomiasis testing, master the relevant theoretical and operational skills, understand the performance of the instruments, and be licensed to work as needed. (50 points)	1. At least 3 years of experience in schistosomiasis testing. (20 points)	
			2. Testing personnel should be familiar with the norms, standards, and methods of testing, master the relevant theoretical and operational skills, and complete the relevant training before starting work. (30 points)	

(continued)

Table 13.5 (continued)

Indicators	Score (points)	Review components	Scoring criteria	Score
Conditions for examinations	140	1. The area of the laboratory room meets the requirements of the test item applied for. (50 points)	1. Separation of laboratory and office (20 points)	
			2. Relatively independent laboratory partitioning, with separate areas for pathogenicity testing, immunology testing, and storage of diagnostic reagents and test samples (30 points)	
		2. Clean environment in the laboratory and reasonable arrangement of instruments, water, electricity, and coal. (30 points)	1. Clean, well-ventilated, and lighting conducive to work (10 points)	
			2. Reasonable arrangement of water and sewage pipes and electrical wiring (10 points)	
			3. Reasonable placement of refrigerators, warming boxes, and various glass instruments (10 points)	
		3. Laboratory biosafety (50 points)	1. Firefighting equipment is fully equipped and within the validity range (10 points)	
			2. Dedicated storage area for reagents, with clear access levels (10 points)	
			3. Double-locked storage of dangerous goods (10 points)	
			4. Special storage equipment for waste liquids and other waste (5 points)	
			5. Ability to dispose of waste as needed, with someone responsible for it (5 points)	
			6. Testing personnel equipped with protective equipment such as white coats and rubber gloves (10 points)	
		4. Sampling equipment should be kept clean and reagent bottle labels written in full (5 points)	1. Clean and uncontaminated sampling equipment (5 points)	
			2. Clearly written labels (5 points)	
		5. Water for experiments is up to standard (5 marks)	Water used for each experimental operation meets requirements and standards (5 points)	

Testing capacity	300	1. Technical documents such as relevant norms, analytical methods, and standards for carrying out schistosomiasis diagnosis should be complete. (30 points)	1. Documents such as technical norms and diagnostic standards for schistosomiasis control (10 points)	
			2. Complete quality system documentation (10 points)	
			3. Diagnostic or technical protocol (10 points)	
		2. Testing should be done in strict accordance with the work procedures. (20 points)	1. Having a work scheme (10 points)	
			2. Having a quality control program (10 points)	
		3. Sample collection and preservation are carried out strictly according to the specifications, and there are corresponding records. (10 points)	1. Sample registration record sheet available (5 points)	
			2. Sample handover records available (5 points)	
		4. In the sample testing, when an accident (water, electricity, etc.) affects the sample determination, there should be clear handling measures. (10 points)	To score appropriately after checking the relevant records	
		5. The format of original records is standardized, records are clear and tidy, and legal units of measurement are used uniformly. (20 points)	A sample of original records will be checked, and points will be deducted on a pro rata basis for any of the following:	
			1. Records completed in pencil (5 points deducted)	
			2. Incorrect method of revision (5 points deducted)	
			3. Failure to use statutory units (5 points deducted)	
			4. Irregular signature (5 points deducted)	
		6. Sound testing files. All kinds of reports should be kept for more than 5 years. (10 points)	The testing file should include business contracts, testing work plans, quality control schemes, raw data, quality control data, and testing reports	
		7. Levels of detection of schistosomiasis diagnostic methods (200 points)	1. IHA full score of 50 points, sample qualitative errors a deduction of 5 points, 2.5 points per sample deducted for qualitative correct validity quantitative errors	
			2. ELISA full score of 50 points, 5 points per sample deducted for sample characterization error, 2.5 points per sample deducted for qualitative correct absorbance quantitative error	
			3. 50 points for nylon silk egg collection and incubation method, 15 points for one incorrectly characterized sample	
			4. 50 points for the modified Kato method, 10 points deducted for each incorrectly characterized sample, 5 points deducted for each film with correct characterization but quantitative egg count <0.8 or 1.2 times the reference value	

(continued)

Table 13.5 (continued)

Indicators	Score (points)	Review components	Scoring criteria	Score
Quality assurance	140	1. There is a special group or dedicated staff responsible for quality assurance. (20 points)	1. A special cabinet for quality management information (10 points)	
			2. Having full-time (or part-time) quality control staff (10 points)	
		2. Strict quality control measures for testing and analysis and assessment records for the whole procedure testing process. (50 points)	1. Each test has an appropriate quality control protocol (10 points)	
			2. A person is responsible for checking and recording the performance of all instruments before and after testing (20 points)	
			3. Quality control staff monitor and assess each step of each test (20 points)	
		3. Data can be reported only after "three reviews" and "three proofs." (30 points)	1. "Three reviews": self-review, mutual review, and quality supervisor review (15 points)	
			2. "Three reviews": self-review, review by quality control staff, and review by the technical director. (15 points)	
		4. A quality control file is in place. (40 points)	1. The quality control file includes records of training and assessment of testing personnel. (10 points)	
			2. The quality control program and quality control data are complete. (20 points)	
			3. There is a quality control assessment procedure specification. (10 points)	
Work system	80	Laboratory management systems and quality assurance systems and documentation are complete. (80 points)	1. Laboratory management system (5 points)	
			2. Instrument and equipment management system (5 points)	
			3. System for the storage of laboratory supplies and management of highly toxic drugs (5 points)	
			4. Test sample management system (5 points)	
			5. Information data confidentiality and audit system (10 points)	
			6. Testing quality assurance work system (10 points)	
			7. Incident analysis and handling reporting system (10 points)	
			8. System for the development, revision, and inspection of the Quality Management Manual (10 points)	
			9. Training system for inspectors (10 points)	
			10. Responsibilities for various categories of personnel (10 points)	
Total	1000			

testing positions, one quality management position, one instrument and equipment management position, and one information management position.

4. System construction. A laboratory quality management system was established, and the development of system documents was completed (including the development of four levels of documents: A, B, C, and D).
5. Work Progress (omitted)
6. Scoring Rules (Table 13.5)

13.8.2 Pilot Program for Symptom Surveillance and Early Warning of Transmission of *Angiostrongylus cantonensis*

13.8.2.1 Proposing the Issue

Angiostrongyliasis is an emerging parasitic disease caused by infection of the larvae of *Angiostrongylus cantonensis* in humans, with eosinophilic meningoencephalitis as the main clinical manifestation. In China, most cases are caused by consumption of raw or undercooked *Pomacea canaliculata* or *Achatina fulica* containing the third-stage larvae of *Angiostrongylus cantonensis*. From 2006 to 2007, a national survey showed that seven provinces (autonomous regions), such as Fujian, were the natural foci of *Angiostrongylus cantonensis*. *Pomacea canaliculata* or *Achatina fulica* were sold in farmers' markets or restaurants in more than 20 counties (cities), among which *Pomacea canaliculata* or *Achatina fulica* infected with *Angiostrongylus cantonensis* were detected in nine counties (cities).

In recent years, there has been a significant increase in the number of reported cases of *Angiostrongylus cantonensis* in China, and several outbreaks have occurred. The expanding distribution of *Pomacea canaliculata* and *Achatina fulica*, the diversification of people's dietary habits, the ease of food transportation, and the lack of knowledge about the prevention and control of *Angiostrongylus cantonensis* have all increased the risk of occurrence of the disease. To explore and establish a sensitive and rapid prevention and control mechanism for *Angiostrongylus cantonensis*, reduce the risk of transmission, effectively reduce the incidence of disease, and control the occurrence of outbreaks, a pilot project on symptom surveillance and early warning of transmission of *Angiostrongylus cantonensis* is being carried out in selected areas. To ensure the implementation quality of the pilot project, the following scheme was developed.

13.8.2.2 Objectives

1. To explore the mechanism for monitoring the symptoms of *Angiostrongylus cantonensis* infection to provide a scientific basis for the formulation of control strategies
2. To understand the incidence of *Angiostrongylus cantonensis* infection and potential food safety problems in the pilot area of Guangzhou and to explore early-warning methods

13.8.2.3 Pilot Areas and Responsible Organizations

1. Principles for determining the pilot areas. The pilot areas are representative of the main types of epidemics and outbreaks of angiostrongyliasis in China; the health institutions responsible for symptom surveillance in the pilot areas have the capacity to treat angiostrongyliasis and are able to undertake and complete the surveillance tasks.
2. Distribution of pilot areas. Based on the distribution of natural epidemic sites and past outbreaks across the country, six areas in Beijing, Fujian Province, and Yunnan Province were selected as pilot areas, including the following three categories:
   - Highly endemic areas: Fuzhou City and Xiamen City in Fujian Province
   - Susceptible areas of natural foci with reported outbreaks: Kunming City, Dali Autonomous Prefecture, and Jinghong City in Yunnan Province
   - Nonnatural foci with outbreaks reported: Beijing

13.8.2.4 Organization of Pilot Surveillance

Sixteen healthcare institutions in the pilot areas were selected to be responsible for symptom surveillance and vector surveillance.

13.8.2.5 Components and Methods of Pilot Work

1. Symptom Surveillance for *Angiostrongylus cantonensis*
   (a) Definition of surveillance cases: A surveillance case is a case with clinical symptoms of meningoencephalitis and a recent history of consumption of snails (mainly freshwater and terrestrial snails) or mollusks such as slugs.
   (b) Surveillance components
      - Confirmed diagnosis: After the medical and health institutions responsible for surveillance discover a indicator case, they should confirm the diagnosis based on epidemiological history, clinical symptoms, and laboratory test results with reference to the following diagnostic criteria.
        Suspected cases: Recent (usually within 1 month) ingestion of or contact with mollusks such as *Pomacea canaliculata* or *Achatina fulica*, slugs or freshwater shrimps, crabs, fish, or frogs accompanied by clinical symptoms of meningoencephalitis such as headache, nociceptive hypersensitivity, and neck stiffness.
        Clinical cases: Suspected cases are consistent with any of four items: eosinophilia in peripheral blood, eosinophilia in cerebrospinal fluid, positive antibody or circulating antigen of *Angiostrongylus cantonensis* in serum or cerebrospinal fluid, and effective treatment targeting specific pathogens.
        Confirmed cases: Clinically diagnosed cases with larvae of *Angiostrongylus cantonensis* found in the cerebrospinal fluid, in the eye, etc.
      - Collection of medical records and specimens: The healthcare institution responsible for surveillance should keep a complete record of the relevant information of the case and retain the serum.

  - Reporting of surveillance cases: Medical and health institutions responsible for surveillance should report symptomatic surveillance cases through the "*Angiostrongylus cantonensis* Surveillance and Reporting System" of NIPD at China CDC (www.ipd.org.cn/angio/) within 24 h after they are detected.
  - Notification of the epidemic: Surveillance health institutions should notify the local CDC within 24 h after finding a case through symptomatic surveillance for *Angiostrongylus cantonensis*, and the CDCs in pilot areas should notify local symptomatic surveillance health institutions within 24 h when they find snails infected with *Angiostrongylus cantonensis* through vector surveillance.
  - Epidemic verification and case investigation: When a local CDC finds a reported indicator case or receives notification of a case from a surveillance health institution, it should verify the reported case and conduct a case investigation in accordance with the requirements of the "Angiostrongyliasis cantonensis Case Investigation Form". The medical and health institutions responsible for symptom surveillance will assist in the case investigation, trace the source of the food, expand the testing in markets and food establishments based on the information from the case investigation, and report the test results to the provincial-level CDC (the YIPD in Yunnan Province) and NIPD at China CDC.

   (c) Surveillance period: Year-round

2. Edible Snail Surveillance
   (a) Surveillance of Snail Infections
  - Scope: According to the local situation, one large aquatic product or farmers' market (hereinafter referred to as the market) and three large catering places (including food stalls) (hereinafter referred to as catering places) that sell edible snails such as *Pomacea canaliculata* or *Achatina fulica* were selected in cities, urban-rural fringe areas, and rural areas as fixed surveillance sites. If the number of stalls in the same market or dining place is 5 or less, all stalls should be checked. The number of snails to check in each stall should be at least 50, and in stalls with fewer than 50 snails, all snails should be checked. If there are more than 5 stalls, 5 stalls will be randomly selected, with the number of snails in each stall to be checked being at least 50 and all snails inspected in stalls with fewer than 50.
  - Methodology: Sampled snails are tested, and the results are summarized in a "Vector Surveillance Survey Registration Form".
  - Monitoring time: During the peak snail selling season (May–October) every year, sampling is carried out twice a month with an interval of 2 weeks. When infected snails are found, samples are taken and tested for 5 consecutive days.
  - Reporting of infected snails: When a snail infected with *Angiostrongylus cantonensis* is found during vector surveillance by a CDC in the pilot

area, it should be reported to the provincial CDC within 24 h. The provincial CDC should report to NIPD within 24 h.

(b) Snail Sales Surveillance
   - Scope: The scope of surveillance is consistent with the monitoring of snail infections.
   - Methodology: The owners of each stall and catering place in the market are interviewed to determine the sales volume of snails, and the "Survey Summary Table of the Sales Volume of *Pomacea canaliculata* in the Market/Catering Place" is filled out.
   - Monitoring time: May to October during the peak season of edible snail sales each year, with survey statistics provided once a month at the same time each month.

(c) Early warning of epidemics: The pilot areas must establish an early-warning mechanism for epidemics, especially for mass epidemics, which must be reported in a timely manner, carry out epidemic tracking and investigation, analyze epidemic trends, and take timely prevention and control measures.

When a local CDC confirms a hospital-reported case of local infection (the patient has consumed snails purchased from the local market during the incubation period), it should increase the surveillance of edible snails in markets and catering places and expand the scope and frequency of testing when appropriate. When the number of reported local infection cases increases to five or more and *Angiostrongylus cantonensis* infection is detected in edible snails sold in markets or catering places, the local authorities should be advised to issue a warning in due course, and the edible snails in markets and catering places should be treated accordingly and banned from being sold.

13.8.2.6 Data Collection, Analysis, and Feedback

Municipal CDCs and medical institutions that undertake the pilot work are responsible for data collection and management. Data from the pilot work should be submitted to the provincial CDCs by the end of January of the following year, the provincial CDCs should submit the data and surveillance reports to the NIPD by the end of February, and the NIPD should submit the summary data and the annual pilot work report to China CDC and Bureau of Disease Prevention and Control by the end of March.

13.8.2.7 Quality Control

NIPD provides technical training to the operational staff of the provincial CDCs and medical institutions that carry out the pilot work, and the provincial CDCs provide technical guidance to the operational staff of municipal CDCs and medical institutions. The Fujian CDC, YIPD, and Beijing CDC are responsible for the supervision of the pilot work in their provinces.

13.8.2.8 Working Mechanisms and Responsibilities

Under the leadership of the Bureau of Disease Prevention and Control of the NHC and the health department (bureau) of each pilot province (municipality directly under the central government), CDCs at all levels are specifically responsible for the organization and implementation of the pilot.

NIPD at China CDC is responsible for the development of the pilot program, training, technical support, and data analysis, and provincial CDCs are responsible for the coordination, guidance, and quality control of municipal CDCs and medical institutions in the pilot areas. Municipal CDCs and medical institutions in the pilot areas are responsible for the implementation of the pilot project and timely reporting of data.

13.8.3 Technical Program for Surveillance Sites of Soil-transmitted Parasitic Diseases

13.8.3.1 Proposing the Issue

In 2005, the Ministry of Health announced the results of a national survey on the status of important human parasitic diseases, which showed that the national infection rate of soil-transmitted diseases was 19.56% and the number of infected people was estimated to be 129 million, indicating that the infection rate of soil-transmitted helminths in China is still high. This is not compatible with the current requirements for building a harmonious society in China. To grasp the epidemiological dynamics and patterns of soil-transmitted diseases in a timely manner and to provide a scientific basis for formulating strategies and measures for the prevention and control of soil-transmitted diseases and evaluating the effectiveness of prevention and control, it is necessary to carry out planned, continuous, and systematic soil-transmitted helminth surveillance on a nationwide scale. In accordance with the National Plan for the Prevention and Control of Key Parasitic Diseases 2006–2015, the National Program for the Surveillance of Soil-Transmitted Parasitic Diseases was formulated as follows.

13.8.3.2 Objectives of Surveillance

To understand the epidemiological dynamics and influencing factors of soil-transmitted helminthiasis, grasp the epidemiological patterns, and predict the epidemiological trends, to provide a scientific basis for formulating prevention and control countermeasures and evaluate their effectiveness, and to provide a relevant evaluation basis for the establishment of national and local healthy cities (towns).

13.8.3.3 Definition of Surveillance Cases

The surveillance cases in this scheme are cases of soil-transmitted helminths when the eggs or bodies of hookworms, roundworms, whipworms, and pinworms are found during the pathogenic examination of soil-transmitted helminths in the population.

13.8.3.4 Surveillance Components

1. Surveillance Sites
   (a) Determination of Principles

- Surveillance sites are categorized according to the national prevalence of soil-transmitted helminth infections.
- The rate of soil-transmitted helminth infection in the population at the surveillance sites should be equivalent to the average infection level of the population in the province, autonomous region, or municipality directly under the central government.
- The counties (cities and districts) where the surveillance sites are located have a certain working basis and can take the initiative to cooperate and undertake certain surveillance tasks.

(b) Distribution of surveillance sites: Twenty-two surveillance sites were set up nationwide in accordance with the principles of categorization.

- A total of ten surveillance sites were selected in areas with infection rates greater than 20%: including Hainan, Guizhou, Sichuan, Guangxi, Hunan, Chongqing, Fujian, Hubei, Jiangxi, and Yunnan.
- A total of seven surveillance sites were selected in areas with infection rates of 5% to 20%: including Anhui, Guangdong, Shandong, Gansu, Qinghai, Henan, and Jiangsu.
- A total of five surveillance sites were selected in areas where the infection rate was less than 5%: including Heilongjiang, Shanghai, Hebei, Shanxi, and Beijing.

(c) Procedures for Determining Surveillance Sites

- The surveillance site should be located in an administrative village or a number of natural villages with a total population of at least 1500 people overall and at least 1000 people being examined each time.
- Surveillance sites are selected by provincial CDCs or institutes of parasitic disease control and reported to China CDC for the record.
- Once a surveillance site has been identified, it will not change for 5 years.

(d) Surveillance Components and Methods

- Population-Based Infection Surveillance
  Components: hookworms, roundworms, whipworms, and pinworms; other species detected are also registered.
  Targets: Permanent residents aged 3 years or older.
  Methods: A modified Kato thick smear method (one fecal triple test) is used from September to November each year, and a transparent adhesive paper anal swab is added to detect pinworm eggs in children aged 3 to 12 years.
  Indicators: The infection rate and degree of infection are calculated separately for hookworms, roundworms, and whipworms, and the infection rate is calculated for pinworms.
- Surveillance of Human Roundworm Egg Contamination in Soil
  Soil sampling sites: Ten households were randomly selected at each surveillance site, and soil samples were collected from each household in four types of locations: vegetable garden, toilet perimeter, courtyard, and kitchen.

Sampling size: 1 sample per household per type of location, 30 g per sample.

Methods: Microscopic examination of fertilized or unfertilized roundworm eggs and soil culture methods were used to distinguish between dead and live fertilized roundworm eggs.

- Investigation of relevant factors: Natural and social factors, such as geographical environment, temperature, humidity, rainfall, crops, economic level, health status, and control measures, were assessed at surveillance sites.

(e) Surveillance of roundworm infection in children under 14 years of age in cities (towns). Areas that have been declared national and local healthy cities (towns) are required to conduct surveillance of roundworm infection in children under 14 years of age. Other areas are required to carry out surveillance in accordance with local parasitic disease control plans and needs.

- Sample size: For cities with a total population of 500,000 or more, the number of people surveyed should be at least 1500; for cities with a total population under 500,000, the number of people surveyed should be at least 1000.
- Sampling method: Two schools in urban areas and 2 schools in suburban areas were randomly selected for roundworm infection surveillance of students in all grades up to 14 years old, with at least 100 students surveyed in each grade.
- Examination components: Roundworm eggs.
- Examination method: Stool specimens were examined for roundworm infection using the modified Kato thick smear method (three tests per specimen).
- Indicator: Roundworm infection rate and degree of infection.

13.8.3.5 Data Collection, Analysis, and Feedback

1. Data Collection, Analysis, and Feedback at Surveillance Sites
   (a) China CDC is responsible for establishing and distributing unified database software for surveillance site surveys.
   (b) Provincial CDCs are responsible for the collection and entry of data from the surveillance sites and for reporting the database and annual surveillance report to China CDC by January 15 of the following year.
   (c) China CDC is responsible for summarizing and analyzing the data from all surveillance sites nationwide and then returning the results to each province, autonomous region, and municipality directly under the central government by the end of March of the following year.
2. Collection of data on surveillance of roundworm infection in children under 14 years of age. In cities (towns) where surveillance of roundworm infection in children under 14 years of age is carried out, the local CDC is responsible for collecting, entering, and managing the surveillance data and reporting the results to the local health administration department and the National Patriotic Health

Campaign Committee as evidence for declaring national and local healthy cities (towns).

13.8.3.6 Quality Control

1. Training. The NIPD is responsible for the training of provincial-level operational personnel who are tasked with surveillance work. Each province, autonomous region, and municipality directly under the central government is responsible for the technical training of surveillance sites in its own region.
2. Supervision and inspection. China CDC is responsible for organizing the supervision and inspection of the surveillance work, including the completion of the work plan, the training of personnel, and the implementation of technical specifications.
3. Quality control of microscopy. NIPD is responsible for organizing the reinspection of the microscopic examination results of the surveillance sites. At least 30% of positive smears (or all smears if there are fewer than ten positive smears) should be sampled for each surveillance site, and the false-positive rate should not exceed 10%; at least 5% of negative smears should be sampled, and the false-negative rate should not exceed 10%. NIPD at China CDC should provide timely feedback and publish the results of microscopic quality control at the surveillance sites.

13.8.3.7 Organizational Leadership and Responsibilities

1. The Ministry of Health is responsible for the organizational leadership of the national surveillance work and formulation of the surveillance program. Provincial health administrative departments lead the surveillance in their jurisdictions, arrange the required funds, and ensure the implementation of the surveillance.
2. China CDC is responsible for the organization and implementation of the work at the surveillance sites nationwide and provides a certain amount of funding for surveillance. NIPD is responsible for undertaking technical guidance, training, quality control, data compilation, and feedback on the work of the surveillance sites.
3. CDCs in the provinces, autonomous regions, and municipalities directly under the central government where the surveillance sites located undertake specific surveillance tasks.
4. The city (district)- and county-level CDCs where the surveillance sites located are responsible for assisting in the surveillance process.

Appendixes

Please refer to the appendixes by scanning the following QR code.

X.-N. Zhou (ed.), *Surveillance and Management of Parasitic Diseases in China*,
https://doi.org/10.1007/978-981-97-1241-0

The manufacturer's authorised representative in the EU is Springer
Nature Customer Service Centre GmbH, Europaplatz 3, 69115 Heidelberg,
Germany. If you have any concerns regarding our products, please
contact ProductSafety@springernature.com

Printed and bound by CPI Group (UK) Ltd, Croydon, CR0 4YY
01/10/2026
02225857-0009